Analyzing Sustainability in Peripheral, Ultra-Peripheral, and Low-Density Regions

Rui Alexandre Castanho
WSB University, Poland

A volume in the Practice, Progress, and Proficiency in Sustainability (PPPS) Book Series

IGI Global
PUBLISHER of TIMELY KNOWLEDGE

Published in the United States of America by
IGI Global
Engineering Science Reference (an imprint of IGI Global)
701 E. Chocolate Avenue
Hershey PA, USA 17033
Tel: 717-533-8845
Fax: 717-533-8661
E-mail: cust@igi-global.com
Web site: http://www.igi-global.com

Library of Congress Cataloging-in-Publication Data

Names: Castanho, Rui Alexandre, editor.
Title: Analyzing sustainability in peripheral, ultra-peripheral, and low-density regions / Rui Alexandre Castanho, editor.
Description: Hershey, PA : Engineering Science Reference, [2023] | Includes bibliographical references and index. | Summary: "This book presents chapters that analyze the dynamics and patterns of cross-border strategies and policies in peripheral, ultra-peripheral, and low-density regions regarding sustainability and the related issues that may influence them"-- Provided by publisher.
Identifiers: LCCN 2022014007 (print) | LCCN 2022014008 (ebook) | ISBN 9781668445488 (h/c) | ISBN 9781668445495 (s/c) | ISBN 9781668445501 (ebook)
Subjects: LCSH: Municipal engineering. | City planning. | Sustainable urban development. | Urban ecology (Biology) | Borderlands.
Classification: LCC TD160 .A53 2023 (print) | LCC TD160 (ebook) | DDC 307.1/216--dc23/eng/20220527
LC record available at https://lccn.loc.gov/2022014007
LC ebook record available at https://lccn.loc.gov/2022014008

This book is published in the IGI Global book series Practice, Progress, and Proficiency in Sustainability (PPPS) (ISSN: 2330-3271; eISSN: 2330-328X)

British Cataloguing in Publication Data
A Cataloguing in Publication record for this book is available from the British Library.

Practice, Progress, and Proficiency in Sustainability (PPPS) Book Series

Ayman Batisha
International Sustainability Institute, Egypt

ISSN:2330-3271
EISSN:2330-328X

Mission

In a world where traditional business practices are reconsidered and economic activity is performed in a global context, new areas of economic developments are recognized as the key enablers of wealth and income production. This knowledge of information technologies provides infrastructures, systems, and services towards sustainable development.

The **Practices, Progress, and Proficiency in Sustainability (PPPS) Book Series** focuses on the local and global challenges, business opportunities, and societal needs surrounding international collaboration and sustainable development of technology. This series brings together academics, researchers, entrepreneurs, policy makers and government officers aiming to contribute to the progress and proficiency in sustainability.

Coverage

- Global Business
- Eco-Innovation
- Knowledge clusters
- Intellectual Capital
- Green Technology
- Sustainable Development
- Socio-Economic
- ICT and knowledge for development
- E-Development
- Strategic Management of IT

IGI Global is currently accepting manuscripts for publication within this series. To submit a proposal for a volume in this series, please contact our Acquisition Editors at Acquisitions@igi-global.com or visit: http://www.igi-global.com/publish/.

The Practice, Progress, and Proficiency in Sustainability (PPPS) Book Series (ISSN 2330-3271) is published by IGI Global, 701 E. Chocolate Avenue, Hershey, PA 17033-1240, USA, www.igi-global.com. This series is composed of titles available for purchase individually; each title is edited to be contextually exclusive from any other title within the series. For pricing and ordering information please visit http://www.igi-global.com/book-series/practice-progress-proficiency-sustainability/73810. Postmaster: Send all address changes to above address.

Titles in this Series

For a list of additional titles in this series, please visit: www.igi-global.com/book-series

Eco-Friendly and Agile Energy Strategies and Policy Development
Mir Sayed Shah Danish (University of the Ryukyus, Japan) and Tomonobu Senjyu (University of the Ryukyus, Japan)
Engineering Science Reference • © 2022 • 335pp • H/C (ISBN: 9781799895022) • US $240.00

Handbook of Research on Global Institutional Roles for Inclusive Development
Neeta Baporikar (Namibia University of Science and Technology, Namibia & University of Pune, India)
Information Science Reference • © 2022 • 398pp • H/C (ISBN: 9781668424483) • US $270.00

Handbook of Research on Building Inclusive Global Knowledge Societies for Sustainable Development
Cristina Raluca Gh. Popescu (University of Bucharest, Romania & The Bucharest University of Economic Studies, Romania)
Information Science Reference • © 2022 • 405pp • H/C (ISBN: 9781668451090) • US $295.00

Handbook of Research on Principles and Practices for Orchards Management
Waleed Fouad Abobatta (Horticulture Research Institute, Agriculture Research Center, Egypt) Ahmed Farag (Central Laboratory for Agricultural Climate, Agriculture Research Center, Egypt) and Mohamed Abdel-Raheem (National Research Center, Egypt)
Engineering Science Reference • © 2022 • 400pp • H/C (ISBN: 9781668424230) • US $295.00

Handbook of Research on Algae as a Sustainable Solution for Food, Energy, and the Environment
Mostafa M. El-Sheekh (Tanta University, Egypt) Norhayati Abdullah (Universiti Teknologi Malaysia, Malaysia) and Imran Ahmad (Universiti Teknologi Malaysia, Malaysia)
Engineering Science Reference • © 2022 • 795pp • H/C (ISBN: 9781668424384) • US $325.00

Socio-Economic Sustainability in the Post-Pandemic Era
Jozef Oleński (High School Technology and Economics, Jaroslaw, Poland) Jeffrey Sachs (Columbia University, USA) Masayuki Susai (Nagasaki University, Japan) Yannis Tsekouras (University of Macedonia, Greece) and Arjan Gjonça (London School of Economics and Political Science, UK)
Information Science Reference • © 2022 • 315pp • H/C (ISBN: 9781799897606) • US $215.00

Climate Change, World Consequences, and the Sustainable Development Goals for 2030
Ana Pego (Nova University of Lisbon, Portugal)
Engineering Science Reference • © 2022 • 335pp • H/C (ISBN: 9781668448298) • US $250.00

IGI Global
PUBLISHER of TIMELY KNOWLEDGE

701 East Chocolate Avenue, Hershey, PA 17033, USA
Tel: 717-533-8845 x100 • Fax: 717-533-8661
E-Mail: cust@igi-global.com • www.igi-global.com

Editorial Advisory Board

Table of Contents

Detailed Table of Contents

European borderlands characteristics are determined by the interaction of differences in geography along with various conflicts in demographic and socio-economic factors. These regions are more isolated than the other areas due to the barriers that restrain the opportunity to interact with cross-border areas. Most cases of border areas are isolated in their own geography. During the last decade, the European Union has designed and implemented several integration procedures to accelerate the transform process of borderlands from primarily peripheral regions into interesting spots for sustainability growth. Addressing accessibility and mobility issues in peripheral regions requires a comprehensive view of the factors and indicators at multiple scales and levels. Mobility-oriented accessibility planning approaches miss some inherent aspects related to spatial and socio-economic circumstances. When moving from theory to practice, analyzing how mobility is addressed by current strategies has resulted in emerging some challenges and inconsistencies in transport systems and infrastructure.

The city of Madrid is the capital of Spain, and the urban sprawl has been important and continuous over the years. There has been a spillover effect since the use of urban land has increased in the municipalities around the capital. Among other factors, the price of housing is increasingly high in Madrid. Thus, in the municipalities near the capital, the population makes trips to work but then returns to their municipality to reside, where the housing price is lower. Satellite images through Land Cover Corine have made it possible to quantify the hectares destined for urban use in 1990, 2000, 2006, 2012, and 2018 through a geographic information system (GIS). It has been possible to identify those municipalities of the autonomous community that have grown better as continuous urban fabric or discontinuous urban fabric. Therefore, it has been possible to identify municipalities most influenced by the capital.

This study aims to examine the flood propensity of the main watercourse of Porto da Cruz drainage basin and, if relevant, to propose two methodologies to alleviate the impacts (i.e., detention basin sizing and riverbed roughness coefficient adjustment). Geomorphological data were obtained from the watershed characterization process and used through the SIG ArcGIS software for the flood propensity assessment and then for the calculation of the expected peak flow rate for a return period of 100 years through the Gumbel Distribution. Subsequently, the drainage capacity of the river mouth was verified using the

Manning-Strickler equation in order to establish whether the river mouth of the watershed has the capacity to drain the entire volume of rainwater in a severe flood event. In conclusion, the results demonstrate that the river mouth of the Porto da Cruz watershed does not have the capacity to drain the rain flow for the predetermined return period; thus, the detention basin was sized through the Dutch method and the simplified triangular hydrograph method.

Extreme precipitation has not only detrimental effects on ecosystems and social and economic sectors, but it is a natural hazard that can trigger floods or soil erosion. This study tries to analyze the extreme rainfalls on different geomorphological units and geographical regions of Chile. For this, data from 87 meteorological stations latitudinally and altitudinally distributed and covering a long period (1980–2018) were used. Results showed that precipitation concentration displays an exponential curve where 30% of the rainiest days were concentrated in only 10% of days with precipitation, proving high irregularity. The decisive weight on annual precipitation falls on a few rainy days with very high rainfall amounts. For return periods > 100 years, extreme events of daily precipitation could reach 109 mm and 305 mm in Northern and Southern Andes Mountains, respectively, while in Northern and Southern Central Depression, their values could be 70 mm and 170 mm, respectively.

Reservoirs play a fundamental role in the hydrological planning of the central valley of Chile as they provide water for human and animal consumption, energy generation, and crop irrigation, especially during the summer season. In agriculture, reservoirs represent a significant source to keep the food security standard for more than half of the population of the country. The water management plans need complete records of their volume to calculate rules of operation or future scenarios; however, currently, these time series include gaps that do not allow better analysis, which increases uncertainty. To address this, the authors test a methodology to assess Sentinel 2 imagery through normalized difference water index (NDWI). The results correctly represent the temporality and seasonality of reservoir dynamics; however, the magnitude of the changes is not well represented when the reservoir is delivering water. This research allows more data-based planning of water resources in the central zone, contributing to better decision-making and more efficient water management.

Food crisis and hunger issues will challenge many countries due to the effects of climate change and global warming. When considering the truth that the Mediterranean Basin is the region that has been challenged by climate change and invasive alien species much more than others in the world, food insecurity is the most critical issue. As achieving food security, Mediterranean countries should keep the balance between the population and food sources in the related region. Accordingly, determining indicators and variables guiding policy makers to develop sustainable strategies for food security will be important cases for researchers. This study aims to review the sustainability level of the agricultural food market among Mediterranean countries. It is thought to give a holistic view for agricultural food markets in the Mediterranean Basin. This study focuses on examining some selected indicators for agricultural food markets among Mediterranean countries by showing recent descriptive statistics for agricultural products as crops and cereals.

For many decades the world witnessed mass displacements. Migration can be either voluntary or forced depending on the reasons. Mass migration can be also due to climate changes that harden the living conditions. Since there is a bidirectional nexus between migration and environmental degradation, the dramatic increase of refugees, asylum seekers, and immigrants greatly affect the environment and conservation efforts. Globally, the movement of people has caused different types of ecosystem changes including deforestation, water, and air pollution, as well as increased waste. Furthermore, as a two-edged sword for conservation, urbanization and growing population cause increasing per capita demand for energy, goods, and services. In this chapter, migration and sudden-onset and slow-onset hazards, the correlation between migration and environmental degradation, environmental and economic effects on the destination, as well as the role of government during migration processes will be analyzed.

The purpose of local environmental programs of municipalities is to involve the entire local population in the solution of environmental issues that are not provided for in the state environmental programs or in addition to them. These programs include measures to maintain the existing ecological balance in the local environment; greening and landscaping of the municipality; collection, transportation, neutralization, processing, and other issues in this area; protection of water, air, and land from all kinds of pollution together with neighboring municipalities, implementation of environmental measures, and other measures of local importance.

The national literature on forest ecosystem valuation is scarce and little is known about how important the valuation of forest ecosystem services and their internalization in low density regions of Portugal are. Hence, there is a need for technicians, academics, and researchers to mitigate this knowledge gap through further research in this area. The chapter is a literature review with the objective of systematizing and synthesizing the knowledge produced in the period between 1992 and 2021 with regard to estimates of the economic value of forest ecosystem services in Portugal as well as finding evidence that relates the mechanisms of internalization of externalities in the sustainable development of low-density regions. A meta-regression was estimated, and the results indicate 220 international dollar/hectare/year in 2019 (190 euros/hectare/year) for forest ecosystem services in Portugal. Payment mechanisms for non-market forest ecosystem are still at an embryonic stage, which does not allow an accurate measurement of their real contribution to the sustainability of low-density regions.

Natural resources are of vital importance in any territory. This is the case of the ultra-peripheral and low-density region such as the island of Palma, where the vegetation is enormously rich and varied, and also together with the water plays a transcendental role in the sustainability of the island. For this reason, the quality of the vegetation and its spatial distribution on the island were analysed. The water content in plants and soils was also determined. These objectives could be met using images from the Landsat satellite and the management of this raster information with a geographic information system. From the results obtained according to the spatial distribution, contrasts were identified between areas of the island where there was vegetation with great vigorousness and abundant water content, concerning others with vegetation of worse quality and lower water content.

Most countries have considered waste management as a high priority since it is a problem of environmental pollution and resource depletion. The main goal of implementing environmental policies is to find or create ways to manage waste beneficial for social, environmental, and economic aspects. In the face of the COVID-19 pandemic, the amount of waste generated got out of control, and therefore, the traditional way

Foreword

This monograph *Analyzing Sustainability in Peripheral, Ultra-Peripheral, and Low–Density Regions* is published with the understanding of serving several purposes and audiences. Within those, it is possible to name some as a research agenda on the subject of peripheral territories and the consequent cooperation in critical border areas around the world, particularly in the Latin American Region and the Caribbean living in "paradoxical poverty" with the enormous biological, mineral, tourist and cultural diversity wealth, with a development model that privileges the extraction of resources in a mercantilist vision of the linear economy.

Contextually, this book delves into policy recommendations conducive to developing an ecologically sustainable economy. In this regard, this book can be read by the entire public (academics, political decision-makers, government agencies, non-governmental organizations, and the general multidisciplinary public) interested in an environmental development model, socially and economically responsible and supportive.

This monograph is a textbook or reference text that I recommend as compelling reading in Political Science, particularly in Environmental Science. It even serves as the basis for undergraduate courses.

As an academic text, it is aimed at researchers in the field of population settlements in accordance with the Future We Want and the Sustainable Development Goals, for the design and application of public policies related to the conservation of matter and energy, for the creation, planning, development, and improvement of low-density peripheral and ultra-peripheral population settlements.

Since this field of research is inherently multidisciplinary, the potential academic audience is broad, including economists, ecologists, conservation biologists, engineers, public policy professionals, geographers, anthropologists, sociologists, and others.

This book presents a methodology for planning and decision-making related to urban land use planning in such a way as to ensure habitability conditions in European cities and towns, promoting their sustainability considering the population impact on the quality of the environment and its components.

I mainly think that the styles and perspectives in this monograph, organized in sixteen chapters, represent the pluralistic nature of sustainable development, in which the Preface by Professor Rui Alexandre Castanho synthesizes these holistic perspectives and serves as a guide to the rest of the book.

It is my most fervent wish that this text becomes a global reference text and that it contributes to further expanding the circle of academics who share the dream and philosophy of sustainable development. With this sincere desire, I publicly congratulate the Editors for this significant publication.

I conclude by pointing out that "Even if renewed systems are implemented for environmental improvement, problems will not be solved unless humans transform themselves."

Sincerely,

Edwin A. Vegas Gallo
Peruvian University of Sciences and Informatics, Peru

Preface

In fact, advances in studies and research associated with the peripheral territories (borderlands) and the subsequent cross-border cooperation (CBC) have been increased and introduced all over the globe (Castanho, 2017; Gómez *et al.*, 2021). Such advances essentially affect the cross-border strategies and policies, processes of border cooperation, and several complex border movements (Vulevic *et al.*, 2020). Moreover, similar scenarios are encountered in ultra-peripheral (islands) and remote territories and low-density regions (Castanho, Couto, and Santos, 2021; Pimentel *et al.*, 2022; Sousa *et al.*, 2022). Common denominators, such as the limited land, water resources, and tourism overexploitation, among many other factors, make these specific territories essential case studies concerning their governance and sustainable development and growth (Spangenberg, 1995; Loures and Crawford, 2008; Štverková, and Pohludka, 2021; Batista *et al.*, 2022).

In this sense, looking retrospectively, it is possible to understand that the notion of 'sustainable development' has generated a lot of debate and interest in the policy and decision-makers circles in the last few decades. Today, such sustainable development is no longer a choice but a necessity for us all (Durán Sosa, Castanho, and Couto, 2022; Pamela Pérez *et al.*, 2022). Thus, it becomes simple to comprehend that if we look to prosper as a society, and probably as a specie, there is no alternative instead of opting for a typology based on sustainable development and growth!

Contextually, this monograph, *Analyzing Sustainability in Peripheral, Ultra-Peripheral, and Low-Density Regions*, investigates activities, processes, and behaviors in light of the new challenges and the desired sustainable development and growth model. Also, it explores the dynamics and patterns ongoing in the peripheral, ultra-peripheral, and low-density regions regarding sustainability and the issues that may influence it. Covering topics such as glamping tourism, vegetation quality, and territorial cohesion, this premier reference source is an essential resource for government officials, business executives and managers, community leaders, environmentalists, researchers, and academicians. Thereby, this publication aims to provide a considerable scientific contribution towards sustainability.

ORGANIZATION OF THE BOOK

This monograph is organized into 16 chapters. A short description of each of the chapters follows:

Chapter 1 provides an interesting discussion about the territorial dimension of policies related to territorial cohesion, peripherality, and accessibility. Moreover, the author uses case studies to contextualize this problem better.

Chapter 2 examines the effect of European Projects in Border Regions by analyzing the efficient management of natural and cultural heritage resources for the sustainable development of cross-border tourism.

Chapter 3 focuses on the accessibility and mobility issues in peripheral regions through a comprehensive view of the factors and indicators at multiple scales and levels. Contextually, the authors discuss how current strategies to address mobility have resulted in emerging challenges and inconsistencies in the transport system and infrastructure.

Chapter 4 addresses the problems related to the urban sprawl in major cities. In this specific case, the authors studied the Spanish Capital City (Madrid) and quantified the hectares destined for urban land use in 1990, 2000, 2006, 2012, and 2018 through Geographic Information System (GIS) tools.

Chapter 5 explores the hot topic of climate change, particularly the flood propensity of the main watercourse of a municipality located in an ultra-peripheral territory (Madeira Island, Portugal).

Chapter 6 continues emphasizing the climate crisis with the analysis of extreme precipitation events. This chapter works over different geomorphological units and regions of Chile where extreme rainfalls occur frequently.

Chapter 7 follows the same line as the previous contributions. Nevertheless, in this case, the authors address the issue of Water resources in Chile. This work enables us to comprehend the hydrological planning of the central valley of Chile and how the reservoirs play a vital role in water supply for human and animal consumption, energy generation, and crop irrigation, especially during the dry season.

Chapter 8 highlights the food crisis and hunger issues that several nations are facing and will face shortly due to the effects of climate change and global warming. In this sense, the study aims to review the sustainability level of the agricultural food market among Mediterranean territories and provide us with a holistic viewpoint of the agricultural food market in the Mediterranean Basin.

Chapter 9 centers on the bidirectional nexus between migration and environmental degradation. So, the authors bring us to a fascinating discussion about migration and sudden-onset and slow-onset hazards, the correlation between migration and environmental degradation, environmental and economic effects on the destination, and the role of government during migration processes.

Chapter 10 studies the purpose of local environmental programs of municipalities to involve the entire local population in solving environmental issues that are not provided for in the state environmental programs. Therefore, the author uses a case study of the Azerbaijan territory to address this issue from a practical approach.

Chapter 11, through the literature review examination, allows us to expand our knowledge on the valuation of forest ecosystem services. In this regard, the authors focused on evidence that relates to the mechanisms of internalization of externalities in the sustainable development of low-density regions - as is the case of Portugal Mainland.

Chapter 12 measures the vegetation quality on Palma island - Canary Archipelago. Throughout this study, it is possible to identify some of the significant contrasts in the spatial distribution of the island vegetation. Besides, this chapter shows where vegetation is located with great vigorousness and abundant water content, concerning others with vegetation of worse quality and lower water content.

Chapter 13 reminds us, once more, about the so-relevant problem of environmental pollution and resource depletion. This chapter analyzes the strategies, regulations, and measures imposed by the governments of Japan, the United Kingdom, Mexico, the United States, and China to compare between them their way of controlling waste and the effects that the pandemic has left regarding the increase in waste and their way of managing them.

Chapter 14 brings a disruptive topic (in this monograph scope) about social sustainability and employee management. Through this work, the authors analyzed the productivity and participation relationship in the performance evaluation through a comparative study of Poland and Azerbaijan companies.

Chapter 15 approaches the topic of tourism massification in major urban centers. Contextually, the authors focused on analyzing the second home tourism market in South Africa.

Chapter 16 deals with issues associated with sustainable tourism. This study reflects the tourism activity in areas exposed to the Northern Lights. In fact, such considerations already exist and tend to develop further through sustainable glamping accommodations in low-density areas.

Rui Alexandre Castanho
WSB University, Poland

REFERENCES

Batista, M., Couto, G., Castanho, R. A., Sousa, Á., Pimentel, P., & Barreto Carvalho, C. (2022). The Rural and Nature Tourism Development Potential in Islands. *Sustainability*, *2022*(14), 5289. doi:10.3390u14095289

Castanho, R. A. (2017). *Sustainable Urban Planning in Transboundary Areas: Analysis of Critical Factors for Territorial Success* [Doctoral Thesis]. University of Extremadura (UEx), Department of Vegetal Biology, Ecology and Earth Sciences.

Castanho, R.A., Couto, G., & Santos, R. (2021). *Introductory Chapter: Rural Tourism as a Catalyst for Sustainable Regional Development of Peripheral Territories*. doi:10.5772/intechopen.96651

Durán Sosa, R., Castanho, R. A., & Couto, G. (2022). *The Challenge of drought in Costa Rica: A Preliminary Research with Economic Implications. WSEAS Transactions on Business and Economics*. doi:10.37394/23207.2022.19.89

Gómez, J. M. N., Velarde, J. G., Gallardo, J. M., Almonte, J. M. J., Aliseda, J. M., & Fernández, J. C. (2021). The Most Meridional Border in Europe. Demographic and Environmental Changes. In R. A. Castanho, G. Couto, & R. Santos (Eds.), Peripheral Territories, Tourism, and Regional Development. IntechOpen. doi:10.5772/intechopen.97566

Loures, L., & Crawford, P. (2008). Democracy in progress: Using public participation in post-industrial landscape (re)-development. *WSEAS Transactions on Environment and Development*, *4*, 794–803.

Pamela Pérez, L., Castanho, R. A., Calvo Martinez, S., & Morales Pachón, A. (2022). *Influence of Inappropriate Basic Sanitation and Lack of Access to Drinking Water in the Community Development of Azama, Otavalo Canton. WSEAS Transactions on Environment and Development*. doi:10.37394/232015.2022.18.45

Pimentel, P., Vulevic, A., Couto, G., Behradfar, A., Naranjo Gómez, J. M., & Castanho, R. A. (2022). Maritime Transportation Dynamics in the Azores Region: Analyzing the Period 1998-2019. *Infrastructures*, *2022*(7), 21. doi:10.3390/infrastructures7020021

Sousa, À., Castanho, R. A., Couto, G., & Pimentel, P. (2022). Pos-Covid Tourism Planning: Based on the Azores Residents' Perceptions About the Development of Regional Tourism. *European Planning Studies*, 1–23. Advance online publication. doi:10.1080/09654313.2022.2079375

Spangenberg, J. (1995). Towards Sustainable Europe. A Study from the Wuppertal Institute for Friends of the Earth Europe, Luton, UK.

Štverková, H., & Pohludka, M. (2021). Sustainable entrepreneurship in small and medium-sized enterprises in the Czech Republic. In *Proceedings of the 14th International Conference Strategic Management and its Support by Information Systems 2021*. VŠB-Technical University of Ostrava.

Vulevic, A., Obradovic, V., Castanho, R.A., & Djordjevic, D. (2020). Cross-Border Cooperation (CBC) in a Multi-Level Governance System in Southeastern Europe Territories: How to Manage Territorial Governance Processes in Serbia-Romania Border Space. In *Cross-Border Cooperation (CBC) Strategies for Sustainable Development*. IGI Global. doi:10.4018/978-1-7998-2513-5.ch004

Chapter 1
Territorial Cohesion:
Territoriality, Peripherality, Accessibility

Ana Vulevic
https://orcid.org/0000-0001-5150-4477
Institute of Transportation CIP, Belgrade, Serbia

ABSTRACT

This chapter contributes to the discussion of the territorial dimension of policies related to territorial cohesion, peripherality, and accessibility. The authors analyze which components and methodology should be considered at the academic level to quantify or measure territorial cohesion trends. The notions of territoriality, peripherality, and accessibility are briefly discussed. The results make it possible to identify problems so that peripheral regions can be further explored. Economic and regional inequalities of the Portuguese territory are shown through indicators related to their cohesion policy. Peripheralization continues to deepen the gap between the center and the periphery, as well as coastal and border regions. The analysis reveals the poor results of the cohesion policies undertaken, although the territoriality of cohesion policies has been attempted and emphasized through many projects over the EU's cohesion policy.

INTRODUCTION

Territorial cohesion in a case of Portugal is analyzed through INTERCO indicators of territorial cohesion presented according to territorial objectives. The INTERCO Report improved four main analytical dimensions of the concept of territorial cohesion, based on six main territorial objectives. Development dimension: (i) stable local economies ensuring global competitiveness; (ii) innovative territories; (iii) fair access to services, market, and jobs; (iv) inclusion and quality of life; (v) attractive regions of high ecological values and substantial territorial capital; and (vi) integrated polycentric territorial development presented through the INTERCO territorial cohesion indicators by territorial objective in case of Portugal. When analyzing territorial cohesion in a given territory of Portugal, the chapter attempts to answer the question: Which components should be taken into account when analyzing the territorial

DOI: 10.4018/978-1-6684-4548-8.ch001

cohesion? The analysis revealed a general picture of Portugal with clearly defined centers and peripheries. In accordance with '*...Faludi (2009: 24) reminds us that defining territorial cohesion should not be a priority because, in the end, these definitions depend on who gives them, when, and with which purpose. Conversely, one can argue that the lack of a proper definition will only contribute to the persistent fuzziness associated with the concept of territorial cohesion, gradually reducing its political and scientific importance*' (Madeiros 2016). The results make it possible to identify problems so that peripheral regions can be further explored. Economic and regional inequalities of the Portuguese territory are shown through indicators related to their cohesion policy. Peripheralization continues to deepen the gap between the center and the periphery, as well as coastal and border regions. Selected indicators show social and economic inequalities, identifying peripheral regions, combining location, demographic, social, economic, cultural, political and natural factors, which are the result of unsustainability of the European project and highlights problems in implementing territorial cohesion policies (Medeiros, 2013; Naranjo, Castanho & Vulevic, 2021).

Discussion of territorial cohesion in the chapter on the results obtained on the basis of the INTERCO indicators in order to achieve teritorial balance in Portugal. The notions of territoriality, peripherality and accessibility are analyzed due to their relevance for territorial cohesion. The aim of the chapter is also to emphasize the relevance of territoriality, peripherality and accessibility in the adoption of new regional and national policy strategies or the adoption of new policies and decisions. The chapter has four subtitle introductory of chapter, approach cohesion by terittorial objectives, presenting indicators and the results by objectives and sucessing conclusions.

TERRITORIAL COHESION, TERRITORIALITY, PERIFERALITY AND ACCESSIBILITY

Territorial cohesion is a multi-dimensional and dynamic concept, susceptible to different

interpretations. Taking its roots in spatial policy development, it has aroused the interest of several actors, from researcher to policy makers and planners. For a long time, debates on what was called European space or territory, spatial or territorial dimension, took place in informal meetings or under the auspices of Council of Europe (INTERCO, 2013) although it took over previous insights on territorial cohesion, it did not propose anyclear definition of it but had a wide and integrated approach, with balanced andsustainable development at its centre. Territorial cohesion is seen as a means of achieving it, by '*transforming diversity into an asset* ' (CEC, 2008; INTERCO Final Report, 2013, Faludi, 2004; 2006). The key challenge identified is '*to ensure a balanced and sustainable territorial development of the EU as whole, strengthening its economic competitiveness and capacity for growth while respecting the need to preserve its natural assets and ensuring social cohesion*' (INTERCO FINAL REPORT, 2013). Thus, territorial cohesion has the difficult task to build '*bridges between economic effectiveness, social cohesion and ecological balance*' (INTERCO FINAL REPORT, 2013). But in real we have increased territorialization through territorial cohesion policies and in same time we do not have good result about territorial cohesion in border areas and periferal regions (Medeiros 2013b; 2014a; 2016b). According to Medeiros (2016): '*the only significant difference between the terms 'territory' and 'region' is the fact that a regional perspective is, as the notion implies,specifically dedicated to the analysis of a certain or several regions. In other words, territory is a more general geographic term, which can cover several scales of analysis, such as the urban, the local, the regional, the national, and the European levels.*'

'Transport accessibility is one of the crucial inequalities in peripheral territories.' Velaga et al. (2012). *'Accessibility is the main product'* of a transport system. It determines the locational advantage of an area (i.e. in ESPON a region, a city or a corridor) relative to all areas (including itself). Indicators of accessibility measure the benefits households and firms in an area enjoy from the existence and use of the transport infrastructure relevant for their area. The important role of transport infrastructure (i.e. networks and transport services) for spatial development in its most simplified form implies that areas with better access to the locations of input materials and markets will, ceteris paribus, be more productive, more competitive and hence more successful than more remote and isolated areas (ESPON TRACC 2013, Vulevic 2016; Castanho et al., 2017; Castanho et al., 2019; Vulevic et al., 2020; Naranjo, Castanho & Vulevic, 2021). Also, consequences of missing modes of transport and accessibility in remote and peripheral regions result in depopulation. The proposed methodology applied on the Portugese case use NUTS2 for few indicators and NUTS3, from 1990 to 2010 (INTERCO, 2013).

APPROACH COHESION BY TERITTORIAL OBJECTIVES (INDICATORS AND RESULTS)

INTERCO methodology and indicators used at several territorial scales of analysis (from European to local). In this chapter we present analyses on NUTS 3 level.

Territorial objective 1. Strong local economies ensuring global competitiveness

Four indicators are proposed as territorial cohesion indicators under this objective:

- GDP per capita in PPS
- Overall unemployment rate
- Old age dependency ratio
- Labor productivity

These indicators are dedicated to measure the overall economic output of all activities (GDP per capita in PPS), the quality of the regional labour markets (unemployment rate), the labour market age structure (old age dependency ratio), and the competitiveness of a region compared to global market (labour productivity) (INTERCO, 2013).

Regarding GDP per capita measured in Purchasing Power Standards (PPS) there is a strong concentration of GDP per capita in capital city of the regions and in the ports and big agglomerations(Porto and Lisboa). It is increasing in Porto and Lisboa from 1998. and in 2008. we have values for Porto 20400 and for the Lisboa 32200. Values arefrom 5000 to 10000 for border regions and higher in coastal regions.

Generally a decrease of unemployment rates over total population is desired, especially attention needs to be paid to decrease unemployment in rural areas or areas with specific geographical handicaps (islands or border regions). Significant increases in unemployment rates, during 2006-2009, unemployment rates again increased due to the economic crisis.Values are partly as high in south and north part of Portugal.

Old age dependency ratio. Generally the indicator illustrates that the size of the working-class population in Portugal. Regions have extremely high variations up to 30 percentage points especialy in border area.

Table 1. Territorial objectives 1. Strong local economies

	NUTS 3 (2006)	GDP per capita measured in Purchasing Power Standards (1997)	GDP per capita measured in Purchasing Power Standards (PPS) (2008)	Old age dependency ratio (2004)	Old age dependency ratio (2009)	Ratio of unemployed people in relation to overall work force (1999)	Ratio of unemployed people in relation to overall work force(2009)
PT111	Minho-Lima	8300	12500	31,20	31,80	3,60	11,00
PT112	Cávado	9400	14800	17,68	18,52	3,30	11,00
PT113	Ave	10200	14100	17,27	18,52	5,70	11,00
PT114	Grande Porto	14600	20400	20,02	22,22	5,60	11,00
PT115	Tâmega	6400	10500	17,76	18,20	3,60	11,00
PT116	Entre Douro e Vouga	11300	15400	19,45	21,29	2,20	11,00
PT117	Douro	6800	11600	30,47	30,74	5,20	11,00
PT118	Alto Trás-os-Montes	7200	12100	36,26	37,63	4,60	11,00
PT150	Algarve	13800	21500	28,02	28,98	4,80	10,30
PT161	Baixo Vouga	12900	17900	23,94	25,59	2,00	6,90
PT162	Baixo Mondego	12100	18500	28,84	30,87	2,50	6,90
PT163	Pinhal Litoral	13500	19700	25,09	27,17	1,60	6,90
PT164	Pinhal Interior Norte	7700	11900	38,61	38,29	2,10	6,90
PT165	Dâo-Lafôes	8500	14400	30,72	31,64	3,40	6,90
PT166	Pinhal Interior Sul	7100	12100	52,80	51,66	2,90	6,90
PT167	Serra da Estrela	5700	9800	36,67	36,36	3,20	6,90
PT168	Beira Interior Norte	7400	12800	41,50	40,52	1,80	6,90
PT169	Beira Interior Sul	10600	16500	46,27	44,98	3,30	6,90
PT16A	Cova da Beira	8400	13000	33,70	34,62	3,20	6,90
PT16B	Oeste	11100	15500	27,07	28,29	2,80	6,90
PT16C	Médio Tejo	10700	15700	32,48	33,35	2,90	6,90
PT171	Grande Lisboa	20300	32200	24,17	26,55	5,80	9,80
PT172	Península de Setúbal	11000	14300	21,31	24,10	6,90	9,80
PT181	Alentejo Litoral	16700	24400	34,72	36,51	8,60	10,50
PT182	Alto Alentejo	10400	15900	42,56	41,63	6,20	10,50
PT183	Alentejo Central	10900	16600	37,05	37,57	4,90	10,50
PT184	Baixo Alentejo	10800	18700	38,02	36,66	8,90	10,50
PT185	Lezíria do Tejo	12100	17400	30,59	32,21	5,70	10,50
PT200	Região Autónoma dos Açores	10200	18300	18,75	18,04	3,40	6,70
PT300	Região Autónoma da Madeira	11500	25800	19,43	18,78	2,80	7,60

Source: INTERCO, 2013

Labour productivity NUTS 2 level only Portugal have levels slightly below EU average (76-100). Big disparities among NUTS regions in Europe for GDP per capita and unemployment rates still remained, while disparities for the old age dependency ration is lowest.

Table 2. Territorial objectives 1. Strong local economies

	NUTS 2	Labour productivity in industry and services (2007)
PT11	Norte	60,02
PT15	Algarve	68,53
PT16	Centro (P)	65,39
PT17	Lisboa	89,37
PT18	Alentejo	75,53
PT20	Região Autónoma dos Açores	66,69
PT30	Região Autónoma da Madeira	86,39

Source: INTERCO, 2013

Territorial objective 2. Innovative territories

Two indicators are proposed as territorial cohesion indicators under this objective:

- Intramural R&D expenditures
- Employment rate 20-64

Table 3. Territorial objectives 2. Innovative territories

	NUTS2	Total intramural R&D expenditures (2007)	Employment rate 20-64 (percentage of population in working age (active population, 25-64 years) with tertiary education on total population aged 25-64.) (2010)
PT11	Norte	1,01	68,9
PT15	Algarve	0,37	72,8
PT16	Centro (P)	1,06	75,2
PT17	Lisboa	1,76	71,1
PT18	Alentejo	0,66	70,3
PT20	Região Autónoma dos Açores	0,43	70,8
PT30	Região Autónoma da Madeira	0,30	71,9

Source: INTERCO, 2013

Total intramural R&D expenditures measured in % of regional GDP. This indicator measures the future orientation of the regional economy in terms of investments in R&D. This support to innovation capacity is considered as a key driver of regional growth (INTERCO,2013). Percentages are generally lower in Portugal regions from 0.08 in South regions until 1.50% in small regional variations in the other part of Portugal.

Indicator **Employment rate** measures the actual participation of working age population in economic regional activities and in producing net added value. High employment rates reflect a vital quality of the labour markets which constitutes a favourable context for innovative territories. Employment rate 20-64 years This indicator measures the actual participation of working age population in economic regional activities and in producing net added value and values for Portugal are 66-84% . High employment rates reflect a vital quality of the labour markets which constitutes a favourable context for innovative territories.

On the contrary, convergence trends for one indicator do not necessarily imply the same development trends for another indicator for over the last decade. As in the whole of Europe this has led to convergence in higher education but to widening gaps in employment rates.

Territorial objective 3. Fair access to services, market and jobs

Potencial accessibility indicators measures market potential and locational advantages of a region. To benefit equally from these development potential is essential for territorial cohesion (INTERCO, 2013).

All NUTS 3 in Portugal have low values (24-40) of potencial accessibility by road as outside of core regions, accessibility by road decreases towards regions located outside the core. New transport infrastructures built between 2001 and 2006 significantly improved road networks improved relative position some regions but the remote regions are futher inaccessible.

Below average accessibility by rail is in all Portugal regions, but slice high in central part. Regions with major airport hubs and their surroundings clearly appear as those regions with highest accessibilities. In most cases these are the capital city regions, plus selected other agglomerations.

"Regions with less than 50% of European average should catch up faster." (INTERCO, 2013.)

Territorial objective 4. Inclusion and Quality of Life

Household income. This indicator measures the welfare of residence population in a region and reflect the level of poverty (INTERCO, 2013). The disposable household income is highest in 5,001-10,001. Regions in the north of Portugal with less than 10,000 EUR mean disposable household income should catch up faster.

Life expectancy. This indicator represents a proxy for the overall quality of the health-care system in a region. It tells us about healthiness of living environment and together with ageing index it allows to assess social policies projections and risk of exclusion (INTERCO, 2013). Life expectancy should at least be stable; decreases should be avoided, regions with expectancies of less than 75 years should catch up faster Portugal has values from 79-80 by regions.

Gender imbalances. This indicator measures differences in the gender composition of a society(INTERCO, 2013). Portugal have higher overrepresentations of women, exept in some south regions.

Table 4. Territorial objectives 3. Access to services

	NUTS 3 (1999)	Potential accessibility by air	Potential accessibility by rail	Potential accessibility by road
PT111	Minho-Lima	70	24	33
PT112	Cávado	76	27	39
PT113	Ave	90	28	40
PT114	Grande Porto	96	27	40
PT115	Tâmega	71	27	39
PT116	Entre Douro e Vouga	74	26	37
PT117	Douro	57	26	39
PT118	Alto Trás-os-Montes	30	26	33
PT121	Baixo Vouga	60	25	36
PT122	Baixo Mondego	42	24	36
PT123	Pinhal Litoral	44	22	32
PT124	Pinhal Interior Norte	38	23	35
PT125	Dâo-Lafôes	51	25	37
PT126	Pinhal Interior Sul	36	23	34
PT127	Serra da Estrela	40	26	34
PT128	Beira Interior Norte	35	26	35
PT129	Beira Interior Sul	29	24	33
PT12A	Cova da Beira	34	24	35
PT131	Oeste	59	21	30
PT132	Grande Lisboa	100	22	36
PT133	Península de Setúbal	78	21	34
PT134	Médio Tejo	45	23	33
PT135	Leziria do Tejo	69	22	35
PT141	Alentejo Litoral	49	16	24
PT142	Alto Alentejo	36	22	33
PT143	Alentejo Central	43	20	31
PT144	Baixo Alentejo	41	18	27
PT15	Algarve	85	16	25

Source: INTERCO, 2013

Differences in female-male unemployment rates. This indicator measures the female participation rate in the economy, and thus the overall quality of labour markets of an inclusive society (INTERCO, 2013). For all regions in Portugal values are 0 -5 for women what means unbalanced unemployment rates across sex.

Aging index. This indicator measures the balance of the age structure of the society (INTERCO, 2013). Portugal have unbalanced age structure and present overaging societies. Values by regions are from 1 in coastal regions until 4 in central and border regions.

Table 5. Territorial objectives 4. Quality of life

	NUTS 2	Average life expectancy at birth in years (2002)	Average life expectancy at birth in years (2008)	Disposable hosuehold income in Euro (1996)	Disposable household income in Euro (2007)	Difference in unemployment rates for women and men (1999)	Difference in unemployment rates for women and men (2010)
PT11	Norte	77,6	79,9	6578,4	9470,3	1,3	4,1
PT15	Algarve	76,5	78,5	8182,7	12082,0	2,8	0,1
PT16	Centro (P)	77,8	79,7	6900,7	10275,2	0,9	1,8
PT17	Lisboa	77,2	79,4	8999,7	14373,8	-0,2	0,1
PT18	Alentejo	77,0	78,5	7127,0	10586,4	6,1	3,8
PT20	Região Autónoma dos Açores	73,4	76,4	6503,6	10739,6	3,6	0,4
PT30	Região Autónoma da Madeira	73,3	75,3	6968,4	11350,9	2,0	-2,4

Source: INTERCO, 2013

Territorial objective 5. High ecological values and strong territorial capital

"These indicators are dedicated to measure the emissions and soil sealing resulting from human behaviour, as well as the general vulnerability of regions to climate change as outcome of human behaviour and adaptation capacities on the one hand, and their climatic, topographic, geological and biological conditions on the other hand" (INTERCO, 2013).

Number of days with ozone exceedances values are more than 140. Generally, the maximum daily 8-hour average is higher the farther south a region is located.

Soil sealing per capita. This indicator measures the degree of de-coupling of economic/demographic development and land take (INTERCO, 2013). In Portugal regons show the largest internal disparities and hot aspots of soil sealing per capita. Highes values are in central and south regions.

Territorial objective 6. Polycentric territorial development

In Portugal related population potentials is above average for coastal regions and below European average for continental.Net migration rate is positive. Cooperation intensity and degree is middle to high for Portugal.

Table 6. Territorial objectives 4. Quality of life

	NUTS 3	Ageing index (2004)	Ageing index (2009)	Gender imbalances (2000)	Gender imbalances (2009)
PT111	Minho-Lima	1,43	1,58	1,05	1,12
PT112	Cávado	0,65	0,76	1,07	1,07
PT113	Ave	0,67	0,82	1,06	1,05
PT114	Grande Porto	0,86	0,99	1,02	1,09
PT115	Tâmega	0,61	0,71	1,07	1,03
PT116	Entre Douro e Vouga	0,80	0,97	1,11	1,05
PT117	Douro	1,36	1,54	1,05	1,07
PT118	Alto Trás-os-Montes	1,84	2,16	1,08	1,07
PT150	Algarve	1,27	1,24	1,08	1,00
PT161	Baixo Vouga	1,00	1,15	1,05	1,06
PT162	Baixo Mondego	1,42	1,54	1,08	1,10
PT163	Pinhal Litoral	1,05	1,18	1,11	1,04
PT164	Pinhal Interior Norte	1,72	1,82	1,08	1,08
PT165	Dão-Lafões	1,33	1,48	1,09	1,08
PT166	Pinhal Interior Sul	2,71	2,93	1,04	1,11
PT167	Serra da Estrela	1,89	2,15	1,07	1,14
PT168	Beira Interior Norte	2,01	2,17	1,10	1,10
PT169	Beira Interior Sul	2,36	2,36	1,05	1,09
PT16A	Cova da Beira	1,63	1,82	1,00	1,07
PT16B	Oeste	1,18	1,24	1,07	1,04
PT16C	Médio Tejo	1,50	1,61	1,06	1,06
PT171	Grande Lisboa	1,08	1,11	1,03	1,09
PT172	Península de Setúbal	0,95	1,01	1,06	1,05
PT181	Alentejo Litoral	1,79	1,91	1,03	1,00
PT182	Alto Alentejo	2,02	2,08	1,12	1,07
PT183	Alentejo Central	1,72	1,78	1,05	1,05
PT184	Baixo Alentejo	1,77	1,76	1,07	1,03
PT185	Lezíria do Tejo	1,45	1,48	1,06	1,05
PT200	Região Autónoma dos Açores	0,62	0,66	1,02	1,02
PT300	Região Autónoma da Madeira	0,71	0,74	1,07	1,11

Source: INTERCO, 2013

Table 7. Territorial objectives 5. High ecological values

	NUTS3 2006	Number of days with ozone exceedances(2008)	Yearly particular matter (PM10) concentration (2009)	Soil sealing per inhabitant (2006)
PT111	Minho-Lima	8	12	257
PT112	Cávado	10	14	237
PT113	Ave	11	14	227
PT114	Grande Porto	13	18	172
PT115	Tâmega	11	14	171
PT116	Entre Douro e Vouga	9	18	153
PT117	Douro	2	11	202
PT118	Alto Trás-os-Montes	0	11	287
PT150	Algarve	11	12	355
PT161	Baixo Vouga	10	15	297
PT162	Baixo Mondego	7	13	264
PT163	Pinhal Litoral	7	13	369
PT164	Pinhal Interior Norte	4	13	230
PT165	Dâo-Lafôes	3	12	223
PT166	Pinhal Interior Sul	4	12	711
PT167	Serra da Estrela	0	11	221
PT168	Beira Interior Norte	0	11	432
PT169	Beira Interior Sul	0	12	559
PT16A	Cova da Beira	0	11	251
PT16B	Oeste	6	15	456
PT16C	Médio Tejo	5	14	409
PT171	Grande Lisboa	8	18	134
PT172	Península de Setúbal	9	20	260
PT181	Alentejo Litoral	8	18	703
PT182	Alto Alentejo	2	13	350
PT183	Alentejo Central	3	14	422
PT184	Baixo Alentejo	4	14	566
PT185	Lezíria do Tejo	6	16	509
	Região Autónoma dos Açores		0	198
	Região Autónoma da Madeira		4	177

Source: INTERCO, 2013

Table 8. Territorial objective 6. Polycentric territorial development

	NUTS 2 (2006)	Cooperation intensity (Number of INTERREG IIIc projects a region is involved in) 2008	Cooperation degree (Number of collaborating regions in INTEREG IIIc projects) 2008
PT1	Continente	116	
PT11	Norte	33	117
PT15	Algarve	7	21
PT16	Centro (P)	27	92
PT17	Lisboa	32	71
PT18	Alentejo	17	45
PT2	Região Autónoma dos Açores	3	41
PT20	Região Autónoma dos Açores	3	41
PT3	Região Autónoma da Madeira	5	17
PT30	Região Autónoma da Madeira	5	17

Source: INTERCO, 2013

CONCLUSION

Analyzing the selected indicators, it can be concluded that there are unbalanced territorial trends, socially and economically, with clearly defined centers and peripheries, although there has been EU financial support to this territory in the last few decades, related to cohesion policies. Grande Lisboa imposes itself with best values of all parameters while on the other side of the region border region Beira and Tamega have the worst results of all parameters. This indicates that the implementation of cohesion policies has not yielded positive results in Portugal. Peripherality seen as regional inequality and accessibility as key of inequality are typical in Portugal regions. In border and north regions are analysis at the NUTS3 level certainly gives a general picture. For a more detailed analysis it is necessary to apply parameters at the local level that are not always available. Moreover, Velaga et al. (2012) state that '*Transport accessibility is one of the crucial inequalities in peripheral territories*', as is the case of a large part of the Portuguese territory. Also, inadequate transport and accessibility possibilities are claimed to have a depopulating outcome in peripheral regions (Vulevic 2016; Castanho et al., 2017; Castanho et al., 2019; Vulevic et al., 2020; Naranjo, Castanho & Vulevic, 2021). Furthermore, its insouciance and inaccessibility may direct the social segregation of a portion of society (Vulevic et al. 2019; Naranjo, Castanho & Vulevic, 2021).

We can say that the Portuguese territory has failed to transform on the territorial level which is a major component of territorial cohesion. It did not have enough capacity to transform and achieve a „dynamic parameter of cohesion" (Faludi, 2006). Although state territoriality has been the category used at the national level and EU officials in their quest to design and implement territorial cohesion policy, there are analysts who believe that the emphasis could be on other levels of territoriality and allow reshuffling regardless of state sovereignty (Faludi, 2006, 2010, 2013, 2016c, 2016d, 2019; Sassen, 2013).

The present analysis indicates that further qualitative research is needed in the Portuguese peripheral regions to clarify the ways in which policy makers should react, and how much they are aware of the current situation and how they can change the current polarization.

Table 9. Territorial objectives 6. Polycentric development

	NUTS 3 (2006)	Net migration rate (Annual average net migration rate per thousand inhabitants for the time period 2001-2007)	Population potential within 50 km airline distance (2008)
PT111	Minho-Lima	5,45	200,49
PT112	Cávado	4,05	338,67
PT113	Ave	2,40	365,48
PT114	Grande Porto	1,68	364,81
PT115	Tâmega	0,04	356,14
PT116	Entre Douro e Vouga	4,88	306,61
PT117	Douro	-1,29	93,47
PT118	Alto Trás-os-Montes	2,77	40,40
PT150	Algarve	15,22	37,37
PT161	Baixo Vouga	6,40	162,63
PT162	Baixo Mondego	0,68	113,14
PT163	Pinhal Litoral	9,72	102,03
PT164	Pinhal Interior Norte	6,93	95,29
PT165	Dâo-Lafôes	6,88	108,75
PT166	Pinhal Interior Sul	0,64	53,97
PT167	Serra da Estrela	4,77	63,41
PT168	Beira Interior Norte	2,54	35,14
PT169	Beira Interior Sul	2,30	25,51
PT16A	Cova da Beira	2,18	43,30
PT16B	Oeste	10,43	216,93
PT16C	Médio Tejo	6,34	84,14
PT171	Grande Lisboa	3,36	318,03
PT172	Península de Setúbal	10,70	306,10
PT181	Alentejo Litoral	2,54	27,26
PT182	Alto Alentejo	0,50	29,33
PT183	Alentejo Central	3,88	28,94
PT184	Baixo Alentejo	1,93	19,21
PT185	Lezíria do Tejo	7,23	154,13
PT200	Região Autónoma dos Açores	2,10	0,00
PT300	Região Autónoma da Madeira	2,63	0,00

Source: INTERCO, 2013

Figure 1.
Source: Eurostat (2010), NUTS - Nomenclature of territorial units for statistics

REFERENCES

Castanho, R., Vulevic, A., Fernández, J.P., Naranjo, G., & Loures, L. (2017). Accessibility andconnectivity—Movement between cities, as a critical factor to achieve success on cross-border cooperation (CBC) projects. A European analysis. Sustain. *Cities Soc., 2*, 181–190.

Castanho, R. A. (2019). Identifying Processes of Smart Planning, Governance andManagement in European Border Cities. Learning from City-to-City Cooperation (C2C). *Sustainability*, *2019*(11), 5476. doi:10.3390u11195476

Castanho, R. A. (2020). The Relevance of Political Engagement and Transparency in Cross-Cooperation (CBC) Environments. Analyzing Border cities in Europe. Lex localis-Journal of Local Self-Government, 18(3), 487–502. doi:10.4335/18.3.487-502(2020)

Castanho, R. A., Vulevic, A., Naranjo Gómez, J., Cabezas, J., Fernández-Pozo, L., Loures, L., & Kurowska-Pysz, J. (2019). *Political Commitment and Transparency as a Critical Factor to Achieve Territorial Cohesion and Sustainable Growth. European Cross-Border Projects and Strategies*. Regional Science Policy and Practice.

CEC (Commission of European Communities). (2008). *Green Paper on territorial cohesion—turning territorial diversity into strength*. Communication from the Commission to the Council, the European Parliament, the Committee of the Regions and the European Economic and Social Committee. CEC.

ESPON. TRACC. (2013). *Transport accessibility at regional/local scale and patterns in Europe*. ESPON, Final Report - Indicator Factsheets.

ESPON. INTERCO. (2013). *Indicators of territorial cohesion - Final Report*. Available at: https://ise.unige.ch/isdd/IMG/pdf/interco_#nal-report_part-a_executive-summary.pdf

Eurostat. (2010). *NUTS - Nomenclature of territorial units for statistics*. http://epp.eurostat.ec.europa.eu/portal/page/portal/nuts_nomenclature/introduction

Faludi, A. (2004). Territorial Cohesion: Old (French) Wine in New Bottles? *Urban Studies (Edinburgh, Scotland)*, *41*(7), 1349–1363. doi:10.1080/0042098042000214833

Faludi, A. (2006). From European spatial developmentto territorial cohesion policy. *Regional Studies*, *40*(6), 667–678. doi:10.1080/00343400600868937

Faludi, A. (2009). *Territorial cohesion under the looking glass: Synthesis paper about the history of the concept and policy background to territorial cohesion*. Retrieved from https://www.researchgate. net/profile/Andreas_Faludi/contributions

Faludi, A. (2010). *Cohesion, Coherence, Cooperation:European Spatial Planning Coming of Age?* Routledge. doi:10.4324/9780203842324

Faludi, A. (2013). Territorial cohesion, territorialism, territoriality and soft planning: A critical review. Environment and Planning A. *Economy and Space*, *45*(6), 1302–1317. doi:10.1068/a45299

Faludi, A. (2016a). European integration and the Territorial – Administrative Complex. *Geografiska Annaler. Series B, Human Geography*, *98*(1), 71–80. doi:10.1111/geob.12090

Faludi, A. (2016b), The territoriality of Cohesion policy. In S. Piattoni & L. Polverari (Eds.), Handbook on Cohesion Policy in the EU (pp. 491-505). Edward Elgar. doi:10.4337/9781784715670.00048

Faludi, A. (2016c). EU territorial cohesion, a contradiction in terms. *Planning Theory & Practice*, *17*(2), 302–313. doi:10.1080/14649357.2016.1154657

Faludi, A. (2016d). The Poverty of Territorialism: Revisiting European Spatial Planning. *disP – The Planning Review, 52*(3), 73-81. doi:10.1080/02513625.2016.1235886

Faludi, A. (2019). New Horizons: Beyond territorialism. *Europa XXI*, *36*, 35–44. doi:10.7163/Eu21.2019.36.3

Medeiros, E. (2013a). Assessing territorial impacts of the EU Cohesion Policy: The Portuguese case. *European Planning Studies*, *22*(9), 1960–1988. doi:10.1080/09654313.2013.813910

Medeiros, E. (2013b). Euro-Meso-Macro:The new regions in Iberian and European Space. *Regional Studies*, *47*(8), 249–1266. doi:10.1080/00343404.2011.602336

Medeiros, E. (2014a). The Europeanization of Spatial Planning processes in Portugal within the EU Cohesion policy Strategies (1989-2013). *Geographyand Spatial Planning Journal*, *6*, 201–222. doi:10.17127/got/2014.6.012

Medeiros, E. (2016a). Territorial Cohesion: An EU concept. *European Journal of Spatial Development*, 60. Available from: http://www.nordregio.se/Global/EJSD/Refereed articles/refereed60.pdf

Medeiros, E. (2016b). EU Cohesion Policy in Spain. *Regional Studies.* doi:10.1080/00343404.2016.1187719

Naranjo, J. M., Castanho, R. A., & Vulevic, A. (2021). Analyzing Transportation Logistics and Infrastructure Sustainability in the Iberian Peninsula: The Case of Portugal Mainland. *European Planning Studies*, 1–23. Advance online publication. doi:10.1080/09654313.2021.2014789

Sassen, S. (2013). When territory deborders territoriality. *Territory, Politics, Governance*, *1*(1), 21–45. doi:10.1080/21622671.2013.769895

Velaga, N. R., Nelson, J. D., Wright, S. D., & Farrington, J. H. (2012). The Potential Role of Flexible Transport Services in Enhancing Rural Public Transport Provision. *Journal of Public Transportation*, *15*(1), 111–131. doi:10.5038/2375-0901.15.1.7

Vulevic, A. (2016). *Accessibility concepts and indicators in transportation strategic planning issues: theoreticalframework and literature review*. Logistics & Sustainable Transport. doi:10.1515/jlst-2016-0006

Vulevic, A., Castanho, R. A., Naranjo Gómez, J. M., Loures, L., Cabezas, J., Fernández-Pozo, L., & Martín Gallardo, J. (2020). Accessibility Dynamics and Regional Cross-Border Cooperation (CBC) Perspectives in the Portuguese—Spanish Borderland. *Sustainability*, *12*(5), 1978. doi:10.3390u12051978

Chapter 2
The Impact of European Union Projects on the Development of the Border Region

Malgorzata Waniek
WSB University, Poland

Rui Alexandre Castanho
https://orcid.org/0000-0003-1882-4801
WSB University, Poland

ABSTRACT

The current work aims to examine the efficient management of natural and cultural heritage resources for the sustainable development of cross-border tourism. Therefore, the authors of the chapter examine what measures should be taken to ensure that the cultural, natural, and residential space in the programme area can be used in a sustainable way in the long term. After a detailed analysis of the selected region, measures to protect and manage natural and cultural resources were considered. The research is expected to increase the expected number of visitors to supported natural and cultural heritage sites and supported monuments, to reduce resource consumption and pollution especially in tourist sites, and to combat the environmental impact of transport.

INTRODUCTION

This work deals with selected issues related to the genesis and functioning of the European Union in the context of experiences and challenges and economic and social opportunities and threats.

The first part deals with the history of European integration. The unification of European countries was a consequence of changes in the economic and political situation after the Second World War. European integration results from the Europeanisation of national policies and the division of common transnational competencies. Nevertheless, it is difficult to say what influenced the integration process to

DOI: 10.4018/978-1-6684-4548-8.ch002

a greater extent, whether it was economic or political change. Today, the situation in Europe is changing dynamically, and it is not easy to predict how it will develop in the coming years.

The subsequent section discusses the rules, funding programs, and possibilities of obtaining funds from the European Union. Over 76% of the EU budget is managed in partnership with national and regional authorities through a system of "*shared management*", mainly through 5 significant funds - the Structural & Investment Funds. Collectively, these help to implement the Europe 2020 strategy.

The final part describes the cooperation program, which addresses the most critical cross-border challenges of implementing the Europe 2020 strategy in the German-Austrian border region. Strategically, the program will promote better integration of research and development capacities on both sides of the border to exploit the region's economic potential fully. At the same time, it will address environmental issues and the growing challenges of climate change, particularly climate and flood risk management. Activities under these two main priorities will be complemented by efforts to promote closer cooperation on legal and administrative matters to create a fully integrated border region.

The texts presented in this publication show the effects of implementing cross-border projects in the EU environment using already existing knowledge and containing new solutions - taking into account quite narrowly specialized knowledge acquired during undergraduate studies and professional practice. The monograph is a voice in the discussion on Cross-Border Cooperation in the European Union at the level of its structures, finances, or security aspects; it describes, diagnoses, and interprets it both as a whole and in various fragments or layers.

The Origins of the European Union

The term "integration", derived from the Latin *integratio,* means the process of combining, merging smaller elements into larger wholes. An important issue is also the disappearance of the distinctness of the pars during this process, their merging and creating larger structures. (Marszałek-Kawa, 2003). Integration can also be defined as cooperation within the new institutions to achieve unity. It is based on a comprehensive structure of common bodies and their broad powers in the relation to the Member States. Integration can take place within the framework of various economic, political, social or institutional models, and the choice of a given model is made primarily by the states. Some researchers also note that, in parallel with the integration process, there is a (slow) process of diminishing the role of the state in favour of both international and local and regional authorities (Rodowski and Zendarowski, 2002).

Of the many existing definitions of integration, it is noteworthy that international economic integration " *(…) is a permanent institutional link and economic process which takes place within (…) relatively state-inspired (…) and is inspired by the countries (and by deepening the international division of labour and economic cooperation) leading to the maximisation of Member States national incomes and to the economic development of the entire economic community thus created*" (Rutkowska, 1998). This definition underlines the role of governments in stimulation economic integration processes and, as practice shows, the creation of macro-regional economic integration structures, this role is usually predominant (Rutkowska, 1998). Integration can cover different spheres of social life, although in the historical dimension it mainly concerns inter-national integration, which boils down to the political and/or economic integration of countries. Economic integration of countries is not always linked to their political integration. In both cases, accession countries are guided by motives (benefits) and interests. Any integration must be based on an established institutional and formal-legal basis. The main features of inter-state integration are the internalisation of economic and political life, the conscious activity of integration actors in forming and regulating lasting and deep integration links, and the processes of bringing nations together (Czachór and Graś, 2006; Vulevic *et al.*, 2020).

In connection with the process of formation of macro-regional integration groups in the aftermath of the Second World War, the development of the theory of integration accelerated also in this period. This is relatively young theory, which makes it understandable that the very definitions of integration are formulated by various authors in different ways and emphasize particular aspects of the phenomenon with different intensity (Rutkowska, 1998). The experience of the processes of creating macro - regional integration groups allowed for generalizations, which today are the basic elements of the integration theory. They relate both to the groupings themselves and to the effects of their creation and functioning on Member States and third countries.

The development of universal principles has been accompanied by the strengthening of regional integration groups. It turned out that the tasks that are today carried out by regional groups, universal international organisations were not able to fulfil. The basic needs for the security and development of the state and nations are not met on a regional rather than global scale and therefore more importance is nowadays attached to regional rather than global institutions. (Kuźniar, 1996). The reason for international regional integration are both global (mainly economic) and regional (mainly political and military) (Dynia, 2003). The source of regional integration is the ever-growing interests of countries and their needs, which require new and more effective forms of cooperation between them. The rapid growth in the number of regional economic groupings after the Second World War and the even more dynamic development of forms of regional cooperation after the end of the Cold War are proof of this thesis (Dynia, 2003).

The Marshal, on the basis of an analysis of the political premises of international integration, came to the following conclusion: "*Only the highly developed neighbouring countries can create effective integration and this happens when the changes taking place in the world economy, which are unfavourable to these countries, force such integration as a form of adjustment, or maybe even opposition to these changes, or actually their negative effect. International regional integration is therefore (...) a function of the changes taking place in the world economy*" (Marszałek, 1996). European integration can be considered from a historical, territorial and thematic perspective. Integration processes have emerged on the old continent throughout history and have been based on various concepts of "European integration", constantly enriched and enriched by new explorations and aspirations (Wojtaszczyk, 2006). Already in the ancient world one can see the beginnings of European unification. In Ancient Greece, merging city-states were formed, with the principle of respecting common solution (Malinowski 2002). Unity was also fostered by universalistic solutions, which were at the inks within the Roman Empire, such as Roman law or later early - medieval political concepts adopted in the Frankish state. Therefore, according to researchers, the contemporary concept of "Europe" has a broad meaning and is not limited to a typically geographical term. Europe was united by individual rulers as well as by culture, civilization, religion, universal law or the establishment of cities under German law, and a little later the Union of Hanseatic cities (Malinowski, 2002 Castanho, 2019; 2020).

For the European idea, there were no national or even continental borders. As a result, it is not possible today to draw the boundaries of Europe both in the area of common citizens and the impact of European culture. However, given the impact of globalization and population migration, Europe's knowledge is diminished by the interpenetration of other civilizations and the achievements of other continents.

The term European integration, which emerged after the Second World War, was initially used to describe the process taking place between the countries of Western Europe with the aim of creating an association, union or federation of states. (Małuszyńska and Gruchman, 2005). The global rationale for starting this process was economic conditions. The countries of Western Europe wanted to regain their lost (until now dominant) position in the world, both economic, political and military. Integration was a form of adaptation of the merging countries to changing, unfavourable external conditions. Elements that were conducive to cooperation between the countries of the region were: their neighbourhood, reduction of their own economic potential and the internal market, as well as the already developed various ties between them - mainly: economic, trade, technological but also political and cultural (Dynia, 2003).

The political premise of integration, on the other hand, is the desire to build a system that could protect Europe from armed conflicts and the new balance of power that emerged on the continent after World War II. The division of Europe into two opposing blocks began a period of the Cold War, leading to the separation of two integration processes.

On the one hand, it included countries in the Soviet Union's sphere of influence and on the other the democratic countries of Western Europe *"(…) for which integration was to become a way to create an effective system of international security and above all, to overstep the expansion of communism in the countries of Western Europe, to the nation of the USSR and the emerging block of socialist countries*" (Dynia, 2003).

The origins of European integration organisations are linked to three major events. The first was the European Congress, which took place in 1948 in The Hague (Hague Congress). He brought together many eminent representatives of political, economic, scientific and cultural life, and the central figure in his deliberations was British Prime Minister Winston Churchill (Rutkowska,1998). The congress from 7-10 May un the Gothic Hall of the Dutch Parliament, organised at the inspiration of the European Union of Federalists, brought together 750 deputies, politicians, cultural people and activists of pro-European movements from 19 European countries. The Congress was held in three session: the economic, political and social, (Czachór and Graś, 2006) as well as the Committee on Culture, chaired by the Spanish writer Salvador Madariaga, assisted by Denis de Rougemont of Swiss origin and a leading promoter of regionalisation. He was opposed to Charles de Gaulle's concept of building a Europe of sovereign nation states in the form of a "Europe of Fatherlands". All Congress participants were unanimous in their support of building a united Europe.

The results of the congress can be seen in the creation of the European Movement (1948), the creation of the Council of Europe (1949) the establishment of the European Cultural Centre (1950), of which Denis de Rougemont was appointed director. On the other hand, the establishment of a special Committee on Culture, acting on an equal footing with the other committees - i.e., the Political, Economic and Social Committee, is a testimony to the fact that creators of the vision of a future integrated Europe treated the importance of cultures for the unification process on an equal footing with political and economic issues. The director of the European Cultural Centre, Denis de Rougemont, commented on the situation as follows: "*True culture is not an ornament, a mere luxury, or a collection of specific values that do not concern people on the street. It is born out of the awareness of life, out of the eternal need to explore the meaning of existence, to increase the power of man over things. She created the size of Europe*" (Czachór and Graś, 2006).

The second major event in the history of European integration was the public announcement of the Schuman Declaration, which took place on 9 May 1950. It was then that Jean Monnet persuaded Robert Schuman, France's Minister of Foreign Affairs, to put forward the idea of placing the development of the French and West German steel and coal industries under the supervision of the High Authority, which later became the main body of the European Coal and Steel Community (ECSC). This was an impulse to start the European integration process. Jean Monnet believed that European integration and the building of a transnational federation structure could be gradually implemented from the integration of economic sectors to political integration. Its concept came to fruition when the ECSC was created (Czachór and Graś, 2006).

Finally, the last of the very important events is the Messina Conference held in June 1995 with the Foreign Ministers of France, Germany, Italy, Belgium, the Netherlands and Luxembourg. The decisions taken during the meeting triggered the process of broadening and strengthening the economic integration of the ECSC countries. The nuclear energy sector, and the expert report prepared under the chairmanship of Paul-Henri Spaak (Belgian Minister of Foreign Affairs) was then used as a programming basis in the negotiation preceding the conclusion of the Treaties of Rome (i.e. the Treaties establishing the European Economic Community and the European Atomic Energy Community) in March 1957.

The initiative to establish the ECSC was launched in 1950 by Jean Monnet, referring to the inter-war integration projects of his compatriot Aristide Brianda. His vision was to gradually subject the competences of individual countries to supernational control. In April 1950, Monnet presented to the French government a proposal to subordinate French and German coal and steel production to a joint administrative body, with the possibility of including other Western European countries in the initiative. The basis for sectoral integration within the ECSC was to be hard coal, which at the time satisfied as much as 90% of the Energy needs of the founding states "The Sixth".

At the meeting of the French Government on 9 May 1950, the Minister for Foreign Affairs – Robert Schuman, presented a plan for the creation of an ECSC (Schuman Plan), based on a draft drawn up on 15 April 1950 by Monnet and his friends: R. Reuter, P. Uri and S. Hirsch. He preceded his declaration with the following words: *"(…) There are no vain words, there is a bold act, a constitutional act. France has acted and the consequences can be enormous. We hope so. It has taken action primarily to promote peace. For peace to have a real Chance, you need Europe first. Exactly five years after the surrender of Germany, France is taking the first decisive act of European construction and incorporation Germany into it. This should completely change European conditions. This change will allow for other, previously impossible, joint action. Europe will be born of this, a strongly united and well-built Europe. A Europe where the standard of living will rise thanks to the clustering of production, the expansion of markets which will lead to lower prices (…)"* (Marszałek-Kawa, 2003). On the same day, during a press conference at the French Ministry of Foregin Affairs, Schuman proposed that French and German coal and steel production be controlled by a supranational High Authority.

According to Monnet and Schuman, the ECSC control of coal and steel production could be the first step towards a federation of the European economy in the future, and in the joint production of the two raw materials they saw a measure to eliminate the issues in Franco-West German relations (Łastawski, 2006) The Schuman Plan referred to the idea of European solidarity and contained proposals for the unification of European activities, complementing the functioning of the Council of Europe, the organisation for European Economic Cooperation and the Western Union.

Both the Schuman Plan and the Monnet Memorial were the direct reason for convening a conference in Paris in June 1950 with representatives of six governments: France, the Federal Republic of Germany (Germany), Italy, Belgium, the Netherlands and Luxembourg (Marszałek-Kawa, 2003). This meeting culminated in the signing of the Treaty establishing the European Coal and Steel Community on 18 April 1951 (Marszałek-Kawa, 2003). This Treaty, called the Paris Treaty, entered into force on 23 July 1952, establishing an organisation with the aim of creating a common market for the coal, iron and steel of the Member States, modernising their economies and having a coordinated impact on the prices of their products (Dynia, 2003). In accordance with Article 97 The Paris Treaty was concluded for a period of 50 years (and on 23 July 2002 the European Community - the former European Economic Community - EEC - took over the tasks carried out by that Community) (Marszałek-Kawa, 2003). The Paris Treaty consists of 100 articles, three annexes, three protocols and introductory provisions. According to its provisions, the founding states shall establish a Coal and Steel Community based on the common market, as well as common objectives and institutions (Article 1). The Community's objective is to contribute to economic development, employment and the raising of the standard of living in the Member States through the establishment of the common market (under the conditions set out in Article 4). This is to be done in accordance with the direction of development of the overall economy of these countries (Rutkowska,1998).

The creation of the ECSC was the first stage of European integration and initiated a new relationship with Western Germany, emerging from post-war isolation. The ECSC was the first international organisation to which governments delegated specific powers, until now reserved exclusively to the highest authorities of the state. Its structure is based on the principle of division of powers between the appointed bodies - i.e., the High Authority, the Council of Ministers, the Joint Assembly and the Court of Justice. The main supranational body has become the High Authority, whose prerogatives, like those of other bodies (especially the Court), can be described as a kind of novelty. The competences of the new institutions have in fact limited the sovereignty of the Member States (Dynia, 2003). According to the provisions of the Paris Treaty, the High Authority is an executive body whose main task was to establish and implement Community law. It consisted of eight members, elected by common accord of the governments of the Member States, and a ninth one, which was appointed by eight (which was to emphasise the supranational character of the Community) (Marszałek-Kawa, 2003).

Another body, the Special Council of Ministers, was tasked with taking action to harmonise the actions of the High Authority and the governments of the Member States.

The General Assembly was a political control body with 68 members (18 from France, Germany and Italy each and 24 from Belgium, the Netherlands and Luxembourg each). It had the power to pass a vote of no confidence in the High Authority, which resulted in its resignation. Members of the General Assembly came from a nomination made by the national parliaments. However, they could also be selected through direct elections. The last of the ECSC bodies, the Court of Justice, ensured respect for and application of the Treaty provisions. It also dealt with complaints that concerned the annulment of the decision or complaints that the High Authority had failed to act. The Court also ruled on cases for compensation for losses incurred in the implementation of the Treaty as a result of an error by a Community institution. It consisted of seven members from the nomination of governments, as well as one of the co-operations (Marszałek-Kawa, 2003).

The creation of the ECSC has brought numerous economic benefits. Awareness of the possibilities to finalise so many integration concepts and plans can be seen as a positive effect and an incentive for Member States to continue their efforts to deepen economic integration. It was also the first ever organisation of a supranational nature, as until it was established, it had not been possible to set up an international institution with the competences on the basis of which it would be able to take over the powers reserved to state authority (Marszałek-Kawa, 2003).

The European Funds Programs

Funding is the act of providing resources to finance a need, program, or project. While this is usually in the form of money, it can also take the form of effort or time from an organization or company. Generally, this word is used when a firm uses its internal reserves to satisfy its necessity for cash, while the term financing is used when the firm acquires capital from external sources. Funds can be allocated for either short-term or long-term purposes.

Funding is managed according to strict rules to ensure there is tight control over how funds are used and that the money is spent in a transparent, accountable manner. As a group, the 27 EU Commissioners have the ultimate political responsibility for ensuring that EU funds are spent properly. But because most of the funding is managed within the beneficiary countries, responsibility for conducting checks and annual audits lies with national governments. Over 76% of the EU budget is managed in partnership with national and regional authorities through a system of "shared management", largely through 5 big funds - the Structural & Investment Funds. Collectively, these help to implement the Europe 2020 strategy (Zamparini and Villani-Lubelli 2019):

I. European Regional Development Fund (ERDF) – regional and urban development,
II. European Social Fund (ESF) – social inclusion and good governance,
III. Cohesion Fund (CF) – economic convergence by less-developed regions,
IV. European Agricultural Fund for Rural Development (EAFRD),
V. European Maritime and Fisheries Fund (EMFF).

Other funds are managed directly by the EU. These are provided in the form of:

1. Grants for specific projects in relation to EU policies, usually following a public announcement known as a 'call for proposals'. Part of the funding comes from the EU, part from other sources.
2. Contracts issued by EU institutions to buy in services, goods or works they need for their operations – like studies, training, conference organisation, IT equipment. Contracts are awarded through calls for tender.

Recipients of tenders, grants, or development aid ear-marked for non-EU countries are published online. The EU provides funding for a broad range of projects and programmes covering areas such as:

I. regional & urban development,
II. employment & social inclusion,
III. agriculture & rural development,
IV. maritime & fisheries policies,
V. research & innovation,
VI. humanitarian aid.

European funds programms:

1. COSME Programme (GROW.A.1 - Publication date: 15/04/2015 "COSME - Leaflet EN") - programme for the Competitiveness of Small and Medium-Sized Enterprises (SMEs). It runs from 2014 to 2020 with a planned budget of €2.3bn.
2. InnovFin Programme (https://www.eib.org/de/products/blending/innovfin/index.htm) - offers suitable financing instruments for research and innovation - for small and medium-sized enterprises (SMEs) as well as for large companies or research institutes.
3. Creative Europe (Regulation (EU) No 1295/2013 of the European Parliament and of the Council of 11 December 2013 establishing the Creative Europe Programme (2014 to 2020) - is a European Union program for the cultural and creative sector for the period 2014-2020 amounting to EUR 1.46 billion.
4. The program was approved by the European Parliament on November 19, 2013 and adopted by the European Council on December 3, 2013. It entered into force on January 1, 2014.
5. Programme for Employment and Social Innovation (EaSI) (REGULATION (EU) No 1296/2013 OF THE EUROPEAN PARLIAMENT AND OF THE COUNCIL of 11 December 2013 s. 238) - is a financing instrument at EU level to promote a high level of quality and sustainable employment, guaranteeing adequate and decent social protection, combating social exclusion and poverty and improving working conditions.
6. European Structural and Investment Funds (ESI funds) (REGULATION (EU) No 1296/2013 OF THE EUROPEAN PARLIAMENT AND OF THE COUNCIL of 11 December 2013 s. 241) - are used to support development in a comprehensive way by investing for instance in businesses, research and development, infrastructure, employment and training, agriculture, forestry and fisheries development, with the overall objective to improve the quality of life of EU citizens.
7. European Investment Bank (http://www.eib.org) - Since its foundation in 1958, the European Investment Bank has had the task of "contributing to the balanced and smooth development of the internal market in the interest of the Union" with its own capital, and in this way operates "economic policy through lending".

Cross-Border Cooperation

European Cross-Border Cooperation, known as Interreg A, supports cooperation between NUTS III regions from at least two different Member States lying directly on the borders or adjacent to them. It aims to tackle common challenges identified jointly in the border regions and to exploit the untapped growth potential in border areas, while enhancing the cooperation process for the purposes of the overall harmonious development of the Union. (https://ec.europa.eu/regional_policy/en/policy/cooperation/european-territorial/cross-border/).

Cross Border Cooperation (CBC) is a key element of the EU policy towards its neighbours. It supports sustainable development along the EU's external borders, helps reducing differences in living standards and addressing common challenges across these borders. It was first recognised as such in the European Neighbourhood and Partnership Instrument (ENPI) regulation for the period 2007-2013. This was confirmed for the period 2014-2020 in the European Neighbourhood Instrument (https://ec.europa.eu/regional_policy/en/policy/what/glossary/e/european-neighbourhood-investment) (ENI) regulation adopted in March 2014.

CBC promotes cooperation between EU countries and neighbourhood countries sharing a land border or sea crossing (Castanho, 2017). Funding can also be provided for a programme between several EU and neighbourhood countries which, for example, are part of the same sea basin.

The project describe the Cooperation programme addresses the most important cross-border challenges which are linked to the implementation of Europe 2020 in the German-Austrian border region. Strategically, the programme will promote the better integration of research and development capacities on both sides of the border to exploit the full economic potential of the region. At the same time, it will address environmental issues and increasing challenges linked to climate change, in particular climate and flood risk management. Actions in these two main priorities will be complemented by efforts to foster closer co-operation in legal and administrative questions to create a fully integrated border region.

To achieve these strategic objectives, the programme aims at increasing the number of cross-border business clusters and research networks, to reduce the areas at risk of flooding through common structures, to promote the joint management of environmental protection areas, to intensify cross-border interaction in the tourism sector, and to further improve cross-border governance structures.

INTERREG Österreich – Bayern

The INTERREG V-A Program Austria - Bavaria 2014 – 2020 (https://www.interreg-bayaut.net/) is one of 60 cross-border funding programs within the European Territorial Cooperation (ETZ). The special feature of cross-border ETC programs is that the funding priorities are adapted to the needs and potential of the participating regions. The program has been developed in partnership between the regions participating in the program (Free State of Bavaria, Vorarlberg, Tyrol, Salzburg and Upper Austria) since the end of 2012. After negotiations with the Eüropean Commission, it was approved as the third cross-border funding program in Europe on December 3, 2014.

A total of around € 9 billion is available at European level for all ETZ programs 2014 - 2020 from funds from the European Fund for Regional Development, and around € 54.5 million for the INTERREG program Austria-Bavaria 2014 – 2020. The program is divided thematically into 3 priorities (funding priorities) with a total of 7 specific goals, to which each project must contribute in order to be eligible. This contribution is measured, among other things, using defined result and output indicators.

Priority 1: Broadening and improving cross-border research, development and innovation capacities (https://www.interregbayaut.net/programm/programminhalte/). The program region has a high potential of research and development institutions, especially in the central rooms. By promoting cross-border capacities, peripheral and near-border areas are to be strengthened in this regard. Central to this is the connection to program area-specific strengths, such as: information and communication technologies, efficient production technologies, mechatronics, automation, robotics, energy management, energy efficiency and renewable energies, life sciences (life sciences; especially biotechnology and systems biology), material sciences and materials technology, renewable raw materials (including biofuels), electromobility, logistics, tourism and leisure, as well as timber research and timber management.

In addition, compared to large companies, SMEs are still rather small in R&D activities. The potential of small and medium-sized enterprises can be promoted and their integration into the innovation system intensified, in particular through a cross-border technology transfer and the establishment of cross-border networks.

There are two specific goals.

Specific goal 1: Establishment and expansion of joint, cross-border research, development and innovation capacities in the university sector and at competence centers.

The existing research capacities are to be expanded further, structurally strengthened through cross-border cooperation, thereby ensuring the region's competitiveness in Europe and internationally.

Joint research capacities, infrastructures and facilities should provide impulses for the regional economy on site. Measures and activities have the aim of contributing to the development of program space-specific R&I excellence and thus making a contribution to employment in the R&D area. The result indicator accordingly measures the "expansion of the workforce in the area of research and development".

Exemplary measures:

1. Development and development of research facilities and structures that support the shared use of capacities in the R&D area;
2. Establishment of incubators as well as investments in the construction, further development and expansion of the range of competence centers;
3. Expansion and development of university and non-university institutions - i.e., user centers;
4. Supporting cooperation between research institutions to develop and strengthen region-specific strengths and research fields.

Target groups:

University and non-university research and technology facilities and their transfer points.

Output indicators:

1. Number of scientists working in improved research infrastructure,
2. Number of research institutions participating in cross-border, transnational or interregional research projects,
3. Number of lighthouse projects implemented.

Specific goal 2: Increase in company-related research, development and innovation activities by strengthening suitable support structures. Research and development in the corporate sector should be improved in the entire program area, but especially in peripheral regions. This should be achieved through suitable cooperation of the existing R&D resources on both sides of the border. The aim is to strengthen the competitiveness of SMEs in particular, but also of business and research as a whole. The "increase in the number of cross-border clusters and other networks" was defined as the result indicator.

Exemplary measures:

1. Development and implementation of (industry-specific) exchange and qualification programs for specialists for the purpose of exchanging experiences and establishing or consolidating cooperation structures; the exchange can take place both between companies and between companies and research institutions;
2. Mobilization of company-based and market-relevant innovations with the support of national / regional research institutions and technology centers, competence centers and science parks;

3. Promotion of innovations and applied research and development, which lead to the use of new processes and which make products and services marketable for new demand structures;
4. Innovative technical and organizational solutions to promote the environmental association;
5. Cooperative research projects between science and applied research with the increased involvement of SMEs in those areas that are part of regional and national strategies and of cross-border importance;
6. Development and further development of cross-border clusters to make regional strengths visible as well as support for companies in integrating into clusters;
7. Establishing industry-specific cooperation platforms and structures that help promote the transfer of innovation and technology;
8. Establishment of cross-border knowledge platforms / knowledge clusters on thematic focus topics (also in connection with the business clusters) by and for companies;
9. Development of common data / information systems and other ICT networks / platforms / programs for knowledge transfer.

Target groups:

Companies, especially SMEs, research institutions and research groups, cluster organizations, technology transfer points, legal interest groups, competence centers.

Output indicators:

1. Number of companies working with research institutions,
2. Number of companies involved in networking activities,
3. Number of companies / organizations that implement process or product innovations,
4. Number of companies receiving support.

Priority 2: Preserving and protecting the environment and promoting resource efficiency (https://www.interreg-bayaut.net/programm/programminhalte/).

The natural and cultural heritage is a central factor for the social and regional identity of a region. In the program region in particular, however, it is also of great importance for its economic development. This potential must be used in an ecologically and economically sustainable manner and thus increase the attractiveness of the region. The program area is also an ecologically highly sensitive region in view of the topographical conditions. Special care must therefore be taken to preserve sensitive natural areas, to deal with the consequences of climate change and to manage cross-border nature reserves.

There are three specific goals.

Specific goal 1: Valuation of natural and cultural heritage with a view to sustainable tourism development. Through cross-border measures, coordinated measures are to be taken to ensure in the long term that cultural, natural and living spaces in the program area can be used in a sustainable manner. Natural tourism and leisure activities are particularly important, as they contribute to resource and energy efficiency. It should be emphasized here:

1. the sustainable conservation and management of natural and cultural resources,

2. the reduction of resource consumption and pollution at tourist destinations as well,
3. combating the environmental impact of transport.

The planned measures are intended to contribute to a more even seasonal distribution of tourism flows. Accordingly, the result indicator measures the "increase in the share of overnight stays in the off-season in the total overnight stays of a year".

Exemplary measures:

1. Establishing and intensifying cross-border cooperations and networks with the aim of making cultural and natural heritage accessible – i.e., guest guidance centers, passage, restoration of cultural assets, nature experience centers, are just a few examples;
2. Development of (common) sustainable tourism concepts especially for national parks, nature parks and other protected areas;
3. Investing activities for the tourist valuation of protected areas and material as well as intangible cultural assets - i.e., themed paths, projects in the field of building culture, shelters, information centers, environmentally friendly development, among several other;
4. Networking of tourist activities in gentle and sustainable tourism, especially in the areas of nature, culture, health, nature experience, accessibility;
5. Preparatory studies and analyzes as a basis and preparation for the implementation of projects that have the theme of sustainable regional development in the field of tourism including cultural and natural heritage.

Target groups:

Municipalities / local authorities, associations, special-purpose associations, tourism institutions, planning authorities, educational institutions, cultural institutions.

Output indicators:

Increase in expected number of visitors to supported natural and cultural heritage sites and supported sights. Number of concepts developed in the field of sustainable tourism / protection of cultural and natural heritage.

Specific goal 2: Improvement of biodiversity through cross-border management structures of protected areas, biodiversity partnerships, species and soil protection projects.

The aim is to preserve biodiversity in the ecologically very sensitive program region or to create a favorable state - both in the Natura 2000 areas and beyond in other program-relevant protected areas. A key expected result is the establishment of common, cross-border management structures. This is primarily intended to ensure the preservation of biodiversity, but also to prevent conflicts of use. In particular, the in part scarce space resources in the Alpine program areas as well as the increase in permanently sealed areas require special care.

The added value of the cross-border initiatives to improve biodiversity is intended in particular to expand the area of the European protected areas in the program area. This is measured in the result indicator as "*increasing the area of the protected European protected areas*".

Exemplary measures:

1. Development of cross-border management and support structures for protected areas as well as development of related concepts for networking the protected areas;
2. Development of monitoring and information systems as well as specialized databases – i.e., area monitoring;
3. Activities to improve knowledge of ecosystems and ecosystem services in the region, measures to impart knowledge and raise awareness;
4. Habitat, soil and species conservation projects in cross-border areas with the aim of promoting the natural dynamics of biodiversity in protected areas, in suitable parts of the Alps as well as in biosphere reserves, in natural forest reserves and in intact, renatured and renaturalized areas;
5. Development of joint projects in the fields of water management, nature experience and protection, agriculture, forestry, including their implementation.

Target groups:

Public and private institutions, regional authorities (municipalities, states, districts, are jus a few examples), special-purpose associations, companies, interest groups.

Output indicators:

1. Number of protected areas under joint management,
2. Number of species and soil protection concepts implemented.

Specific goal 3: Protection of the habitat and biodiversity by building and expanding the green infrastructure.

Healthy ecosystems provide protection against extreme weather and natural disasters, which occur in the program area due to climate changes with increasing frequency. The increased susceptibility to floods, landslides, avalanches and storms in the program room is to be countered by gentle measures within the framework of the development of a green infrastructure.

A cross-border exchange of information and best practices in the face of common challenges is sought. The aim is to establish green infrastructures across borders in order to safeguard biological diversity and to protect endangered settlement areas from natural hazards. The result indicator mainly covers the area of flood protection. Accordingly, the contribution to the *"reduction of water stretches with potentially significant flood risk"* is measured.

Exemplary measures:

1. Common concepts and measures for water development, for structuring feeder and main waters and cross-border water retention measures;
2. Cross-border detection of hazards at local level - Creation of common regional hazard zone plans;
3. Development and establishment of a cross-border risk and natural hazard management;
4. Development of joint cross-border civil protection measures: Development of common cross-border strategies in the field of civil protection, development of disaster management systems, investment construction measures.

Target groups:

Public and private institutions, regional authorities – i.e., municipalities, states, districts; special-purpose associations, companies, interest groups.

Output indicators:

Number of protective measures in the area of green infrastructure.

Priority 3: Promotion of cooperation in legal and administrative matters and cooperation between citizens and institutions (https://www.interreg-bayaut.net/programm/programminhalte/).

Different political, administrative and legal framework conditions as well as different regional organizational structures still make cross-border cooperation difficult. The establishment and further development of cross-border cooperation structures and projects are seen as the heart of cross-border programs and are intended to help to remove these barriers.

There are two specific goals.

Specific goal 1: Strengthening cross-border structures to support regional governance and an instrument to promote cross-border initiatives and projects involving civil society.

Despite many years of experience in the development and implementation of cross-border projects, there are still differences at the border in all areas of life from kindergarten to school, vocational training and the world of work to health care and care for the elderly, two different systems collide at the state borders. In these everyday matters, people mostly only move within their own country, the border is not crossed for this.

The aim is to overcome this border by promoting cross-border people-to-people projects and other small projects and by supporting those organizations that coordinate and implement these projects on site. The result indicator measures the intensity of cross-border governance as "*broadening of the number of cross-border governance structures*".

Exemplary measures:

1. Creating platforms for exchanging experiences and coordinating cross-border initiatives;
2. Establishment of information, service and advice centers for cross-border cooperation for the general public;
3. Improving cross-border cooperation through information exchange and mutual coordination;
4. Establishment and development of support structures for cross-border projects.

Target groups:

Civil society institutions, educational, social and care institutions, rescue and civil protection organizations, transport institutions, advocacy groups.

Output indicators:

Number of project partners involved in small projects

Specific goal 2: Establishment and intensification of long-term and structural cross-border cooperation to strengthen social and economic integration and to remove administrative and legal barriers.

The aim is to support the establishment and strengthening of the structural and thus long-term cooperation of institutions beyond the project duration and funding period. Accordingly, the aim is to further reduce obstacles that exist due to the different administrative and legal circumstances, and thus to promote long-term growth in the region. The result indicator accordingly measures the "*increase in cooperation intensity*" using a cross-border survey.

Exemplary measures:

1. Support for long-term university cooperation / research / personnel cooperation, e.g. for the development and strengthening of region-specific strength and research fields;
2. Measures for area-saving settlement development (including pilot projects, implementation planning) based on existing concepts;
3. Cooperation between social and health organizations in the area of designing offers, qualifying employees, coordinating structures, and others;
4. Cooperation between emergency services and technical and rescue-related disaster protection, for example in the area of training, material management, logistics, adaptation of the legal framework, development of joint operational plans, implementation of exercises, among many others;
5. Cooperation of interest groups – i.e., Chamber of Commerce, Chamber of Commerce, Chamber of Crafts, unions, are just a few examples);
6. Cross-border cooperation between local and regional authorities to develop strategies, plans, processes and common services;
7. Cooperation in the field of education and lifelong learning through institutional cooperation for the purpose of coordinating formal and non-formal education / qualification (horizontal and vertical cooperation between various educational institutions, also in the field of youth work and adult education as well as environmental education, cooperation between schools and business) as well as development and implementation of pilot projects;
8. Measures to improve the environmental association – i.e. pilot measures in the area of passenger information, tariff information, best price calculations and ticketing in public transport, cross-border tariff models and transport associations, multimodal cross-border traffic plans).

Target groups:

Civil society institutions, educational, social and care institutions, rescue and disaster management organizations, transport institutions, advocacy groups.

Output indicators:

1. Number of institutional long-term cooperation partnerships,
2. Number of pilot actions implemented under long-term cooperation partnerships.

In terms of coordination, the following funding programmes are of importance for the programme area of the INTERREG Austria-Bavaria programme: the cross-border ETC programmes Austria-Czech Republic, Alpenrhein-Bodensee-Hochrhein, Italy-Austria and Bavaria-Czech Republic; the transnational programmes Central Europe, the Danube Region Programme and the Alpine Space Programme and the IWB programmes in Austria and Bavaria. Furthermore, there are links to the EAFRD programmes as well as the LIFE programme and Horizon 2020.

CONCLUSION

In economic and political terms, for a long time, European integration has been an attempt to develop a new political and legal order in Europe and to give the united continent a high position on the international scene. It started in the 1950s. In the 1990s, the integration process could not achieve its current level of development if it were carried out only through intergovernmental cooperation between the Member States. As a result, the originally economic-oriented process has increasingly extended to other areas, which today cover both foreign and domestic policy. Member States have been faced with the difficult task of choosing the political structure for the EU, but also with several crises, such as the European immigration crisis and the problem with the agreement on the reception and relocation of ears or the fight against terrorism and the need to protect European borders.

The Interreg V-A program Austria-Bavaria 2014-2020 was developed from 2012 in a broad partnership process and approved by the European Commission on December 3, 2014. The program has EUR 54.5 million in ERDF funding for 2014-2020. The program is divided thematically - without technical assistance - into three priority axes with a total of seven specific goals, as shown in the following figure with their respective ERDF equipment and their current rate of utilization.

In the meantime, 63 large projects (including 4 technical assistance projects) and 120 small projects have been approved, of which only a few have been completed. This means that the impact evaluation can primarily show the impact paths that have been embarked on and the impacts triggered thereby. The following figure shows the current distribution (as of January 2019) of the approved projects (left picture) and the ERDF funds used thereby (right picture) both per priority axis and per specific goal. The distribution corresponds well to the program's intention to allocate funds in SZ 1 in a concentrated manner to a few projects and, for example, in SZ 7 to promote a more significant number of projects at manageable volumes.

The overall picture is relatively balanced if you consider the regional participation in the previous program implementation. Of the 41 NUTS-3 regions in the program area (counties in Germany, districts in Austria), only eight have shown no participation. These are exclusively regions on the outer edge of the program area. The lead partners of the approved large-scale projects are also spread across many program regions.

One hundred forty different institutions (as of January 2019) are now involved in the large projects approved by all seven SZs to date, and almost 280 different institutions are involved in the small projects (as of April 2019). The diversity of the project partners is, therefore, quite high. The intensive involvement of universities and other public research institutions is interesting, in line with the program's focus on strengthening research and innovation capacities, and shows that these comparatively new target groups were easily reached. In addition to many municipalities, private companies and tourism organizations are particularly well represented. All in all, the sample reached by institutions suggests that the paradigm shift in the program from broad support to a largely thematically focused focus has been implemented successfully.

Following art. 56 section 3 VO (EU) No 1303/2013, an impact assessment to demonstrate the program's impact must be carried out during the 2014-2020 funding period. The assessment report on the "Impact Assessment of the INTERREG V-A Austria-Bavaria 2014-2020 Program" was prepared on behalf of the administrative body by the Institute of System Management and Public Management at the University of St. Gallen.

The results generally show a very positive picture of the implementation of the INTERREG V-A program to date and provide further optimization options to strengthen the program's effectiveness further. Therefore, the implementation and impact experience of the currently launched program should also be effectively used in the programming process of the INTERREG VI-A Bavaria-Austria 2021-2027 program.

In May 2018, the European Commission presented its vision to continue European Territorial Cooperation (ETC). On this basis, it could be considered to start programming work for the future Interreg VI-A Austria-Bavaria program. In addition to European specifications, the experience gained from the program also makes an essential contribution to programming to configure the future program efficiently and effectively. An impact assessment prepares these experiments. It is based on the program intervention logic with defined product and result indicators for each specific goal. The indicators are constantly documented.

REFERENCES

Barcik, J., & Wentkowska, A. (2014). *Prawo Unii Europejskiej, (European Union law).* C.H. Beck.

Betka, T. (2011). *Wspólnoty Europejskie na arenie międzynarodowej w okresie Zimnej Wojny* [European Communities in the international arena during the Cold War]. Analyses European Union.

Bokajło, W., & Dziubik, K. (Eds.). (2003). Unia Europejska. Leksykon integracji [European Union. Lexicon of Integration]. Europa Publishing House.

Borucki, M. (1996). *Historia powszechna (Common history) 1945-1995.* Mada.

Castanho, R. A. (2017). *Sustainable Urban Planning in Transboundary Areas: Analysis of Critical Factors for Territorial Success* [Doctoral Thesis]. University of Extremadura (UEx), Department of Vegetal Biology, Ecology, and Earth Sciences.

Castanho, R. A. (2019). Identifying Processes of Smart Planning, Governance and Management in European Border Cities. Learning from City-to-City Cooperation (C2C). *Sustainability, 2019*(11), 5476. doi:10.3390u11195476

Castanho, R.A. (2020). The Relevance of Political Engagement and Transparency in Cross-Border Cooperation (CBC) Environments. Analyzing Border cities in Europe. *Lex localis - Journal of Local Self-Government, 18*(3), 487 – 502. doi:10.4335/18.3.487-502(2020)

Czachór, Z., & Graś, A. (2006). Vademecum Europa od A do Z [Vademecum Europa from A to Z]. Vizja Press & It.

Dynia, E. (2003). *Integracja Europejska [European integration].* LexisNexis.

Gacka-Asiewicz, A. (2016). *Prawo Unii Europejskiej w pigułce* [European Union law in a nutshell]. Academic Press.

Jednolity Akt Europejski – podpisany 17 lutego 1986 r. (Single European Act - signed on 17 February 1986) (Luksemburg) oraz 28 lutego 1986 r. (Haga) (wszedł w życie 1 lipca 1987 r.) (Dz.U. z 2004 r. Nr 90, poz. 864/5).

Kowalski, J., & Ślusarczyk, Z. (2006). *Unia Europejska. Proces integracji europejskiej i zarys problematyki instytucjonalno-prawnej [European Union. The process of European integration and the outline of institutional and legal issues].* Polskie Wydawnictwo Prawnicze.

Kuligowski, R. (2014). *Unia Europejska po Traktacie z Lizbony, Wydanie II – poprawione i rozszerzone* [The European Union after the Treaty of Lisbon, Edition II - Revised and extended]. Academic Press.

Kuźniar, R. (1996). Porządek międzynarodowy –faza nierównowagi [International order - phase of imbalance, International affairs]. *Sprawy międzynarodowe, 2.*

Łastawski, K. (2006). *Historia integracji europejskiej [History of European integration].* Wydawnictwo Adam Marszałek.

Malinowski, L. (2002). Unia Europejska: geneza, współczesność [European Union: genesis, modernity]. SIG Management School.

Małuszyńska, E., & Gruchman, B. (Eds.). (2005). *Kompendium wiedzy o Unii Europejskiej [Compendium of knowledge on the European Union].* PWN.

Marszałek, A. (1996). Członkostwo krajów Europy Środkowo-Wschodniej we Wspólnotach Europejskich, a przesłanki międzynarodowej integracji [Membership of Central and Eastern European countries in the European Communities and the rationale for international integration]. In J. Fiszer (Ed.), Państwa narodowew euroatlantyckich strukturach, Kom [Nation states in Euro-Atlantic structures, Kom]. The Political Sciences Department of PAN.

Marszałek-Kawa, J. (2003). *Procesy integracyjne w Europie po II wojnie światowej [Integration processes in Europe after World War II].* Wydawnictwo Adam Marszałek.

Parzymies, S. (2012). *Unia Europejska od Maastricht do Lizbony. Polityczne aspekty aktywności [European Union from Maastricht to Lisbon. Political aspects of activity].* DIALOG.

Rodowski, A., & Zendarowski, R. (2002). *Integracja Europejska [European integration].* Wydawnictwo UKSW.

Ruszkowski, J., Górnicz, E., & Żurek, M. (2004). *Leksykon integracji europejskiej [Lexicon of European integration].* PWN.

Rutkowska, I. (1998). *Od wspólnot europejskich do Unii Europejskiej [From the European Communities to the European Union].* Wydawnictwo Zachodnio-pomorskiej Szkoły Biznesu w Szczecinie.

Traktat o funkcjonowaniu Unii Europejskiej – tekst skonsolidowany uwzględniający zmiany wprowadzone Traktatem z Lizbony (Dz.U.2004.90.864/2) (Treaty on the Functioning of the European Union - consolidated text taking into account the changes introduced by the Treaty of Lisbon).

Traktat o funkcjonowaniu Unii Europejskiej, (Treaty on the Functioning of the European Union) art. 13.

Traktat o Unii Europejskiej (tzw. Traktat z Maastricht) (The Treaty on European Union (the so-called,, Treaty on European Union"). (Maastricht Treaty)) – podpisany 7 lutego 1992 r. (Dz.U. z 2004 r. Nr 90, poz. 864/30).

Traktat o utworzeniu Europejskiej Wspólnoty Energii Atomowej. (Treaty establishing the European Atomic Energy Community).

Traktat o utworzeniu Europejskiej Wspólnoty Energii Atomowej (tzw. Traktat Rzymski) (Treaty establishing the European Atomic Energy Community) – podpisany 25 marca 1957 r. (Dz.U. z 2004 r. Nr 90, poz. 864/3); J. Kowalski, Z. Ślusarczyk, Unia Europejska (…) (European Union (…)).

Traktat o utworzeniu Europejskiej Wspólnoty Gospodarczej. (Treaty establishing the European Economic Community).

Traktat o utworzeniu Europejskiej Wspólnoty Gospodarczej (tzw. Traktat Rzymski) (The Treaty establishing the European Economic Community (the so-called,, Treaty of Lisbon"). Treaty of Rome)) – podpisany (signed) 25 marca 1957 r. (Dz.U. z 2004 r. Nr 90, poz. 864/2).

Traktat o utworzeniu Europejskiej Wspólnoty Węgla i Stali (tzw. Traktat Paryski) – podpisany 18 kwietnia 1951 r. (wszedł w życie 23 lipca 1952 r.; wygasł 23 lipca 2002 r.) (Treaty on the establishment of the European Coal and Steel Community (the so-called,, Treaty of Lisbon"). Treaty of Paris) - signed on 18 April 1951 (entered into force on 23 July 1952; expired on 23 July 2002).

Traktat ustanawiający jednolitą Radę i jednolitą Komisję Wspólnot Europejskich (tzw. traktat o fuzji) – podpisany 8 kwietnia 1965 r. (Treaty establishing a single Council and a single Commission of the European Communities (the so-called,, Single Council"). the Merger Treaty) - signed on 8 April 1965.

Traktat ustanawiający Konstytucję dla Europy (Treaty establishing a Constitution for Europe) (Dz.Urz. UE, C 310, 16 grudzień 2004).

Traktat z Amsterdamu zmieniający Traktat o Unii Europejskiej, traktaty ustanawiające Wspólnoty Europejskie i niektóre związane z nimi akty (tzw. Traktat Amsterdamski) (The Treaty of Amsterdam amending the Treaty on European Union, the Treaties establishing the European Communities and certain related acts (the so-called,, Amsterdam Treaty")) – podpisany 2 października 1997 r. (wszedł w życie 1 maja 1999 r.) (Dz.U. z 2004 r. Nr 90, poz. 864/31).

Traktat z Lizbony zmieniający Traktat o Unii Europejskiej i Traktat ustanawiający Wspólnotę Europejską podpisany w Lizbonie dnia 13 grudnia 2007 r. (Treaty of Lisbon amending the Treaty on European Union and the Treaty establishing the European Community signed in Lisbon on 13 December 2007.) (Dz.U. 2009 nr 203 poz. 1569).

Traktat z Nicei zmieniający Traktat o Unii Europejskiej, traktaty ustanawiające Wspólnoty Europejskie i niektóre związane z nimi akty (tzw. Traktat Nicejski) –przyjęty 11 grudnia 2000 r. i podpisany 26 lutego 2001 r. (wszedł w życie 1 lutego 2003 r.) (Dz.U. z 2004 r. Nr 90, poz. 864/32). (The Treaty of Nice amending the Treaty on European Union, the Treaties establishing the European Communities and certain related acts (the so-called,, Nice Treaty") Treaty of Nice) - Adopted on 11 December 2000 and signed on 26 February 2001 (entered into force on 1 February 2003) (OJ L 338, 30. 12. 2000, p. 1. U. of 2004, No. 90, item. 864/32).

Tyranowski, J. (1999). *Prawo Europejskie. Zagadnienia instytucjonalne z uwzględnieniem Traktatu Amsterdamskiego [European law. Institutional issues in the light of the Amsterdam Treaty]*. Properium.

Vulevic, A., Obradovic, V., Castanho, R.A., & Djordjevic, D. (2020). Cross-Border Cooperation (CBC) in a Multi-Level Governance System in Southeastern Europe Territories: How to Manage Territorial Governance Processes in Serbia-Romania Border Space. In *Cross-Border Cooperation (CBC) Strategies for Sustainable Development*. IGI Global. doi:10.4018/978-1-7998-2513-5.ch004

Wojtaszczyk, K. (Ed.). (2006). Integracja europejska [European integration]. Poltext.

Chapter 3
Addressing Critical Challenges of Accessibility and Mobility in Peripheral Areas Toward Sustainable Spatial Development and Infrastructure Provision

Arian Behradfar
University of Extremadura, Spain

Rui Alexandre Castanho
https://orcid.org/0000-0003-1882-4801
University of Extremadura, Spain

ABSTRACT

European borderlands characteristics are determined by the interaction of differences in geography along with various conflicts in demographic and socio-economic factors. These regions are more isolated than the other areas due to the barriers that restrain the opportunity to interact with cross-border areas. Most cases of border areas are isolated in their own geography. During the last decade, the European Union has designed and implemented several integration procedures to accelerate the transform process of borderlands from primarily peripheral regions into interesting spots for sustainability growth. Addressing accessibility and mobility issues in peripheral regions requires a comprehensive view of the factors and indicators at multiple scales and levels. Mobility-oriented accessibility planning approaches miss some inherent aspects related to spatial and socio-economic circumstances. When moving from theory to practice, analyzing how mobility is addressed by current strategies has resulted in emerging some challenges and inconsistencies in transport systems and infrastructure.

DOI: 10.4018/978-1-6684-4548-8.ch003

INTRODUCTION

Combination of specific geography, culture, language, socio-economic interactions and conflicts, and a wide range of demographic engagement has formed state cross-borders and borderlands (Kolossov & Scott, 2013). In fact, border regions are considered as a multi-dimensional materiality that consists of particular correlated features and characteristics. These areas also include all kind of natural and artificial limitations along with political interactions (Kmec et al., 2016). As a result, borderland and cross-border region studies have been considered as a significant branch of regional studies over the last two decades (Makkonen & Williams, 2016). In this regard, the concentration of extensive literature review is mainly being triggered by a series of high-impact developments accompanied by an increasing interest in internal and the cross-border isolation barriers (Martín-Uceda & Vicente Rufí, 2021).

Borderlands and cross-border areas are directly in the forward-facing of geographical cohesion of the EU countries and states. However, most of them commonly appear to be out of sorted developed and heterogeneous in the context of transport infrastructure and planning (Christodoulou & Christidis, 2018). These regions are principally noticeable by an overall weak connectivity at local to regional scales, low-density and unbalanced distribution of settlement, fragility in socio-economic structure, and debilitated interrelation among stakeholders, facilities, and services (Mercado, 2002). Specifically, environmental obstacles and inadequacy of capacity in potential regional planning have caused limited inter-regional accessibility for a remarkable part of the EU borderlands (Medeiros et al., 2021).

Nevertheless, the EU has designed and implemented a wide range of integration action plans to meet the demand of accessibility and mobility among border areas over the past decades (Pagu, 2002) (Gkoumas, 2021). These efforts have undergone some conceptual, moral and methodological paradigm shifts resulting in the proposal of different planning approaches that highlight the importance and combination of new values and principles such as regional planning, infrastructure, mobility, accessibility, and sustainability (Bibri et al., 2020). They mainly aim to achieve transformative regional accessibility and mobility toward improvement and opportunities and take place at various levels (Kammerlander et al., 2015).

In this regard, a growing number of border region studies is being triggered by a series of high-impact regional development plans, including the Brexit process in EU (De Ville & Siles-Brügge, 2019), Cross-Border Cooperation (CBC) (Castanho et al., 2017), Cohesion Policy of the European Commission (EC) for 2021–27 (Naldini, 2018), and The Trans-European Network (TEN-T) Policy (Casaca & Marlow, 2007). Furthermore, European Spatial Planning Observation Network (ESPON) (Vitale Brovarone & Cotella, 2020). European Groupings of Territorial Cooperation (EGTC) (Evrard & Engl, 2018), Cohesion Fund (Martín et al., 2019), Single European Transport Area (Wiesenthal et al., 2015), Connecting Europe Facility (CEF) (Vettorazzi, 2018), and Communication Boosting Growth and Cohesion (Hix et al., 2005) are the main supplementary European initiatives related to the improvement of cross-border accessibility and mobility.

The accessibility and mobility planning approaches recommend the development of specific sustainable regional accessibility and mobility plans. These plans require a comprehensive point of view at multi-scale including various impacts that have been taken into account in the context of accessibility and mobility. They also have to contain a long-term and sustainable vision for regional studies and regional science fields (Fatima, 2020). These approaches aim to identify and inverse socio-economic dynamics and environmental issues that have led to the marginalization of European peripheral regions (Basile & Cavallo, 2020). Funding of mentioned plans is identified as the main driver to design the required accomplishments. On the other hand, funding has been identified as the main challenge of accessibility and mobility issues in borderlands (Pécoud & de Guchteneire, 2006). Recently, there is more funding available from various EU and national sources for accessibility and mobility measures (Petrauskiene et al., 2020) (Pinna et al., 2017).

From the planning perspective, accessibility is defined as ease of reaching specific destination while mobility is defined as the ease of movement (Handy, 2005) (Levine & Garb, 2002). Both concepts are intertwined, however they are readily distinguishable. Based on the common statement, we could mention that the more one moves, the more opportunities he can access (Flamm & Kaufmann, 2006). Nevertheless, when destination is close, though great accessibility can be reached, even if mobility is limited or constrained. Moreover, when destination is distant, mobility could be high, notwithstanding the low level of accessibility (Ferreira et al., 2020). This declaration that link both accessibility and mobility together lacks in considering particular inherent characteristics of accessibility, mainly associated with spatial challenges in socio-economic and environmental issues (Bocarejo. Et al., 2012) In this context, borderlands features are very critical due to specific circumstances (Vulevic et al., 2020).

Inevitably, it is essential to accurately analyze the way accessibility and mobility in these regions are considered to address the theoretical and practical challenges along with existing restrictions and inconsistencies toward achievement of sustainable regional planning. In this regard, the main objective of this chapter is to investigate and evaluate the current strategies in accessibility and mobility in European borderlands and peripheral regions. Therefore, the accessibility and mobility challenges in these regions could be effectively reflected.

In fact, accessibility is often reduced to mobility focusing on the most instinctive explanation of the both concepts. It is a fact that the words accessibility and mobility appear with an increasing frequency in the political discourses which could be seen as a positive element in the bridging of the implementation gap in regional planning for peripheral and ultra-peripheral areas (Ferreira & Papa, 2020). However, it may refer to this issue that the current policy agenda has effectively assumed accessibility concerns. In particular, the existing strategies grasped the meaning of accessibility-based planning in the context of sustainability (Le Clercq et al., 2003). In fact, it could be that an un-reflected use of accessibility concepts and measures may create an illusion of change while actually continuing the current practice, focused on issues such as providing more infrastructure that are probably impossible (Silva & Larsson, 2018).

Even so, the contribution of this study is to address the potentialities and shortcomings of current policies and strategies with the main focus on accessibility and mobility improvement based on infrastructure provision throughout the lens of accessibility and mobility points of view. At the first step, the existing strategies regarding the accessibility and mobility measurements will be represented and fully discussed. Afterwards, the study concentrates on how the chosen peripheral areas considered by the existing strategies address mobility as an assessment tool for efficient accessibility measurements. Therefore, the proposed analysis addresses conflicts and critical concerns regarding hinterland strategies and approaches for accessibility and mobility evaluation, specifically in practice. Eventually, we represent some practical suggestions for upcoming studies and projects.

Accessibility and Mobility

In recent decades, literature on accessibility measures has grown and a number of accessibility measurement methods have been developed (Paiva et al., 2021) (Siddiq & Taylor, 2021). In spite of theoretical evidence and the believed added value for planning process, accessibility measures are scarcely put into practice to support mobility management (Silva & Larsson, 2018). In the field of planning, accessibility is defined as the ease with which activities can be reached from a certain place and with a certain system of transport (Liu et al., 2018). The concept generally takes the combination of spatial distribution of potential destinations, the ease of reaching each destination, as well as the magnitude, quality and character of the activities found there (Brussel et al., 2019). This implies that the concept of accessibility is determined by four interdependent components: a transport component (transport system), a land-use component (the magnitude, quality and characteristics of activities found at each destination), a temporal component (availability of activities) and an individual component (needs, abilities and opportunities of individuals) (Järv et al., 2018).

In general, accessibility measurement methods consist of both individual-oriented indicators, mostly characterized by time-space or cost constraints and location-oriented indicators, typically denoted by the possible conditions of specific destinations. In both explications, accessibility is determined by the spatial distribution of the destinations based on ease of reaching along with the features of the accomplishments come upon. In this way, accessibility could be assessed as the multiparty impact of priorities in activities and the spatial analysis of the mobility (Scott & Morency, 2012).

The literature on the subject of proposed context suggestions various theoretical studies and reviews regarding the state of the art of accessibility and mobility (Hoang Pham et al., 2021). In this regard, four main categories of measurement methods could be classified and introduced as infrastructure-based, location-based, individual-based, and utility-based approaches (Vale et al., 2015). The evaluation method that distinctly consider 4 cores of accessibility including transport infrastructure, land use planning, temporal dimension, and individual-oriented allocation is the major differentiation among the mentioned approaches (Melo, 2013).

Infrastructure-based approaches mostly focus on the efficiency of transport infrastructure and consider indicators such as level of service, that concern with the level of congestion and the mobility speed (Melo et al., 2013). Location-based approaches consist of level of accessibility based on spatial distribution of specific activities and services. In general, they operate at macro-level scales (Bryant Jr & Delamater, 2019). Individual-based approaches deal with accessibility measures from the social groups perspective that determine the amount of activities or services that an individual could access within a certain interval of time or distance (Levesque. Et al., 2013) Utility-based approaches that have been mostly taken into account in the field of accessibility studies reflect the profits that individuals could attain while accessing to a set of spatial distribution of services and activities (Zhang et al., 2011) (Qin et al., 2022).

At the same time, mobility comprises numerous features. From the planning view, the concept of mobility sets out criteria, goals and priorities for ability to travel that specifically determines the individual potential movement. It is a qualitative idiom principally relies on the capacity of specific groups of individuals to travel that essentially is contingent of the availableness of various transport modes (Docherty et al., 2018). Besides, it includes interpretations regarding the socio-economic issues, demographic features, and behavioral interactions. Regardless of the complex interconnection with the occurrence and situation of motion, mobility could represent the practical implementation of regional alteration and transformation that is a rather new line of research called transformative mobility (Pullano et al., 2020) (Behradfar & Mohammadi, 2020). As claimed by the sustainable mobility paradigm, it is essential to shift from conventional accessibility planning on the basis of transformative mobility to developing particular strategies that target sustainability-oriented approaches for improvement of accessibility. These strategies play a crucial role in the interpretation of planning purposes (Jordová & Brůhová-Foltýnová, 2021).

In many conventional professional parlances, accessibility and mobility are often used without clear distinction as co-equal regional transportation planning objectives. In contrast, the derived demand framework of borderlands in EU denotes that the former policies in regional planning inherently incorporate the current ones. In other hand, improved regional mobility in general, and developed local automobility in particular are valued in planning terms to the extent that these matters enhance accessibility (Medeiros, 2019). In fact, mobility advantages that is interpreted into accessibility fail to accurately represent transportation policies and planning defects. Although any mobility gain could lead to accessibility improvement, Sustainable accessibility consequently stand for techniques to enable regular communications with no direct dependency on long-distance, fast, and energy-intensive transportation (Butler et al., 2020) (Alyavina et al., 2020).

As we discussed, paradigm shift in planning from mobility-enhancing techniques toward accessibility-enhancing strategies is an essential issue. This procedure mostly consists of improve opportunities for services and activities by extending the range of availability in borderlands (Hersperger et al., 2018). In this regard, a wider regional planning context and innovative ways and means to reconceive transport infrastructure planning and integrate it with land-use planning and socio-economic planning science is required (Hern, 2013). Planning and policy-making for mobility management predominantly deal with resources that are free of any trepidation for ultimate resolution (Landau et al., 2013) Despite the fact that this process for accessibility principally relies on the denouement rather than the implemented strategies (Dhaliwal & Zheng, 2019).

The concept of ‘‘peripherality” is also used in the literature and is synonymous with inaccessibility to economic activity. Peripheral regions generally face poor economic performances and negative net-migration rates. However, it is not always true and some regions can have a high economic performance in spite of low levels of accessibility (Piekermann & Neubauer, 2002) (Vandenbulcke et al. 2009).

In peripheral and ultra-peripheral regions of EU, accessibility and mobility comprise of different characteristics and dimensions from regional contexts due to particular geography and conditions and need to be investigated in a different way (Gómez et al., 2021). Complex and mostly distant services and opportunities, spatio–temporal geographies circumstances, and finally mobility patterns and transport modes are extremely limited in compare with other regions (Castanho et al., 2021).

In these areas, distances are higher, trips are longer and mobility dependence is immoderate. As a result, mobility is a critical element in accessibility (Liu et al., 2021). Spatial mobility is more than connectedness between two distinct places. In fact, it is a constituting measurement of daily transportation (Kwan, 1998). Hence, integrated policies and planning process are required in borderlands more than, any other region that performing at different extent and scale based on various components of accessibility (Wong Villanueva et al., 2020). Then again, mobility is a strategic and critical element to modify and improve accessibility and regional transportation planning (Okraszewska et al., 2018).

MATERIALS AND METHODS

Peripheral Regions in the Strategies

Building the theoretical framework of the research, literature review has underlined the manifestation of various terms in scientific and political consideration such as marginal areas, peripheral areas, inner areas, inner peripheries, inland areas. All terms referred to unbalanced development dynamics and peripheralisation processes (Oppido & Ragozino, 2019).

The main focus regarding the concept of "peripherality", is the central discussion concerns levels of accessibility (Luukkonen, 2010). This spatial denotation has been increasingly associated with socio-economic performances, attached to distances from the main center of any specific activity (Lang, 2015). This conversion defines that areas that are not geographically remote or isolated could be peripheral in terms of dealing with socio-economic networks, underlining the role of lack of connectedness and weakness of interaction (Felzensztein et al., 2013). From a first screening of scientific literature and policy documents, the conditions of marginalization rely on the exclusion from physical, social, economic, institutional and cultural accessibility (Sevelius et al., 2020).

As stated by place-based approaches of regional development action plans promoted by EU cohesion policy and planning, peripheral regions are the territories placed at a remarkable distance from centers represent essential activities and services at various levels and scales (Gutiérrez et al. 2017). In general, they are Low-density and depopulating areas characterized by several interconnected factors. Specific geographic characteristics are likely to influence local demographic aspects which may determine the economic context and transport needs. The interplay of these factors is hardly captured by the existing definitions (Vitale Brovarone & Cotella, 2020). Regions fit in the same classification will demonstration substantial alterations in the context of demography, socio-economic and transport infrastructure. Firm groupings would lead to a failure in taking the exclusivity in low-density and depopulating areas (Prus & Sikora, 2021).

In general, existing strategies applied in peripheral areas focus on identify and implement action plans that could lead to an efficient reversion in marginalization development dynamics. These steps toward better regional planning act based on structural prerequisites for territorial development in order to achieve environmental preservation goals of territory, cultural resources promotion, and exploitation in potentiality of underutilized resources. They also deal with both fundamental issues for development such as health, education and mobility and reinforcement and activation of regional development procedures (Breathnach, 2013).

In this regard, accessibility to vital activities and services is envisioned as a critical prerequisite for imbalanced regional and territorial development in peripheral and ultra-peripheral regions. This context is mostly investigated at lower scale to accomplish better easily spread management of intraregional inequalities. Therefore, current strategies evidently state the theoretical relationship between accessibility to activities and services and development (Gómez et al., 2021). However, the inclusive and comprehensive implementation of existing to achieve sustainability objectives strategy is complicated and determined including multi-level, multi-sectors, multi-actor, and multi-funded procedure. Consequently, strategic governance regarding European borderlands dominance comprises various levels including EU to minor municipalities and stakeholders and actors as of a wide range of sectors (Kurowska-Pysz et al., 2018).

Furthermore, various local action plans are combined with current European strategies. As soon as regions are nominated for additional considerations, an extended procedure of explanation regarding the existing strategies initiates for any distinct project extent including national, regional, sub-regional and local sectors. Local sectors are at the central of the planning progression along with critical impact to main dynamics and trends in the strategic planning. Simultaneously, the principal management is so determinative regional planning is the turning point of fitting together between the others (Kişi, 2019).

Accessibility and Mobility Challenges in Strategies

As we discussed, accessibility planning is a key area of regional transport policy. Indeed, enhanced access to activities and services is a key cornerstone of integrated transport strategies. It properly reflects the multiple interactions among the spatial features of peripheral territories, socio-economic conflicts, and mobility of residents (Malý, 2019).

In peripheral regions, and even more so in remote and isolated territories, the perspective of ensuring suitable rates of accessibility is predominantly challenging. It is not only due to geomorphological and territorial structures, but also related to socio-economic and cultural dynamics that caused accelerating marginalization in compared with central territories representing activities, services and development opportunities. Moreover, various contexts have interweaved with accessibility, including spatial and socio-economic distribution of values, capacities, infrastructures, and environmental resources (Basile & Cavallo, 2020). In order to deal with the accessibility of peripheral territories that are considerably distant from activities, services, it is essential to form a transformation of perspective in current strategies. While peripheral regions comprise a great range of challenges that have been directed over and done with strategies by a growing policy-making and planning advanced at the international stages for the purpose of foster socio-economic and territorial cohesion guidelines (Loukaitou-Sideris, 2013).

Remoteness is certainly a crucial concern in investigating the state of affairs of the accessibility and mobility in peripheral and ultra-peripheral areas. The accessibility and mobility context delivers numerous procedural strategies to determining spatial accessibility and mobility (Stanković et al., 2018). Regardless of the various framework that can be taken into account for accessibility and mobility measurements, peripheral regions typically face lower spatial accessibility level and mobility management. Moreover, accessibility had better to be considered in both physical dimensions and in terms of socio-economic sector (Vitale Brovarone & Cotella, 2020).

In peripheral territories, transport is the main component of accessibility while the other three have been rarely taken into account due to the specific characteristics and unbalanced distance challenges (Martínez et al., 2021). On the other hand, peripheral territories basically embody inadequate transport infrastructure. Consequently, advanced resolutions in providing appropriate accessibility to activities and services is extremely essential (Lebo & Schelling, 2001). Lack of digital literacy is another challenge that these regions are mainly facing due to remoteness and other mentioned elements. Thus, enhancing the accessibility and mobility of these territories relies on performing based on quite a lot of forefronts (Vitale Brovarone & Cotella, 2020). These pole positions include all-purposed themes such as mobility, infrastructure planning, transport services along with aggregation of activities and services, knowledge, and flexible socio-economic interactions in all sectors of territories at regional scale (Rodrigue, 2020).

Throughout highlighting challenges of accessibility and mobility in peripheral regions, the need for developing efficient and comprehensive methods that could represent integrated policy-making and planning packages is critical. While the proposed challenges are specifically associated with issues in territorial levels and at regional scales, the solutions are mostly introduced and implemented based on exceeded local governance (Soliman, 2004). Subsequently, the practical employment of effective resolutions and achievement of accessibility enhancement stalwartly depend on interlacing the adoption of multi-level and multi-sector governance perception to accomplish sustainable accessibility planning and mobility management (Laurenti et al., 2019). The proposed strategy and related specifics and features have been taken into account for EU peripheral territories as a part of EPSON URRUC scheme (Britchenko et al., 2022). Main characteristics and final results of implementation of the mentioned action plan will be additionally introduced and discussed in the following sections of the chapter.

Evidence-Based Accessibility Practices

In general, current strategies of accessibility planning and mobility management in EU peripheral regions that have come to operational stage rely on three main themes: transport planning, transport infrastructure development, and transport services development (Porru et al., 2020). Transport planning process aims to achieve the promoted regional accessibility, confirmed ease of mobility for individuals and opportunities at territorial levels, and integrated perspective of strategic land-use planning (Vulevic et al., 2020). Transport infrastructure development mainly deal with expanding accessibility in both inside and outside of any region by promoting transport infrastructure. This objective is achieved by mean of efficient adaption in transport service access points, info-mobility, and fleet renewal process. Sustainable strategies in tourism management, priorities in services and activities, and conservation schemes of environment and natural resources are also a key element in this subject (Szymańska et al., 2021) Finally, transport services development consists of developing accessibility in the context of specific functional places for both systematic and non-systematic mobility, enhancing territorial cohesion of borderlands based on inner and outer reduced mobility, and improving opportunities and services for slow mobility on pedestrian linkages and shared-use paths (Lebo & Schelling, 2001).

SOLUTIONS AND RECOMMENDATIONS

Current strategies in the context of accessibility and mobility in EU peripheral and ultra-peripheral region represent a number of distinct points of view that are necessary to discuss and essential for prospective issues and policy-making process (Di Bella et al., 2019) (Vendemmia et al., 2021). Although the principles of accessibility planning and mobility management in these territories are in a suitable consistency with theoretical accessibility-based methodologies, a number of fragilities could possibly be appeared in addressing and achieving accessibility and mobility practical objectives (Lanzini & Stocchetti, 2021). Specifically, the extent that existing strategies have put into practice in the projected perspective to identify and fulfil mentioned challenging instabilities directly depends on the accessibility and mobility sections under examination and comparison of criticalities and actions envisioned by project area and addressed in previous guidelines and action plans (Kinigadner & Büttner, 2021).

Additionally, the analysis of existing strategies represents particular constituents for innovative policy-making and planning process (Tiboni et al., 2021) (Wulfhorst et al., 2017). The integration of multi-sector and multi-level of accessibility and mobility domination developments along with various perspectives provide collective action plans including top-down and bottom-up methodologies (Turienzo et al., 2008). While strategies are designed and implemented at regional scales, local sectors play the most significant role in this system. Highlighting local developments as the core axes of accessibility and mobility planning by reconsidering the spatial distribution of activities and services, current strategies involving all range of individuals in adaptation and contribution in provision of the objectives at all governance levels (Rodríguez-Pose, 2008). All the same, the examination of accessibility planning and mobility management in EU borderlands enable us to identify and analyze the possibility of challenges and discrepancies that represent exact ground-breaking assessments. This issue is largely remarkable in the time of technique assortment of current strategies operationalization in comparison with project area strategies (Vitale Brovarone & Cotella, 2020).

Specifically, in transport planning, accessibility is predestined by the strategies as the capability of system while mobility has been considered as a right or even an option (Larsson & Olsson, 2017). Moreover, in the guideline methodologies, accessibility is denoted as a comprehensive concept and related multiple scopes are acknowledged. In practical issues, it turns into a spatial theme that could be determined by proposed methods regardless of the unbalanced opportunities, exist as spatio-temporal dimensions (Petrie & Bevan, 2009). In the context of accessibility and mobility in peripheral and ultra-peripheral territories, multi-modality of transport system come with complex critical issues while public transport provision is associated with almost impenetrable inadequacies and restrictions (Couto et al., 2021).

Straight-forward determinable and transmissible indicators are essential to investigate and evaluate these regions by mean of available data sets. However, this issue sets up a vast theoretical limitation in the strategies, deserting the proposed multiple constituents and elements that determine accessibility and mobility measurements (Martínez et al., 2021). As we discussed, the implemented approaches of accessibility measures would result in diverse outcomes and miscarry in achieving specificities and disparities. Furthermore, nominated region, beneficial derived from resources and development prospects are also associated with the measurement tools of accessibility (Apparicio et al., 2017). The ease of mobility is an inherent characteristic of the accessibility planning strategies (Wang et al., 2012).

However, it is not certainly true that the improvement of accessibility is directly and proficiently depended on mobility enhancement (Yan et al., 2021). In this regard, peripheral and ultra-peripheral regions are assumed to require more mobility in comparison with other territories (Zolin et al., 2020). Moving from theoretical concerns to practical challenges, the accessibility and mobility strategies in these regions necessitate altered integrated actions that would not act as the same they do in other territories (Silva & Larsson, 2018).

An underlying issue that appeared in analysis of the strategies and further in practical observations is the path-dependence, or adherence to situations that are deeply existed in these territories as far as mobility is concerned (Bibri et al., 2020). For instance, dependence on the exterior of the territory is extremely embedded in the attitude of local sectors (Mayer et al., 2016). Nevertheless, dependencies are not considered as a critical concern in the list of challenging issues in perspectives and objectives of accessibility and mobility strategies in peripheral regions (Vitale Brovarone & Cotella, 2020).

In general, the outcomes of accessibility planning and mobility management strategies reflect critical concerns and describe the schedules for objectives as a constellation captivating perspectives on policy-making process (Bibri, 2019). The methodology of proposed strategies implicates a wide range of both actors and sectors in thematic working agendas and joint discussions (Sundberg, 2011).

Furthermore, EU borderlands strategies that have been taken into account to address the efficient incorporation among actions and services could not completely deal with challenges of essential engagements for mobility. They also approving a perspective mainly refers to mobility in the context of accessibility. This perspective includes guidelines and strategies, prescriptive practical framework, acquiescence logical theory, specific descriptions of schemes, restrictions and limitations, innovative and potential alternatives of mobility management and services (Marenin, 2010). Similarly, while aiming to proceed from the theoretical concepts to the applied environment of policies, incorporated perception clarifies through distinct schedules and engagements. In many cases, it has resulted in misplacing vision of prospective interactions and collaborations at regional scales (Petheram et al., 2010).

As we mentioned, a primary issue that has appeared in the context of accessibility and mobility analysis in peripheral territories is the path-dependence, or devotion to situations that are firmly established in these territories as far as mobility is concerned. Dependencies on the outermost regions are extremely embedded in the attitude of stakeholders. Car dependency is not considered as a prior nor critical issue in the list of challenging planning process. Even though, the deficiency in adequate public transport infrastructure and services and extraordinary proportions of commuting based on regular mobility patterns have not been considered as predominantly critical by the strategies (Hidayati et al., 2019)

Deficiency in transport infrastructure and services would results in slow mobility. This issue could be interpreted by mean of several techniques (van Eldijk et al., 2022). Inconsistency between the critical challenges, objectives and required accomplishments stands in the first place (Levine, 2020). When deficient mobility has not been considered as a significant criticality, tourism management is the second most declared objective as long as all territories plan to participate in transport infrastructure and services to achieve sustainable accessibility planning and mobility management (Coppola et al., 2020). The latter topic is principally engaged with tourism industry along with hardly any exceptions, where the required undertakings are the engagement of all groups of local to regional stakeholders to develop accessibility and multi-modality in transport system and moderate car dependencies in peripheral territories (Abdi, 2021).

The demand analysis and evaluation of the desire tools in accessibility planning and mobility management are envisioned by the existing strategies as essential introductory phases for strategic involvement in mobility in EU borderlands (Ye et al., 2021). In the description of the strategies by projective objectives, it relies on the background context. In-depth information of demand analysis in mobility management, comprehensive understanding and extensive-range planning play the key role to highlight multi-level and multi-sector issues peripheral territories (Harma & Gutiérrez, 2010). The increasing interest to invest in planning and management in the context of accessibility depends on various elements (Loucks & van Beek, 2017). In fact, substantial issues reduce the significance of long-term strategic planning in accessibility and mobility. In the other hand, regional administrations in peripheral territories are frequently familiar with focused efforts at micro-scale, in short-term objectives and probably irrelevant conflicts due to inadequate resource management or established availability and capacity (Tennøy et al., 2016). Nevertheless, short-vision strategies comprise of principals for long-term developing policies while the core strategies target the provision of demand receptive transport system of addressing accessibility challenges (Auvinen & Tuominen, 2014).

CONCLUSION

This chapter aims to represent an examination of existing strategies in the context of accessibility planning and mobility management challenges in EU peripheral and ultra-peripheral territories. This analysis reflects the extent and efficiency of current conditions and scopes extend innovative comprehensions and put them into practical projects and implementations. In this regard, the main themes and methodologies, availabilities and priorities, and capability and restrictions of accessibility and mobility in borderlands have been introduced and fully discussed to precisely address accessibility planning and mobility management.

The impact of mentioned implemented strategies could not be completely evaluated while they are being applied (Moullin et al., 2020). It is only possible to investigate the prospective potentials and existing challenges from the conceptual point of view including in-depth policy-making and planning in the context of accessibility and mobility in the first steps of implementation (Silva & Larsson, 2018). However, they have represented a proficient flow of regional development in EU policies while major ones mostly focus on urban and metropolitan regions that are different from borderlands (Olsson et al., 2021).

The acknowledgement of accessibility planning and mobility management in peripheral regions indicates that accessibility could be considered as an alternative way to improve mobility-focused procedures based on cooperative interactions and incorporations. In order to achieve sustainable accessibility and mobility in these territories, not only mobility improvement depends on individuals, but also enhancement in equipping of activities, services and opportunities to ensure strengthening regional services and communities (Vitale Brovarone & Cotella, 2020).

While we advance from theoretical concept to practical implementation, the current strategies in EU borderlands mainly deal with transport system and infrastructure development that do not comprise all components and aspects of accessibility (Gudmundsson & Höjer, 1996). Furthermore, most applied methods are approximate-based in the structure and intentions of the strategies that could only provide short-sighted solutions (Vitale Brovarone & Cotella, 2020). It is essential to fulfill this gap is as learnt from the previous experiences for supplementary policy action to enhance the accessibility level and mobility in peripheral territories in the context of EU cohesion policy.

REFERENCES

Abdi, M. H. (2021). What the newcomers to transit-oriented development is confronted with? Evidence from Iranian policy and planning. *Journal of Transport Geography, 92*.

Akse, R., Thomas, T., & Geurs, K. (2021). Mobility and accessibility paradigms in Dutch policies: An empirical analysis. *Journal of Transport and Land Use*, *14*(1), 1317–1340. doi:10.5198/jtlu.2021.2097

Alyavina, E., Nikitas, A., & Tchouamou Njoya, E. (2020). Mobility as a service and sustainable travel behaviour: A thematic analysis study. *Transportation Research Part F: Traffic Psychology and Behaviour*, *73*, 362–381. doi:10.1016/j.trf.2020.07.004

Apparicio, P., Gelb, J., Dubé, A.-S., Kingham, S., Gauvin, L., & Robitaille, É. (2017). The approaches to measuring the potential spatial access to urban health services revisited: Distance types and aggregation-error issues. *International Journal of Health Geographics*, *16*(1), 16. doi:10.118612942-017-0105-9 PMID:28830461

Auvinen, H., & Tuominen, A. (2014). Future transport systems: Long-term visions and socio-technical transitions. *European Transport Research Review*, *6*(3), 343–354. doi:10.100712544-014-0135-3

Basile, G., & Cavallo, A. (2020). Rural Identity, Authenticity, and Sustainability in Italian Inner Areas. *Sustainability*, *12*(3), 1272. doi:10.3390u12031272

Behradfar, A., & Mohammadi, S. (2020). Spatio-Temporal Understanding and Representation of Transformative Urban Mobility and Trip Patterns, A Review. *J. Civil Eng. Urban.*, *10*(4), 35–41. doi:10.29252cil.2020.jceu6

Bibri, S. E. (2019). On the sustainability of smart and smarter cities in the era of big data: An interdisciplinary and transdisciplinary literature review. *Journal of Big Data*, *6*(1), 25. doi:10.118640537-019-0182-7

Bibri, S. E., Krogstie, J., & Kärrholm, M. (2020). Compact city planning and development: Emerging practices and strategies for achieving the goals of sustainability. *Deve Built Environ*, *4*, 1–2. doi:10.1016/j.dibe.2020.100021

Bocarejo, S. (2012). Transport accessibility and social inequities: A tool for identification of mobility needs and evaluation of transport investments. *Journal of Transport Geography*, *24*(C), 142–154. doi:10.1016/j.jtrangeo.2011.12.004

Breathnach, P. (2013). Regional governance and regional development: Implications of the Action Programme for Effective Local Government. *Administration*, *61*(3), 51–73.

Britchenko, I., Zolotarov, V., Levchenko, Y., & Losonczi, P. (2022). Inter-Territorial Collaboration in the Context of Strengthening Its Economic Security. *IJCSNS International Journal of Computer Science and Network Security.*, *21*(12), 675–683.

Brussel, M., Zuidgeest, M., Pfeffer, K., & van Maarseveen, M. (2019). Access or Accessibility? A Critique of the Urban Transport SDG Indicator. *ISPRS International Journal of Geo-Information*, *8*(2), 67. doi:10.3390/ijgi8020067

Bryant, J. Jr, & Delamater, P. L. (2019). Examination of spatial accessibility at micro- and macro-levels using the enhanced two-step floating catchment area (E2SFCA) method. *Annals of GIS*, *25*(3), 219–229. doi:10.1080/19475683.2019.1641553

Butler, L., Yigitcanlar, T., & Paz, A. (2020). How Can Smart Mobility Innovations Alleviate Transportation Disadvantage? Assembling a Conceptual Framework through a Systematic Review. *Applied Sciences (Basel, Switzerland)*, *10*(18), 6306. doi:10.3390/app10186306

Casaca, A., & Marlow, P. (2007). The Impact of the Trans-European Transport Networks on the Development of Short Sea Shipping. *Maritime Economics & Logistics*, *9*(4), 302–323. doi:10.1057/palgrave.mel.9100184

Castanho, R., Loures, L., Cabezas, J., & Fernández-Pozo, L. (2017). Cross-Border Cooperation (CBC) in Southern Europe—An Iberian Case Study. The Eurocity Elvas-Badajoz. *Sustainability*, *9*(3), 360. doi:10.3390u9030360

Castanho, R. A., Naranjo Gómez, J. M., Vulevic, A., Behradfar, A., & Couto, G. (2021). Assessing Transportation Patterns in the Azores Archipelago. *Infrastructures*, *6*(1), 10. doi:10.3390/infrastructures6010010

Christodoulou, A., & Christidis, P. (2018). *Cross-border transport infrastructure in the EU: A methodology to assess the role of cross-border road networks, EUR 29565 EN*. Publications Office of the European Union.

Coppola, P., Carbone, A., Aveta, C., & Stangherlin, P. (2020). Assessing transport policies for tourist mobility based on accessibility indicators. *European Transport Research Review*, *12*(1), 56. doi:10.118612544-020-00444-4

Couto, G., Castanho, R. A., Santos, C., Pimentel, P., Sousa, Á., Faria, S., & Batista, M. da G. (2021). Guidelines for Tourism Sustainability in Ultra-Peripheral Territories: A Research Based on the Azores Region's Touristic Companies' Analysis. *Sustainability*, *13*(7), 3895. doi:10.3390u13073895

De Ville, F., & Siles-Brügge, G. (2019). The Impact of Brexit on EU Policies. *Politics and Governance*, *7*(3), 1–6. doi:10.17645/pag.v7i3.2129

Dhaliwal, A., & Zheng, G. (2019). Improving accessibility of EPR-insensitive tumor phenotypes using EPR-adaptive strategies: Designing a new perspective in nanomedicine delivery. *Theranostics*, *9*(26), 8091–8108. doi:10.7150/thno.37204 PMID:31754383

Di Bella, A., Petino, G., & Scrofani, L. (2019). The Etna macro-region between peripheralization and innovation: Towards a smart territorial system based on tourism. *Reg Sci Policy Pract*, *11*(3), 493–507. doi:10.1111/rsp3.12176

Docherty, I., Marsden, G., & Anable, J. (2018). The governance of smart mobility. *Transportation Research Part A, Policy and Practice*, *115*, 114–125. doi:10.1016/j.tra.2017.09.012

Evrard, E., & Engl, A. (2018). Taking Stock of the European Grouping of Territorial Cooperation (EGTC): From Policy Formulation to Policy Implementation. In E. Medeiros (Ed.), *European Territorial Cooperation. The Urban Book Series*. Springer. doi:10.1007/978-3-319-74887-0_11

Fatima, K., Moridpour, S., De Gruyter, C., & Saghapour, T. (2020). Elderly Sustainable Mobility: Scientific Paper Review. *Sustainability*, *12*(18), 7319. doi:10.3390u12187319

Felzensztein, C., Gimmon, E., & Aqueveque, C. (2013). Entrepreneurship at the Periphery: Exploring Framework Conditions in Core and Peripheral Locations. *Entrepreneurship Theory and Practice*, *37*(4), 815–835. doi:10.1111/j.1540-6520.2012.00515.x

Ferreira, A., & Papa, E. (2020). Re-enacting the mobility versus accessibility debate: Moving towards collaborative synergies among experts. *Case Studies on Transport Policy.*, *8*(3), 1002–1009. doi:10.1016/j.cstp.2020.04.006

Ferreira, A. F., Akasaka, Y., Greiner de Oliveira Pinheiro, M., & Chang, S. K. J. (2020). Information as the First Attribute of Accessibility: A Method for Assessing the Information Provided by Urban Rail Systems to Tourists with Reduced Mobility. *Sustainability*, *12*(23), 10185. doi:10.3390u122310185

Flamm, M., & Kaufmann, V. (2006). Operationalising the Concept of Motility: A Qualitative Study. *Mobilities*, *1*(2), 167–189. doi:10.1080/17450100600726563

Gkoumas, K., dos Santos, F. L. M., Stepniak, M., & Pekar, F. (2021). Research and Innovation Supporting the European Sustainable and Smart Mobility Strategy: A Technology Perspective from Recent European Union Projects. *Applied Sciences (Basel, Switzerland)*, *11*(24), 1981. doi:10.3390/app112411981

Gómez, J. M. N., Vulevic, A., Couto, G., & Alexandre Castanho, R. (2021). Accessibility in European Peripheral Territories: Analyzing the Portuguese Mainland Connectivity Patterns from 1985 to 2020. *Infrastructures*, *6*(6), 92. doi:10.3390/infrastructures6060092

Gudmundsson, H., & Höjer, M. (1996). Sustainable Development Principles and Their Implications for Transport. *Ecological Economics*, *19*(3), 269–282. doi:10.1016/S0921-8009(96)00045-6

Gutiérrez, J., García-Palomares, J. C., Romanillos, G., & Salas-Olmedo, M. H. (2017). The eruption of Airbnb in tourist cities: Comparing spatial patterns of hotels and peer-to-peer accommodation in Barcelona. *Tourism Management*, *62*, 278–291. doi:10.1016/j.tourman.2017.05.003

Handy, S. (2005). Planning for Accessibility: In Theory and in Practice. In Access to Destinations. Emerald Group Publishing Limited. doi:10.1108/9780080460550-007

Harma, S., & Gutiérrez, J. A. (2010). An evaluation framework for viable business models for m-commerce in the information technology sector. *Electronic Markets*, *20*(1), 33–52. doi:10.100712525-010-0028-9

Hern, J. (2013). Urban landscape sustainability and resilience: The promise and challenges of integrating ecology with urban planning and design. *Landscape Ecology*, *28*(6), 1203–1212. doi:10.100710980-012-9799-z

Hersperger, A. M., Oliveira, E., Pagliarin, S., Palka, G., Verburg, P., Bolliger, J., & Grădinaru, S. (2018). Urban land-use change: The role of strategic spatial planning. *Global Environmental Change*, *51*, 32–42. doi:10.1016/j.gloenvcha.2018.05.001

Hidayati, I., Yamu, C., & Tan, W. (2019). The Emergence of Mobility Inequality in Greater Jakarta, Indonesia: A Socio-Spatial Analysis of Path Dependencies in Transport–Land Use Policies. *Sustainability*, *11*(18), 5115. doi:10.3390u11185115

Hix, S., Noury, A., & Roland, G. (2005). Power to the Parties: Cohesion and Competition in the European Parliament, 1979–2001. *British Journal of Political Science*, *35*(2), 209–234. doi:10.1017/S0007123405000128

Järv, O., Tenkanen, H., Salonen, M., Ahas, R., & Toivonen, T. (2018). Dynamic cities: Location-based accessibility modelling as a function of time. *Applied Geography (Sevenoaks, England)*, *95*, 101–110. doi:10.1016/j.apgeog.2018.04.009

Jean-Paul Rodrigue. (2020). *The Geography of Transport Systems. FIFTH EDITION*. Routledge.

Jordová, R., & Brůhová-Foltýnová, H. (2021). Rise of a New Sustainable Urban Mobility Planning Paradigm in Local Governance: Does the SUMP Make a Difference? *Sustainability*, *13*(11), 5950. doi:10.3390u13115950

Kammerlander, M., Schanes, K., Hartwig, F., Jäger, J., Omann, I., & O'Keeffe, M. (2015). A resource-efficient and sufficient future mobility system for improved well-being in Europe. *Eur J Futures Res*, *3*(1), 8. doi:10.100740309-015-0065-x

Kinigadner, J., & Büttner, B. (2021). How accessibility instruments contribute to a low carbon mobility transition: Lessons from planning practice in the Munich region. *Transport Policy*, *111*, 157–167. doi:10.1016/j.tranpol.2021.07.019

Kişi, N. (2019). A Strategic Approach to Sustainable Tourism Development Using the A'WOT Hybrid Method: A Case Study of Zonguldak, Turkey. *Sustainability*, *11*(4), 964. doi:10.3390u11040964

Kmec, S., Hesse, M., Wille, C., & Reckinger, R. (Eds.). (2016). *Spaces and Identities in Border Regions: Politics - Media – Subjects*. Transcript.

Kolossov, V., & Scott, J. (2013). Selected conceptual issues in border studies. *Belgeo, 1*.

Kurowska-Pysz, J., Castanho, R., & Loures, L. (2018). Sustainable Planning of Cross-Border Cooperation: A Strategy for Alliances in Border Cities. *Sustainability, 10*(5), 1416. doi:10.3390u10051416

Kwan, M.-P. (1998). Space-Time and Integral Measures of Individual Accessibility: A Comparative Analysis Using a Point-based Framework. *Geographical Analysis, 30*(3), 191–216. doi:10.1111/j.1538-4632.1998.tb00396.x

Landau, L. B., Segatti, A., & Misago, J. P. (2013). Planning and participation in cities that move: Identifying obstacles to municipal mobility management. Public Admin. *Dev., 33*, 113–124.

Lang, T. (2015). Socio-economic and political responses to regional polarisation and socio-spatial peripheralisation in Central and Eastern Europe: A research agenda. *Hungarian Geographical Bulletin, 64*(3), 171–185. doi:10.15201/hungeobull.64.3.2

Lanzini, P., & Stocchetti, A. (2021). From techno-centrism to socio-centrism: The evolution of principles for urban sustainable mobility. *International Journal of Sustainable Transportation, 15*(11), 815–825. doi:10.1080/15568318.2020.1827315

Larsson, A., & Olsson, J. (2017). Potentials and limitations for the use of accessibility measures for national transport policy goals in freight transport and logistics: Evidence from Västra Götaland County, Sweden. *REGION, 4*(1), 71–92. doi:10.18335/region.v4i1.172

Laurenti, R., Singh, J., Cotrim, J. M., Toni, M., & Sinha, R. (2019). Characterizing the Sharing Economy State of the Research: A Systematic Map. *Sustainability, 11*(20), 5729. doi:10.3390u11205729

Le Clercq, F., & Bertolini, L. (2003). Achieving Sustainable Accessibility: An Evaluation of Policy Measures theAmsterdam in Area. *Built Environment, 29*(1), 36–47. doi:10.2148/benv.29.1.36.53949

Lebo, J., & Schelling, D. (2001). Design and Appraisal of Rural Transport Infrastructure: Ensuring Basic Access for Rural Communities. World Bank Technical Paper; No. 496. Washington, DC: World Bank.

Levesque, J.F., Harris, M.F., & Russell, G. (2013). Patient-centered access to health care: conceptualising access at the interface of health systems and populations. *Int J Equity Health, 11*, 12-18.

Levine, J. (2020). A century of evolution of the accessibility concept. *Transportation Research Part D: Transport and Environment*.

Levine, J., & Garb, Y. (2002). Congestion pricing's conditional promise: Promotion of accessibility or mobility? *Transport Policy, Elsevier, 9*(3), 179–188. doi:10.1016/S0967-070X(02)00007-0

Liu, Q., Liu, Y., Zhang, C., An, Z., & Zhao, P. (2021). Elderly mobility during the COVID-19 pandemic: A qualitative exploration in Kunming, China. *Journal of Transport Geography, 96*(C).

Liu, R., Chen, Y., Wu, J., Xu, T., Gao, L., & Zhao, X. (2018). Mapping spatial accessibility of public transportation network in an urban area – A case study of Shanghai Hongqiao Transportation Hub. *Transportation Research Part D, Transport and Environment, 59*, 478–495. doi:10.1016/j.trd.2018.01.003

Liu, Y., Cao, X., & Li, T. (2020). Influence of Accessibility on Land Use and Landscape Pattern Based on Mapping Knowledge Domains: Review and Implications. *Journal of Advanced Transportation, 2020*, 7985719–12.

Loucks, D. P., & van Beek, E. (2017). Water Resources Planning and Management: An Overview. In *Water Resource Systems Planning and Management*. Springer. doi:10.1007/978-3-319-44234-1_1

Loukaitou-Sideris, A. (2013). New Rail Hubs along High-Speed Rail Corridor in California: Urban Design Challenges. *Transportation Research Record: Journal of the Transportation Research Board, 2350*(1), 1–8. doi:10.3141/2350-01

Luukkonen, J. (2010). Territorial cohesion policy in the light of peripherality. *The Town Planning Review, 81*(4), 445–466. doi:10.3828/tpr.2010.12

Makkonen, T., & Williams, A. M. (2016). Border region studies: The structure of an "offbeat" field of regional studies. *Regional Studies, Regional Science, 3*(1), 355–367. doi:10.1080/21681376.2016.1209982

Malý, J. (2019). Polycentric Urban Systems and Territorial Cohesion. In E. Medeiros (Ed.), *Territorial Cohesion. The Urban Book Series*. Springer. doi:10.1007/978-3-030-03386-6_4

Marenin, O. (2010). *Challenges for Integrated Border Management in the European Union*. Geneva Centre for the Democratic Control of Armed Forces Occasional Paper No. 17: 28-45.

Martín, R., Martín, M., Fernández, S., Mejía, Z., & Bedriñana, A. (2019). A Spatial Analysis of the Achievements, in Terms of Regional Development, accomplished by the Initial EU-Member Cohesion Fund Beneficiaries Using a Synthetic Indicator. *Sustainability, 11*(8), 2343. doi:10.3390u11082343

Martín-Uceda, J., & Vicente Rufí, J. (2021). Territorial Development and Cross-Border Cooperation: A Review of the Consequences of European INTERREG Policies on the Spanish–French Border (2007–2020). *Sustainability, 13*(21), 12017. doi:10.3390u132112017

Martínez, M., Rojas, C., Condeço-Melhorado, A., & Carrasco, J. A. (2021). Accessibility Indicators for the Geographical Assessment of Transport Planning in a Latin American Metropolitan Area. *Geographies, 1*(2), 124–142. doi:10.3390/geographies1020008

Martínez, M., Rojas, C., Condeço-Melhorado, A., & Carrasco, J. A. (2021). Accessibility Indicators for the Geographical Assessment of Transport Planning in a Latin American Metropolitan Area. *Geographies, 1*(2), 124–142. doi:10.3390/geographies1020008

Mayer, H., Habersetzer, A., & Meili, R. (2016). Rural–Urban Linkages and Sustainable Regional Development: The Role of Entrepreneurs in Linking Peripheries and Centers. *Sustainability, 8*(8), 745. doi:10.3390u8080745

Medeiros, E. (2019). Cross-border transports and cross-border mobility in EU border regions. *Case Studies on Transport Policy., 7*(1), 1–12. doi:10.1016/j.cstp.2018.11.001

Medeiros, E., Ferreira, R., Boijmans, P., Verschelde, N., Spisiak, R., Skoniezki, P., Dietachmair, J., Hurnaus, K., Ebster, M., Madsen, S., Ballaguy, R.-L., Volponi, E., Isinger, E., Voiry, P., Markl-Hummel, L., Harster, P., Sippel, L., Nolte, J., Maarfield, S., ... Berzi, M. (2021). Boosting cross-border regions through better cross-border transport services. The European case. *Case Stud Transp Policy*, *9*(1), 291–301. doi:10.1016/j.cstp.2021.01.006

Melo, P. C., Graham, D. J., & Brage-Ardao, R. (2013). The productivity of transport infrastructure investment: A meta-analysis of empirical evidence. *Regional Science and Urban Economics*, *43*(5), 695–706. doi:10.1016/j.regsciurbeco.2013.05.002

Melo, P. C., Graham, D. J., & Brage-ardao, R. (2013). Theproductivity of transport infrastructure investment: A meta-analysis of empirical evidence. *Regional Science and Urban Economics*, *43*(5), 695–706. doi:10.1016/j.regsciurbeco.2013.05.002

Mercado, R. G. (2002). *Regional Development in the Philippines: A Review of Experience*. State of the Art and Agenda for Research and Action.

Moullin, J. C., Dickson, K. S., Stadnick, N. A., Albers, B., Nilsen, P., Broder-Fingert, S., Mukasa, B., & Aarons, G. A. (2020). Ten recommendations for using implementation frameworks in research and practice. *Implement Sci Commun*, *1*(1), 42. doi:10.118643058-020-00023-7 PMID:32885199

Naldini, A. (2018). Improvements and risks of the proposed evaluation of Cohesion Policy in the 2021–27 period: A personal reflection to open a debate. *Evaluation*, *24*(4), 496–504. doi:10.1177/1356389018804261

Okraszewska, R., Romanowska, A., Wołek, M., Oskarbski, J., Birr, K., & Jamroz, K. (2018). Integration of a Multilevel Transport System Model into Sustainable Urban Mobility Planning. *Sustainability*, *10*(2), 479. doi:10.3390u10020479

Olsson, L. E., Friman, M., & Lättman, K. (2021). Accessibility Barriers and Perceived Accessibility: Implications for Public Transport. *Urban Science*, *5*(3), 63. doi:10.3390/urbansci5030063

Oppido, S., & Ragozino, S. (2019). Unbalanced development and peripheralisation processes: atesting phase to map studies.*AESOP Annual Congress Venice 2019 Planning for Transition*, 3381–3393.

Páez, A., Scott, D. M., & Morency, C. (2012). Measuring accessibility: Positive and normative implementations of various accessibility indicators. *Journal of Transport Geography*, *25*, 141–153. doi:10.1016/j.jtrangeo.2012.03.016

Paiva, D. M. B., Freire, A. P., & de Mattos Fortes, R. P. (2021). Accessibility and Software Engineering Processes: A Systematic Literature Review. *Journal of Systems and Software*, *171*, 110819. doi:10.1016/j.jss.2020.110819

Pécoud, A., & de Guchteneire, P. (2006). International migration, border controls and human rights: Assessing the relevance of a right to mobility. *Journal of Borderlands Studies*, *21*(1), 69–86. doi:10.1080/08865655.2006.9695652

Peña, S., & Durand, F. (2022). Mobility planning in cross-border metropolitan regions: The European and North American experiences. *Territory, Politics, Governance*, *10*(2), 219–236. doi:10.1080/21622671.2020.1769716

Petheram, L., Zander, K. K., Campbell, B. M., High, C., & Stacey, N. (2010). Strange Changes: Indigenous perspectives of climate change and adaptation in NE Arnhem Land (Australia). *Global Environmental Change, 20*(4), 681–692. doi:10.1016/j.gloenvcha.2010.05.002

Petrauskiene, K., Dvarioniene, J., Kaveckis, G., Kliaugaite, D., Chenadec, J., Hehn, L., Pérez, B., Bordi, C., Scavino, G., Vignoli, A., & Erman, M. (2020). Situation Analysis of Policies for Electric Mobility Development: Experience from Five European Regions. *Sustainability, 12*(7), 2935. doi:10.3390u12072935

Petrie, H., & Bevan, N. (2009). *The Evaluation of Accessibility, Usability, and User Experience*. The Universal Access Handbook. doi:10.1201/9781420064995-c20

Pham, D. H., Shimizu, T., & Van Nguyen, T. (2021). A Literature Review on Interactions Between Stakeholders Through Accessibility Indicators Under Mobility as a Service Context. Int. J. *Intelligent Transportation Systems Research, 19*(2), 468–476. doi:10.100713177-021-00257-2

Piekermann, K., & Neubauer, J. (2002). *European Accessibility and Peripherality: Concepts, Models, and Indicators*. Nordregio Working Paper.

Pinna, F., Masala, F., & Garau, C. (2017). Urban Policies and Mobility Trends in Italian Smart Cities. *Sustainability, 9*(4), 494. doi:10.3390u9040494

Porru, S., Misso, F. E., Pani, F. E., & Repetto, C. (2020). Smart mobility and public transport: Opportunities and challenges in rural and urban areas. *Journal of Traffic and Transportation Engineering, 7*(1), 88–97. doi:10.1016/j.jtte.2019.10.002

Prus, P., & Sikora, M. (2021). The Impact of Transport Infrastructure on the Sustainable Development of the Region—Case Study. *Agriculture, 11*(4), 279. doi:10.3390/agriculture11040279

Puga, D. (2002, October 1). European regional policies in light of recent location theories. *Journal of Economic Geography, 2*(4), 373–406. doi:10.1093/jeg/2.4.373

Pullano, G., Valdano, E., Scarpa, N., Rubrichi, S., & Colizza, V. (2020). Evaluating the effect of demographic factors, socioeconomic factors, and risk aversion on mobility during the COVID-19 epidemic in France under lockdown: A population-based study. *The Lancet. Digital Health, 2*(12), e638–e649. doi:10.1016/S2589-7500(20)30243-0 PMID:33163951

Qin, J., Luo, S., Yi, D., Jiang, H., & Zhang, J. (2022). Measuring Cluster-Based Spatial Access to Shopping Stores under Real-Time Travel Time. *Sustainability, 14*(4), 2310. doi:10.3390u14042310

Rodríguez-Pose, A. (2008). The Rise of the "City-region" Concept and its Development Policy Implications. *European Planning Studies, 16*(8), 1025–1046. doi:10.1080/09654310802315567

Sevelius, J. M., Gutierrez-Mock, L., Zamudio-Haas, S., McCree, B., Ngo, A., Jackson, A., Clynes, C., Venegas, L., Salinas, A., Herrera, C., Stein, E., Operario, D., & Gamarel, K. (2020). Research with Marginalized Communities: Challenges to Continuity During the COVID-19 Pandemic. *AIDS and Behavior, 24*(7), 2009–2012. doi:10.100710461-020-02920-3 PMID:32415617

Siddiq & Taylor. (2021). Tools of the Trade? *Journal of the American Planning Association, 87*(4), 497–511. doi:10.1080/01944363.2021.1899036

Silva, C., & Larsson, A. (2018). *Challenges for Accessibility Planning and Research in the context of Sustainable Mobility.* Discussion Paper, International Transport Forum, Paris.

Soliman, A. M. (2004). Regional planning scenarios in South Lebanon: The challenge of rural-urban interactions in the era of liberation and globalization. *Habitat International, 28*(3), 385–408. doi:10.1016/S0197-3975(03)00039-0

Stanković, M., Gladović, P., Popović, V., & Lukovac, V. (2018). Selection Criteria and Assessment of the Impact of Traffic Accessibility on the Development of Suburbs. *Sustainability, 10*(6), 1977. doi:10.3390u10061977

Sundberg, J. (2011). Diabolic *Caminos* in the Desert and Cat Fights on the Río: A Posthumanist Political Ecology of Boundary Enforcement in the United States–Mexico Borderlands. *Annals of the Association of American Geographers, 101*(2), 318–336. doi:10.1080/00045608.2010.538323

Szymańska, E., Panfiluk, E., & Kiryluk, H. (2021). Innovative Solutions for the Development of Sustainable Transport and Improvement of the Tourist Accessibility of Peripheral Areas: The Case of the Białowieża Forest Region. *Sustainability, 13*(4), 2381. doi:10.3390u13042381

Tennøy, A., Hansson, L., Lissandrello, E., & Næss, P. (2016). *How planners' use and non-use of expert knowledge in land use and transport planning affect the goal achievement potential of plans? Experiences from three Scandinavian cities.* Academic Press.

Tiboni, M., Rossetti, S., Vetturi, D., Torrisi, V., Botticini, F., & Schaefer, M. D. (2021). Urban Policies and Planning Approaches for a Safer and Climate Friendlier Mobility in Cities: Strategies, Initiatives and Some Analysis. *Sustainability, 13*(4), 1778. doi:10.3390u13041778

Turienzo, J., Cabanelas, P., & Lampón, J. F. (2022). The Mobility Industry Trends Through the Lens of the Social Analysis: A Multi-Level Perspective Approach. *SAGE Open, 12*(1). Advance online publication. doi:10.1177/21582440211069145

Vale, D. S., Saraiva, M., & Pereira, M. (2015). Active accessibility: A review of operational measures of walking and cycling accessibility. *Journal of Transport and Land Use, 9*(1). Advance online publication. doi:10.5198/jtlu.2015.593

van Eldijk, J., Gil, J., & Marcus, L. (2022). Disentangling barrier effects of transport infrastructure: Synthesising research for the practice of impact assessment. *European Transport Research Review, 14*(1), 1. doi:10.118612544-021-00517-y

Vandenbulcke, G., Steenberghen, T., & Thomas, I. (2009). Mapping accessibility in Belgium: A tool for land-use and transport planning? *Journal of Transport Geography, 17*(1), 39–53. doi:10.1016/j.jtrangeo.2008.04.008

Vendemmia, B., Pucci, P., & Beria, P. (2021). An institutional periphery in discussion. Rethinking the inner areas in Italy. *Applied Geography (Sevenoaks, England), 135*, 102537.

Vettorazzi, S. (2018). Establishing the Connecting Europe Facility 2021-2027. European Parliamentary Research Service.

Vitale Brovarone, E., & Cotella, G. (2020). Improving Rural Accessibility: A Multilayer Approach. *Sustainability*, *12*(7), 2876.

Vulevic, A., Castanho, R. A., Naranjo Gómez, J. M., Loures, L., Cabezas, J., Fernández-Pozo, L., & Martín Gallardo, J. (2020). Accessibility Dynamics and Regional Cross-Border Cooperation (CBC) Perspectives in the Portuguese—Spanish Borderland. *Sustainability*, *12*(5), 1978.

Wang, Z., Han, Q., & De Vries, B. (2012). Land Use/Land Cover and Accessibility: Implications of the Correlations for Land Use and Transport Planning. *Applied Spatial Analysis and Policy*, *12*, 923–940.

Wiesenthal, T., Condeço-Melhorado, A., & Leduc, G. (2015). Innovation in the European transport sector: A review. *Transport Policy, Elsevier*, *42*(C), 86–93.

Wong Villanueva, J. L., Kidokoro, T., & Seta, F. (2020). Cross-Border Integration, Cooperation and Governance: A Systems Approach for Evaluating "Good" Governance in Cross-Border Regions. *Journal of Borderlands Studies*.

Wulfhorst, G., Büttner, B., & Ji, C. (2017). The TUM Accessibility Atlas as a tool for supporting policies of sustainable mobility in metropolitan regions. *Transportation Research Part A: Policy and Practice, Elsevier.*, *104*, 121–136.

Yan, X., Zhao, X., Han, Y., Hentenryck, P. V., & Dillahunt, T. (2021). Mobility-on-demand versus fixed-route transit systems: An evaluation of traveler preferences in low-income communities. *Transportation Research Part A: Policy and Practice, Elsevier.*, *148*, 481–495.

Ye, E. M., Du, J. T., Hansen, P., Ashman, H., Sigala, M., & Huang, S. S. (2021). Understanding roles in collaborative information behaviour: A case of Chinese group travelling. *Information Processing & Management*, *58*(4), 102581.

Zhang, X., Lu, H., & Holt, J. B. (2011). Modeling spatial accessibility to parks: A national study. *International Journal of Health Geographics*, *10*, 31.

Zolin, M. B., Ferretti, P., & Grandi, M. (2020). Sustainability in Peripheral and Ultra-Peripheral Rural Areas through a Multi-Attribute Analysis: The Case of the Italian Insular Region. *Sustainability*, *12*(22), 9380.

KEY TERMS AND DEFINITIONS

Accessibility: Measurements for spatial distribution and disparities of opportunities to access for specific individuals that mainly deal with spatial impedance between demand and supply, and the availability of services.

Accessibility Planning: The process of identifying and investigating the challenges in the access to opportunities and services for specific social groups in a particular region by developing strategies for accessibility improvement.

Borderlands: The land on either side of a border between countries that is mostly far from core of socio-economic flows of territories.

Mobility: Moving individuals from one place to another one within or between regions based on necessity of access to opportunities and services, and displayed preferences for mobility due to transportation efficiency.

Peripheral Regions: Territories on the edge of larger areas that are less developed than the semi-periphery and core-countries due to environmental and socio-economic obstacles resulted in receiving disproportionately small share of global wealth and regional opportunities and services.

Regional Planning: Geo-technology techniques and specific structure for comprehensive development at regional scale through rational transformation of spatio-temporal dimensions that aim to manipulate specific areas in a rational way toward a wider view to achieve better regional integration.

Sustainable Development: The way of unifying communities and society in order that they could be extant in the long term in the context of environmental preservation and socio-economic equity based on present and future imperative concerns.

Transport Infrastructure Provision: The process of setting up transport infrastructure including essential steps toward improving accessibility to specific opportunities and services to make them available for all social groups.

Chapter 4
Assessment of the Urban Land Area in the Municipalities of the Community of Madrid in 1990, 2000, 2006, 2012, and 2018

José Manuel Naranjo Gómez
https://orcid.org/0000-0001-7998-9154
Polytechnic School, University of Extremadura, Spain

José Cabezas
https://orcid.org/0000-0002-7209-3731
University of Extremadura, Spain

José Martín-Gallardo
University of Extremadura, Spain

ABSTRACT

The city of Madrid is the capital of Spain, and the urban sprawl has been important and continuous over the years. There has been a spillover effect since the use of urban land has increased in the municipalities around the capital. Among other factors, the price of housing is increasingly high in Madrid. Thus, in the municipalities near the capital, the population makes trips to work but then returns to their municipality to reside, where the housing price is lower. Satellite images through Land Cover Corine have made it possible to quantify the hectares destined for urban use in 1990, 2000, 2006, 2012, and 2018 through a geographic information system (GIS). It has been possible to identify those municipalities of the autonomous community that have grown better as continuous urban fabric or discontinuous urban fabric. Therefore, it has been possible to identify municipalities most influenced by the capital.

DOI: 10.4018/978-1-6684-4548-8.ch004

INTRODUCTION

The implementation of an increasingly dispersed metropolitan model, and with a lower population density, has been accentuated in Spain since 1990. This urbanizing construction dynamic accelerated in the following years, as it was sustained by the loans offered by banks with low-interest rates to both builders and home buyers. The Spanish economy has been sustained by this model of urban development (Fernández, 2006).

The most inhabited Spanish city with 3,305,408 inhabitants (Spanish Statistical Office [INE], 2021) is its economic and administrative capital, Madrid. This city exerts an enormous influence on the 178 municipalities that are around it and that, together with the Municipality of Madrid, form the Autonomous Community of the same name. Indeed this Autonomous Community has become one of the main sources of urban growth in the European Union (Domingo et al., 2021).

As for the city of Madrid, the urban form has changed rapidly and convulsively, especially since 1990. These changes have been especially evident in the peripheral areas of the city, where in the last twenty years urban development has been characterized by the sprawl model, with more extensive spaces, but discontinuous, of low population density and fragmented. (García-Palomares, 2010).

Regarding the Autonomous Community of Madrid, there has also been a significant increase in the area of land for urban use since the 1990s. Predominantly, this growth has occurred in six rings of municipalities (Díaz-Pacheco and García-Palomares, 2014). These rings are formed by the union of municipalities that are surrounding the central municipality that is Madrid, and which hosts the city of the identical name. Likewise, the municipalities that make up these rings have been influenced by public transport, which allows residents to go to work in the city of Madrid and return to rest in their municipality of residence, that is to say, commuters. In fact, in the 1995 Urban Development Plan, one of the basic lines of action was focused on improving transport infrastructure by establishing a comprehensive system, to promote collective transport and adapt the transport system to the framework of action of the Commission of the European Union (Fernández et al., 1998).

The objective of this work is to evaluate the urban dynamics in the Autonomous Community of Madrid. For this reason, the specific objectives are to locate and quantify the urban land area considered as a continuous or discontinuous urban fabric in 1990, 2000, 2006, 2012, and 2018.

URBAN SPRAWL

Urban sprawl has had a major influence on land-use changes in recent decades, as population pressure on a global scale is increasing. As a result, land uses are changing intensively in many parts of the world. Thus, some of them become urban lands. This phenomenon turns out to be very important for different reasons. Because urbanization produces radical changes in the quantity, type, and spatial distribution patterns of land use (Xu et al., 2016). Therefore, the description and analysis of these changes are key to proposing strategies for territorial development (Díaz-Pacheco et al., 2014).

In Western countries, urban sprawl is generally considered to have various negative effects on urban growth, such as heavy traffic loads, deteriorated inner-city neighbourhoods, and the disappearance of public spaces (Wang et al., 2009). However, real estate developers can make big profits from building residential housing in new urbanized areas. Due to high housing prices and commercial redevelopment projects in city centre areas, through urban renewal, local and regional governments can reap great benefits from the transmission of land-use rights and related taxes (Wang et al., 2001). Thus, urban sprawl is emerging from suburban centres to peripheral areas, and the increase of built urban areas is generally considered as an effect of expansion (Barnes et al., 2012).

Urban growth and low population density are accompanied by increasing discontinuity and spatial fragmentation of suburban areas. Isolated urbanizations extend on the outskirts of cities beyond the supposed city limits (García-Palomares, 2010). In this way, decentralization and re-centralization take place, forming peripheral cities and a multipolar metropolitan structure (Giulano and Small, 1999). Thus, the monocentric city gives way to a polycentric urban system, where proximity to the central city is no longer essential, but it is enough to be close to any of these multiple poles (Filion et al., 1999). In fact, in all metropolitan areas of the world, the recent period of suburbanization is characterized by following new polycentric forms (Dupuy, 1995).

Even though on our planet, lands considered as urban, occupy a small percentage of its surface, their effects on the alteration of the environment are very important. For this reason, it is paramount to understand and explain them, depending on various types of factors, taking into account that urban expansion is characterized by a high degree of complexity (Rocha et al., 2008).

Controversy and Handling Cartographic Data through a GIS

Currently, there is great controversy about urban sprawl. Even, there is no complete and exact definition, even there are no specific measures (Wang et al., 2020). On the one hand, some academics are advocating strategies for more compact cities and controlling expansion (Burton et al.,1996; Johnson, 2001). On the other hand, others are in favour of urban expansion as a result of residential preference (Squires, 2002), and also state that governments should not intervene, since non-optimal distributions of land resources occur (Dowall et al., 1996).

Nonetheless, the ease of handling cartographic data through a GIS is increasing. As well as the management of data obtained by remote sensing techniques to analyze land uses. In addition, these data are increasingly abundant, higher quality, and some of them are in open access. As a result, the number of studies intended for the analysis of urban sprawl through a GIS is also increasing (Díaz-Pacheco and García-Palomares, 2014).

Precisely, studies carried out in the Community of Madrid reveal that the city of Madrid exerts an enormous influence on the municipalities around and that even the functional region of Madrid exceeds its administrative limit (Solís, 2008; Gutiérrez and García, 2007; Córdoba and Morcillo, 2020). It has also been established that within the Autonomous Community, the intense urbanizing activity has had economic and urban connotations between the different municipalities, altering and conditioning their expansive capacities (Naredo, 2010; Daher, 2013; Gaja, 2013). Because the region has been transformed according to the needs of the city of Madrid, becoming a node for the accumulation of technical resources and human resources from the surrounding municipalities (Romero et al., 2015; Carmona and González, 2018).

Other studies even state that there has been an urban sprawl without an elaborated model for optimal development because the appropriate regulatory instruments have not been used. These would have facilitated the realization of a territorial model under which to carry out an adequate urban, economic, and social development (De Terán and Sánchez, 1999; Valenzuela, 2010). The Spatial Planning Guidelines, the Coordinated Action Programmes, and the Physical Environment Management Plan adopted in 1985 aimed to regulate the settlement processes in the territory of the different economic and social activities, of public and private actors. But before 1990 they were replaced by sectoral plans and territorial strategic planning, which did not achieve optimal spatial planning for economic activities. (Carmona and González, 2018). In fact, in 2020 the Autonomous Community was still without strategic territorial planning (Córdoba and Morcillo, 2020).

Urban land area in the municipalities of the Community of Madrid in 1990, 2000, 2006, 2012 and 2018

Information was initially obtained on the land uses recorded in the Corine Land Cover (CLC)) in the European Environment Agency (EEA) platform. This inventory began in 1985 and offers land uses for the years 1990, 2000, 2006, 2012, and 2018, of 44 different classes. Besides the Minimum Mapping Unit (MMU) is 25 hectares (ha) for areal phenomena and a minimum width of 100 for linear phenomena. The information is offered in Shapefile format, and each file contains polygons that evoke in reality the different land uses. The alphanumeric information associated with each of them contains a code that distinguishes the registered land use (Table 1).

The second source of information was the Spanish National Geographic Information Centre (CNIG in Spanish). The administrative limits of the municipalities and the Autonomous Community of Madrid were obtained. The graphical entity used to represent them are also polygons, with the name associated with each of them, as alphanumeric information.

Finally, the third information source was the Spanish National Statistics Institute (INE in Spanish). Where the population was downloaded for the last year analyzed, that is to say, in 2018. The information obtained here is a database.

All information obtained previously was managed through a Geographic Information System (GIS), for this purpose the ArcGis 10.3 software ESRI was used, this software is manufactured by ESRI Company. In addition, Microsoft Office 2021 was also used. Specifically, Microsoft Access 2021 was used.

Firstly, the limits of the Municipal Terms of the Madrid Autonomous Community served as a limit for cutting the land uses obtained from CLC. This intersection of two layers whose graphical entities in both cases are polygons resulted in another layer of polygons. Thus, a polygon was obtained that represented the limit of the municipality and in turn, contained several polygons that evoked, in reality, the land uses. This operation was repeated 5 times, for the years 1990, 2000, 2006, 2012, and 2018. In this way, it was possible to obtain thematic maps according to the different land uses in each year. Subsequently, using the projection Universal Transverse Mercator (UTM) European Terrestrial Reference System 1989 (ETRS89) and zone 30, each of the polygons representative of the land uses in each of the Municipal Terms was measured in hectares.

Secondly, each database file obtained from the GIS for each year analyzed was exported to Microsoft Access 2021. Thus, it was possible to make a union between the database that offered the number of inhabitants in each Municipality and land uses. To achieve this, the code assigned according to the INE to each Municipal Term was used as a common field. Then, it was possible to select those municipalities that had a higher population in 2018 and obtain the amount of urban land resulting from adding the area of each of the polygons whose code CLC was either 111 or 112, corresponding to the continuous or discontinuous urban fabric, for the years analyzed.

Table 1. CLC nomenclature

Level 1	Level 2	Level 3
1 Artificial surfaces	11 Urban fabric	111 Continuous urban fabric 112 Discontinuous urban fabric
	12 Industrial, commercial, and transport units	121 Industrial or commercial units 122 Road and rail networks and associated land 123 Port areas 124 Airports
	13 Mine, dump, and construction sites	131 Mineral extraction sites 132 Dump sites 133 Construction sites
	14 Artificial, non-agricultural vegetated areas	141 Green urban areas 142 Sport and leisure facilities
2 Agricultural areas	21 Arable land	211 Non-irrigated arable land 212 Permanently irrigated land 213 Rice fields
	22 Permanent crops	221 Vineyards 222 Fruit trees and berry plantations 223 Olive groves
	23 Pastures	231 Pastures
	24 Heterogeneous agricultural areas	241 Annual crops associated with permanent crops 242 Complex cultivation patterns 243 Land principally occupied by agriculture, with significant areas of natural vegetation 244 Agro-forestry areas
3 Forest and semi-natural areas	31 Forests	311 Broad-leaved forest 312 Coniferous forest 313 Mixed forest
	32 Scrub and/or herbaceous associations	321 Natural grassland 322 Moors and heathland 323 Sclerophyllous vegetation 324 Transitional woodland-scrub
	33 Open spaces with little or no vegetation	331 Beaches, dunes, sands 332 Bare rocks 333 Sparsely vegetated areas 334 Burnt areas 335 Glaciers and perpetual snow
4 Wetlands	41 Inland wetlands	411 Inland marshes 412 Peat bogs
	42 Marine wetlands	421 Salt marshes 422 Salines 423 Intertidal flats
5 Water bodies	51 Inland waters	511 Water courses 512 Water bodies
	52 Marine waters	521 Coastal lagoons 522 Estuaries 523 Sea and ocean

Source: EEA,2020

Figure 1. Madrid autonomous community municipalities
Source: authors

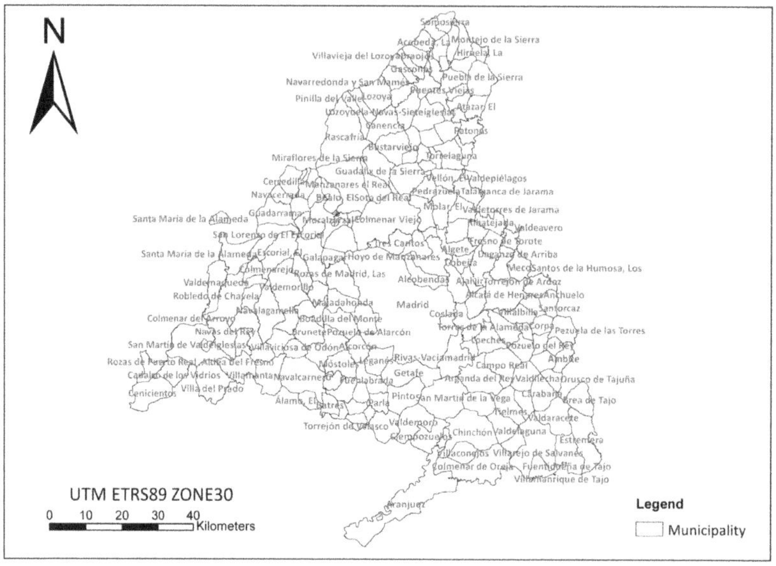

From the information obtained by the intersection in the GIS of the land uses recorded by CLC and the boundaries of each of the municipalities, different thematic maps were obtained. A thematic map for each of the years was analyzed. In this way, thematic maps were obtained on the land uses registered within each municipality for the years 1990, 2000, 2006, 2012, and 2018.

In this regard, the same representation symbology was used for each of the thematic maps obtained. The municipal terms were represented by polygons whose edge was black and transparent filling. Likewise, the same symbology was always used for the representation of the different land-use values according to the CLC codes for level 3. In this way, the different maps obtained in the different years can be compared.

In Figure 2 it can be seen that there is no definite pattern regarding the arrangement of the different land uses in the Madrid Autonomous Community. Although it is true that in a certain way, in the strip that runs from the north, passing through the west and reaching the south, those land uses destined to forests and semi-natural areas, wetlands and waterbodies predominate. All these land uses are represented in shades of green, from the lightest to the darkest. Also, in a parallel strip, but located in the central part the land predominant are artificial surfaces and agricultural areas. Finally, there is another strip located in the southernmost part, where there is an alternation of land uses for agricultural areas and forests, and seminatural areas.

As for the urban land uses continuous or discontinuous urban fabric, represented in two shades of red, they appear centralized in the centre of the central strip. Corresponding its highest concentration to the municipality of Madrid and the municipalities that surround it, especially those municipalities that are located south of the city of Madrid. In the remaining two strips, the urban lands appear in a distributed and disseminated form.

Figure 2. Map of land uses in the Madrid Autonomous Community in 1990
Source: authors

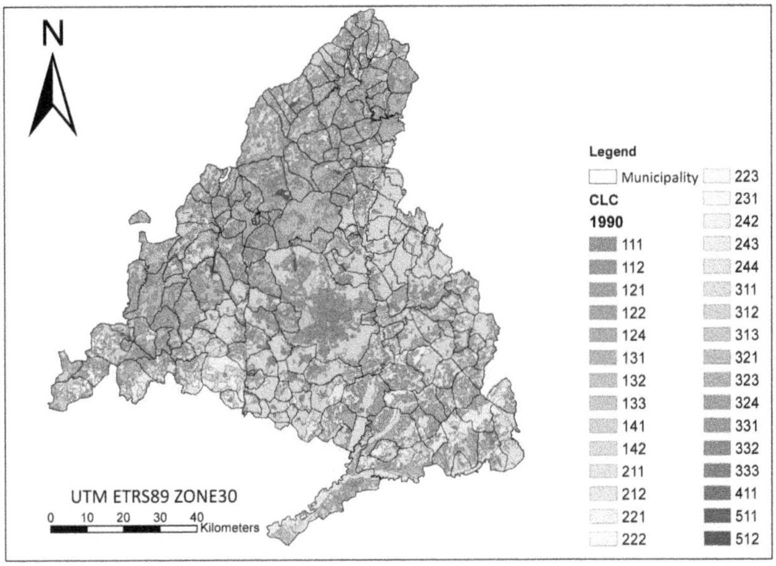

Figure 3. Map of land uses in the Madrid Autonomous Community in 2000
Source: authors

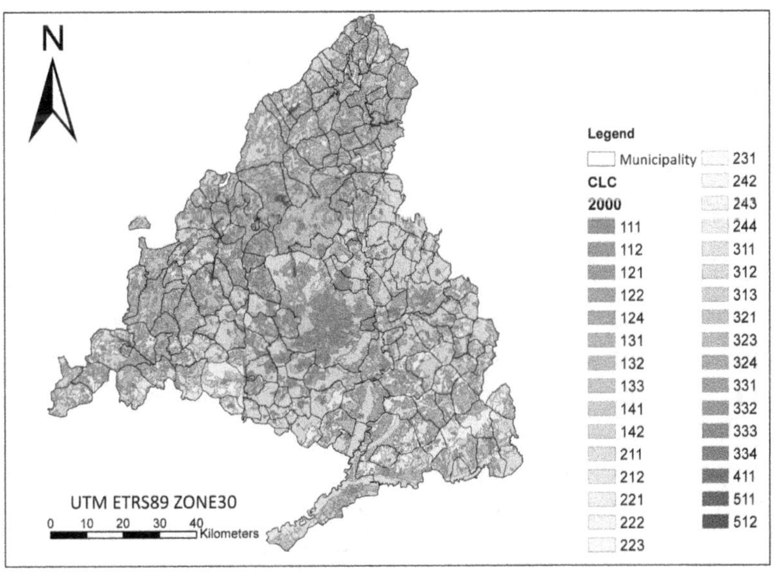

From the comparison of Figures 2 and 3, it can be seen that there has been little variation in terms of the location and extent of land uses in general, and in particular of urban land uses. Nonetheless, in the northernmost strip, there is a greater extension of the continuous or discontinuous urban fabric. Nevertheless, this effect is more pronounced in the central strip. Because in the year 2000 appear more visible urban centres that were previously barely noticeable in the northern part of the strip. Also, in this same central strip in the south, it is possible to realize the increase in the area of settlements located

Figure 4. Map of land uses in the Madrid Autonomous Community in 2006
Source: authors

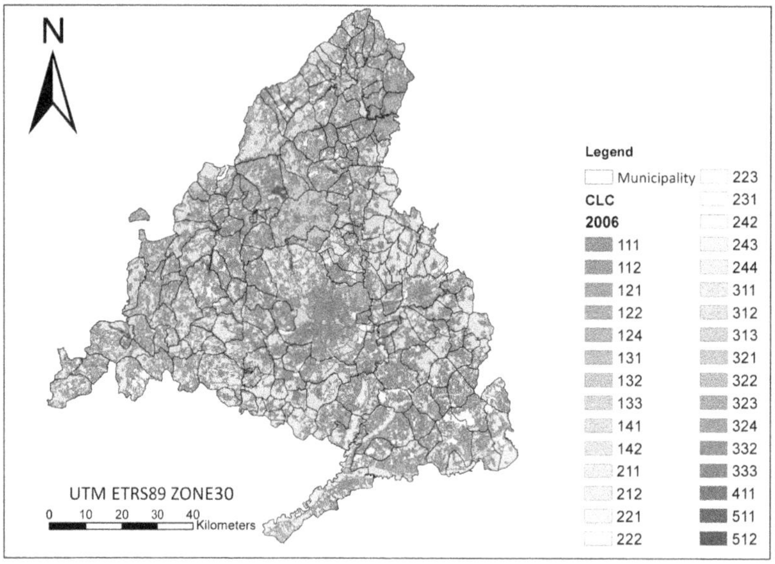

south of the Madrid municipality. Finally, in the southernmost strip, this effect is also appreciated, but it is of lower intensity than in the previous two stripes.

As before, the comparison of figures 3 and 4 again shows that the trend observed between 1990 and 2000 continues to increase. Because urban lands continue to increase in extent. Even, also as before in the central strip, this effect is still more intense. Indeed, it can be perceived that among the urban lands of the Madrid municipality, those located further south seem to join forming a large belt around the capital of the country. As a consequence, between 1990 and 2000 there was urban consolidation of the municipalities forming the belt south of the city of Madrid. Likewise, in the north of the city, it can be observed a great urban expansion. Although, it is not as intense as the one produced to the south.

Again, Figures 4 and 5 allow us to compare the situation and expansion of urban lands in two different temporal sections between 2006 and 2012. Maps (Figures 4 and 5) are not so different. As a consequence, urban expansion between these years has not been as intense as in former years. Although, it can be observed that there is a tendency to the higher surface occupation of the land that is intended for urban use. Due to the fact that those red polygons can be seen more easily. However, the pattern previously defined in previous years is not broken. Certainly, in the south of the city of Madrid urban expansion is still more intense than in the municipalities located to the north.

Finally, Figure 6 offers the most recent map of land uses in the Madrid Autonomous Community, referring to the year 2018. The pattern in a certain way of grouping the land uses marked since 1990 is maintained since there are the three strips well differentiated in terms of the predominance of the different land uses. Also, the largest amount of urban land is clustered in the central strip. Likewise, the comparison of figures 5 and 6 shows that there is still an expansion of urban lands. However, this expansion of urban land is much less intense than that observed in former years. As a result, the most recent urban expansion in the Madrid Autonomous Community around the city of Madrid seems to be decreasing in intensity and if this trend continues there could be a standstill.

Figure 5. Map of land uses in the Madrid Autonomous Community in 2012
Source: authors

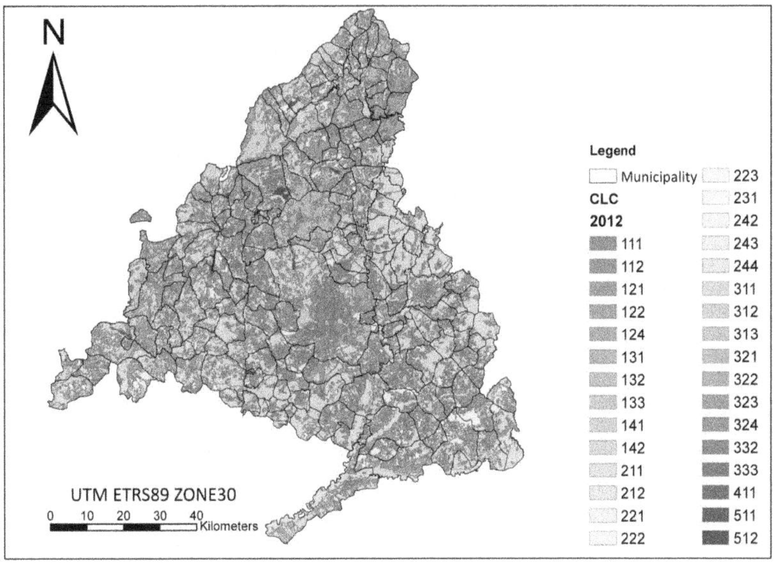

Figure 6. Map of land uses in the Madrid Autonomous Community in 2018

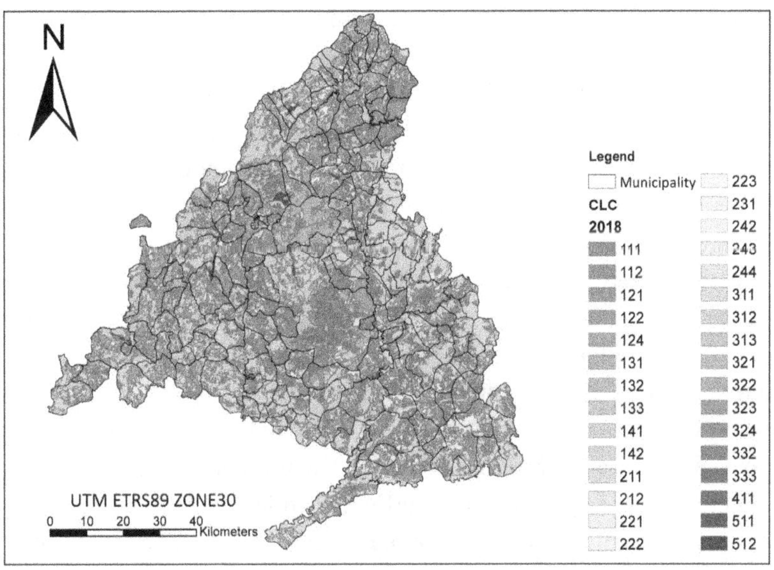

In addition, from the thematic maps previously reviewed it was possible to obtain alphanumeric information regarding the resident population, and urban extension in hectares in each of the years analyzed.

Firstly, of the 178 municipalities in the Community of Madrid, only 19 did not register urban land uses. Braojos, Berzosa del Lozoya, Redonda y San Mamés, Madarcos, La Hieruela, Horcajuelo de la Sierra, Horcajo de la Sierra, Robregordo, Cervera de Buitrago, El Atazar, Puebla de la Sierra, Montejo de la Sierra, Piñuecar-Gandullas, Olmeda de las Fuentes, Prádena del Rincón, La Acebeda, La Serna del Monte, Somosierra, and Robledillo de la Jara.

As for the most populated municipalities, the hectares of registered urban land were:

Table 2. Urban land uses measured in hectares in the most populated municipalities of the Madrid Autonomous Community

Municipality	Population	1990	2000	2006	2012	2018
Madrid	3223334	13524.72	15455.89	14238.99	15025.76	15131.91
Móstoles	207095	499.28	854.66	900.38	967.37	968.24
Alcalá	193751	882.05	1075.35	1078.39	1149.02	1124.29
Fuenlabrada	193586	603.73	758.37	745.70	820.43	848.51
Leganés	188425	527.64	771.96	840.85	953.58	909.83
Getafe	180747	459.84	570.53	873.24	1009.66	1253.80
Alcorcón	169502	314.59	869.42	752.20	828.73	803.87
Torrejón de Ardoz	129729	366.65	435.09	738.66	822.95	872.20
Parla	128256	216.08	287.71	465.48	701.02	632.25
Alcobendas	116037	969.18	1354.77	1064.24	1136.93	1193.96

Source: authors

Table 2 shows that there is a large amount of accumulated population in the municipality of Madrid, where the capital of the same name is located. There is a great difference of inhabitants between the second most inhabited municipality which is Móstoles. Moreover, between Móstoles and the other more populated municipalities, the difference is not so deep and is progressive. All these municipalities would be under the influence of the city of Madrid, and they would be the municipalities where the largest number of people reside who travel to go to work in the capital and then they return home, that is to say, commuters.

Table 2 also shows the hectares of urban land in each of the years analyzed, for each of the municipalities. In this respect, the hectares obtained are the result of adding the hectares referred to discontinuous and continuous urban fabric land. In each of the municipalities, it can be observed that the trend in urban land has been different during the years analyzed (Figure 7).

Figure 7 shows how there has been a continuous increase in the urban land area in the municipalities of Móstoles, Getafe, and Torrejón de Ardoz. There is also a trend of increasing urban land in the municipalities of Alcalá, Fuenlabrada, Leganés, Parla, and Alcobendas. Although, it is not always progressive and continuous. As a consequence, there is a marked influence of the capital, on the most populated municipalities of the Autonomous Community and this trend is growing, except in the municipality of Alcorcón, where urban sprawl appears to have declined between 2012 and 2018.

Figure 7. Hectares of urban land uses in the most inhabited municipalities in the Madrid Autonomous Community
Source: authors

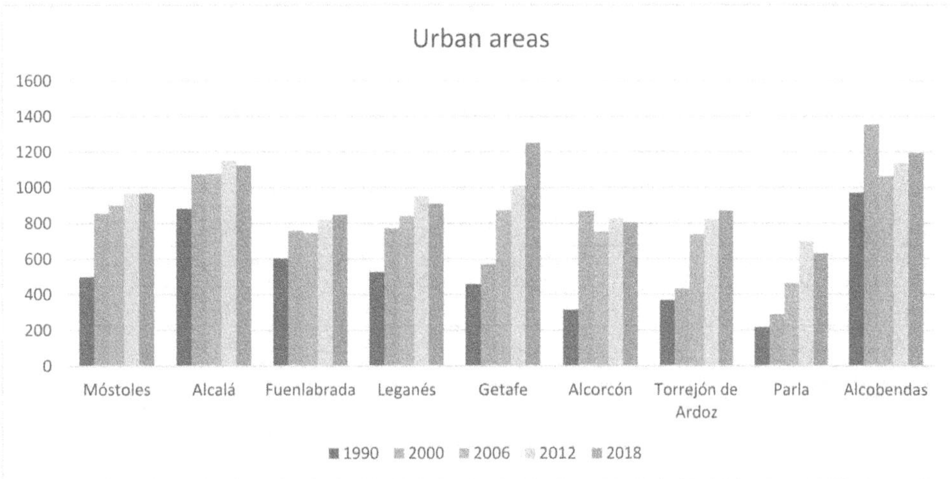

Regarding municipalities whose population is less than 100000 and more than 10000 inhabitants:

Table 3. Urban land uses measured in hectares in municipalities of the Madrid Autonomous Community with a population between 100000 and 10000 inhabitants

Municipality	Population	1990	2000	2006	2012	2018
Rozas de Madrid, Las	95550	1194.01	1591.10	1880.75	1952.40	1955.08
San Sebastián de los Reyes	87724	664.76	865.30	1054.34	1221.64	1170.84
Pozuelo de Alarcón	86172	1146.67	1696.67	1896.30	2091.76	2075.39
Rivas-Vaciamadrid	85893	0.00	496.87	858.41	970.28	1003.65
Coslada	81860	288.22	389.88	415.20	439.97	446.24
Valdemoro	74745	91.26	227.34	441.51	632.58	623.34
Majadahonda	71785	507.04	603.02	764.97	842.27	811.05
Collado Villalba	63074	729.69	854.53	816.94	850.42	851.66
Aranjuez	59037	298.91	373.24	493.03	618.16	641.58
Arganda del Rey	54554	112.36	256.62	371.45	436.72	526.00
Boadilla del Monte	52626	1121.75	1362.90	1398.41	1457.19	1480.71
Pinto	51541	200.15	249.25	305.67	353.86	350.81
Colmenar Viejo	49498	513.56	644.12	609.52	712.60	778.65
Tres Cantos	46750	310.65	500.94	399.20	399.20	417.73
San Fernando de Henares	39466	80.64	222.44	268.15	243.76	295.95
Galapagar	33379	973.17	1097.40	1198.02	1203.69	1217.93
Arroyomolinos	30052	0.00	133.74	211.02	462.18	485.23

continues on following page

Table 3. Continued

Municipality	Population	1990	2000	2006	2012	2018
Navalcarnero	28305	99.48	286.87	299.91	451.11	451.83
Villaviciosa de Odón	27596	745.71	1203.60	1026.06	1026.06	1061.81
Paracuellos de Jarama	24521	273.10	350.46	242.39	485.89	474.29
Ciempozuelos	24087	141.60	187.50	222.09	250.81	255.32
Torrelodones	23361	670.08	775.30	792.20	779.77	791.81
Mejorada del Campo	23241	173.59	309.69	379.15	409.95	398.84
Villanueva de la Cajada	21000	408.00	639.99	669.24	775.19	800.38
Algete	20473	411.89	467.32	525.38	525.38	522.13
Humanes de Madrid	19587	53.11	82.48	170.23	170.23	158.08
San Martín de la Vega	18784	116.16	191.82	304.11	377.06	318.20
San Lorenzo de El Escorial	18088	250.75	336.88	305.92	339.33	412.97
Villanueva del Pardillo	17127	94.48	122.56	333.67	341.38	326.29
Escorial, El	15842	590.81	646.39	807.06	803.70	803.70
Guadarrama	15785	421.75	668.57	704.04	710.08	710.09
Alpedrete	14364	362.93	374.18	487.96	487.96	487.96
Meco	13959	2.93	141.26	191.22	233.05	257.60
Villalbilla	13421	241.06	343.87	375.81	476.45	482.11
San Agustín	13273	74.57	144.22	192.77	242.87	243.64
Moralzarzal	12697	114.57	384.98	386.91	386.91	396.22
Valdemorillo	12280	795.56	858.01	934.18	934.18	934.18
Velilla de San Antonio	12193	26.15	131.09	100.20	100.20	104.61
Brunete	10596	66.65	165.67	279.90	279.90	280.40
Griñón	10178	119.40	178.74	296.67	319.07	300.85
Daganzo de Arriba	10061	0.00	108.65	111.61	131.90	131.90

Source: authors

The population for municipalities between 100000 and 10000 inhabitants decreases progressively (Table 3). In this regard, in 1990 the municipality with the largest urban area was Las Rozas de Madrid, but in 2000, 2006, 2012, and 2018 the municipality is Pozuelo de Alarcón. As a consequence, Las Rozas de Madrid would be more densely populated than Pozuelo de Alarcón. Likewise, the municipality with the least population Daganzo de Arriba is not the municipality that has the least urban area, since this is the municipality Velilla de San Antonio in 2018. In addition, in 1990 no urban area was recorded for 3 municipalities, but from 2000 all municipalities had urban area.

As for the municipalities that have between 10000 and 1000 inhabitants.

Table 4. Urban land uses measured in hectares in municipalities of the Madrid Autonomous Community with a population between 10000 and 1000 inhabitants

Municipality	Population	1990	2000	2006	2012	2018
Álamo, El	9353	121.79	198.59	227.29	221.90	221.89
Colmenarejo	9124	71.01	185.83	281.86	281.86	292.50
Sevilla la Nueva	9095	28.45	56.78	182.77	182.77	204.25
Soto del Real	8694	335.58	454.41	477.23	477.23	477.23
Loeches	8673	0.00	0.00	105.71	136.51	122.49
Molar, El	8666	60.73	60.69	202.29	202.29	196.48
Manzanares el Real	8597	257.91	406.37	368.43	368.43	346.01
Torrejón de la Calzada	8582	0.00	0.00	109.46	140.27	155.73
San Martín de Valdeiglesias	8318	189.12	239.30	247.35	254.64	439.19
Hoyo de Manzanares	8222	382.16	540.33	391.36	391.36	391.36
Colmenar de Oreja	7902	484.43	510.33	557.23	562.69	652.93
Torres de la Alameda	7760	57.04	143.18	148.80	148.80	148.80
Morata de Tajuma	7553	49.10	74.01	121.87	121.87	117.53
Boalo, El	7399	81.82	403.21	418.38	466.17	463.78
Villarejo de Salvanis	7291	157.01	178.28	233.04	233.04	236.44
Cobeña	7280	0.00	65.40	84.11	84.11	141.72
Camarma de Esteruelas	7226	113.75	147.62	224.33	224.33	224.33
Cercedilla	6948	246.21	305.68	302.51	302.51	302.51
Collado Mediano	6781	193.65	307.17	285.32	285.32	285.33
Fuente el Saz de Jarama	6541	34.82	86.00	135.17	135.17	221.61
Villa del Prado	6409	276.02	338.95	365.42	365.42	402.86
Cubas de la Sagra	6310	67.74	67.70	216.81	252.74	219.15
Nuevo Baztán	6154	570.75	587.86	634.65	634.65	626.95
Guadalix de la Sierra	6119	101.84	192.55	148.56	162.74	173.48
Campo Real	6075	35.35	63.72	137.54	146.47	145.77
Miraflores de la Sierra	5897	157.06	197.49	341.59	341.59	329.18
Pedrezuela	5746	95.20	149.80	222.34	222.34	231.20
Becerril de la Sierra	5564	157.27	376.97	343.83	343.83	348.73
Chinchón	5239	104.27	104.19	177.24	177.24	245.38
Moraleja de Enmedio	5021	2.62	72.74	135.37	147.34	129.25
Torrelaguna	4724	0.00	93.75	113.94	113.94	113.94
Ajalvir	4559	0.00	73.40	44.16	44.16	44.16
Valdetorres de Jarama	4384	195.26	278.50	325.95	325.95	320.99
Molinos, Los	4328	204.66	288.89	311.35	311.35	304.99
Torrejón de Velasco	4298	44.89	44.86	97.11	111.27	131.51
Robledo de Chavela	4159	281.94	359.41	467.18	467.18	468.59
Serranillos del Valle	4066	0.00	70.41	124.23	127.86	158.43

continues on following page

Table 4. Continued

Municipality	Population	1990	2000	2006	2012	2018
Valdeolmos-Alalpardo	4039	47.77	47.73	106.38	108.27	108.27
Talamanca de Jarama	3725	16.79	16.94	107.93	122.07	122.07
Casarrubuelos	3707	0.00	0.00	68.81	74.20	61.35
Villaconejos	3363	39.40	39.37	97.55	97.55	65.32
Quijorna	3357	61.38	69.67	121.82	138.68	138.68
Cadalso de los Vidrios	3049	35.90	96.36	284.72	284.72	276.96
Navacerrada	2911	30.70	103.54	174.04	174.04	171.73
Valdilecha	2840	0.00	0.00	54.28	54.28	52.93
Perales de Tajura	2831	0.00	0.00	55.35	55.35	76.38
Navas del Rey	2819	33.98	33.96	151.15	151.15	153.41
Aldea del Fresno	2698	203.48	268.00	130.26	130.26	150.89
Cabrera, La	2613	100.91	253.99	223.42	223.42	223.43
Tielmes	2604	34.73	34.71	90.95	90.95	87.56
Navalagamella	2601	63.55	63.51	78.68	78.68	78.68
Santos de la Humosa, Los	2542	0.00	0.00	97.58	97.58	97.59
Villamanta	2501	30.44	30.42	101.33	101.33	101.34
Bustarviejo	2486	0.17	35.88	95.62	126.85	158.27
Pelayos de la Presa	2475	200.50	242.74	253.26	253.26	252.23
Chapinería	2240	76.73	124.98	149.00	154.14	154.14
Fresno de Torote	2181	173.62	197.37	238.89	238.62	239.74
Venturada	2081	160.84	200.78	209.58	209.58	213.45
Fuentidueña del Tajo	1985	155.59	155.47	241.78	241.78	246.12
Cenicientos	1980	49.02	67.08	93.67	93.67	93.67
Villar del Olmo	1967	146.76	146.65	154.22	154.22	149.15
Vellón, El	1942	29.42	29.40	78.48	78.48	78.48
Carabaña	1905	0.00	0.00	83.70	83.70	85.77
Buitrago del Lozoya	1875	64.41	64.36	83.51	83.51	78.36
Colmenar del Arroyo	1685	96.75	96.70	118.87	118.87	118.87
Rascafrla	1663	0.00	0.00	107.55	107.55	107.55
Batres	1638	137.68	137.59	163.43	163.43	163.43
Belmonte de Tajo	1627	0.00	0.00	52.30	57.84	57.61
Zarzalejo	1620	31.16	31.14	74.94	74.94	76.50
Fresnedillas de la Oliva	1554	3.07	47.74	84.24	84.24	84.24
Villanueva de Perales	1547	0.00	0.00	48.98	48.98	53.26
Valdeavero	1510	77.79	77.73	158.05	158.05	158.05
Villamantilla	1420	0.00	0.00	57.82	56.67	51.20
Navalafuente	1397	41.57	41.54	89.94	89.94	106.08
Titulcia	1304	0.00	0.00	55.21	55.21	48.36

continues on following page

Table 4. Continued

Municipality	Population	1990	2000	2006	2012	2018
Anchuelo	1272	26.55	26.53	43.46	43.46	43.46
Lozoyuela-Navas-Siete iglesias	1266	43.70	43.67	150.82	157.05	157.05
Estremera	1236	0.70	32.66	43.63	43.63	120.95
Orusco de Tajufa	1202	0.00	0.00	54.44	54.44	54.44
Santa Mar	1182	72.92	86.01	96.24	96.24	123.83
Pozuelo del Rey	1120	0.78	0.78	36.34	36.34	36.34

Source: authors

Table 4 shows that the most populated municipality in 2018 is El Álamo, but in 1990, 2000, and 2012 the municipality with the most urban area was Nuevo Batzán, and in 2018 it was the municipality Colmenar de Oreja. The municipality of El Álamo occupies the 26th position in terms of the municipalities with the largest urban area in 2018. Therefore, in the municipality of El Álamo, there is a high population density. On the contrary, in 1990 there is no registered urban area for 17 municipalities, in 2000 for 13, but as of 2006, there is an urban area for all municipalities.

Regarding the least inhabited municipalities with less than 1000 inhabitants.

Table 5. Urban land uses measured in hectares in the municipalities of the Autonomous Community of Madrid with less than 1000 inhabitants

Municipality	Population	1990	2000	2006	2012	2018
Torremocha de Jarama	975	6.67	6.67	114.41	114.41	114.05
Valdemanco	927	0.00	0.00	38.97	38.97	38.97
Valdelaguna	903	0.00	0.00	45.16	45.16	45.82
Santorcaz	850	0.00	0.00	56.30	56.30	56.30
Pezuela de las Torres	821	0.00	0.00	39.84	39.84	39.84
Valdemaqueda	761	79.26	90.48	86.58	86.58	86.58
Berrueco, El	741	0.00	0.00	88.37	88.37	88.37
Cabanillas de la Sierra	712	0.00	27.67	91.26	91.26	91.26
Ribatejada	706	117.25	119.23	111.53	111.53	113.35
Villamanrique de Tajo	699	0.00	0.00	48.42	48.42	48.42
Corpa	697	0.65	0.65	38.76	38.76	38.76
Puentes Viejas	642	0.00	0.00	39.70	39.70	39.70
Valdaracete	626	0.00	0.00	26.91	26.91	26.91
Ambite	601	57.10	57.06	69.63	75.75	105.13
Valdepillagos	574	2.52	5.08	0.02	0.02	25.09
Lozoya	559	36.70	36.67	66.19	66.19	66.20
Rozas de Puerto Real	527	1.94	1.94	36.06	36.06	63.21
Brea de Tajo	525	56.96	56.92	144.16	144.16	144.16

continues on following page

Table 5. Continued

Municipality	Population	1990	2000	2006	2012	2018
Patones	523	33.87	33.85	24.97	24.97	24.97
Valverde de Alcalá	432	0.41	0.41	47.19	47.19	30.25
Canencia	423	0.00	0.00	31.14	31.14	31.14
Garganta de los Montes	348	0.00	0.00	28.25	28.25	28.25
Gargantilla del Lozoya y Pinilla de Buitrago	318	28.30	28.28	44.51	44.51	44.62
Villavieja del Lozoya	258	0.00	0.00	0.00	0.00	46.53
Redue	247	0.90	0.90	0.04	0.04	0.04
Alameda del Valle	195	0.00	0.00	25.58	25.58	25.58
Gascones	188	0.00	0.00	0.27	0.27	2.94
Pinilla del Valle	187	0.00	0.00	0.00	0.00	28.91

Source: authors

The most inhabited municipality is Torremocha de Jarama (table 5), but in 1990 and 2000, the municipality with the largest urban area was Ribatejada, in 2006, 2012, and 2018 it was Brea de Tajo.

However, Torremocha de Jarama occupies the second position in terms of the urban area occupied by the municipalities in 2018. Also, it is noteworthy that in 1990 there is no urban area registered for 15 municipalities, in 2000 for 14, in 2006 and 2012 for 2 municipalities and in 2018 all municipalities have urban areas.

Regarding land use continuous urban fabric:

Table 6. Difference in hectares between 2018 and 1990 of the land classified as continuous urban fabric in the Municipalities of the Madrid Autonomous Community

Municipality	Difference 2018-1990	Municipality	Difference 2018-1990
Alcalá de Henares	259.54	Navalcarnero	2.02
San Martín de la Vega	102.20	Leganés	1.21
Parla	90.17	Fuente el Saz de Jarama	-0.38
Torres de la Alameda	74.84	Guadarrama	-0.51
Velilla de San Antonio	73.58	Pozuelo de Alarcón	-2.42
Ciempozuelos	56.83	Torrejón de Velasco	-5.92
Arganda del Rey	48.90	Pinto	-8.96
Cenicientos	44.64	San Lorenzo de El Escorial	-9.53
Móstoles	44.24	Rozas de Madrid, Las	-10.83
Colmenar de Oreja	39.38	Villaviciosa de Odón	-15.35
Chinchón	33.56	Alcorcón	-18.82
Valdemoro	28.93	Majadahonda	-20.37
Villarejo de Salvanis	27.27	Galapagar	-22.67
Villaconejos	25.92	Humanes de Madrid	-23.65
Valdemorillo	25.26	Torrejón de Ardoz	-29.52

continues on following page

Table 6. Continued

Municipality	Difference 2018-1990	Municipality	Difference 2018-1990
Griñón	19.67	Molar, El	-33.60
Morata de Tajuma	14.84	San Sebastián de los Reyes	-41.46
Campo Real	10.29	Aranjuez	-46.23
Colmenar Viejo	3.51	Getafe	-55.27
Villa del Prado	3.31	Collado Villalba	-55.42
Boadilla del Monte	3.20	Alcobendas	-95.74
Álamo, El	2.77	Fuenlabrada	-96.11

Source: authors

The variation of the area destined to continuous urban fabric between 2018 and 1990, has been experienced in 44 municipalities of the 179 that exist in the Community of Madrid (Table 6), that is, approximately 25% or a quarter of them. Among which there is an increase in 24 municipalities and a decrease in 20 municipalities. Therefore, only 13.40% of the municipalities concentrate an increase in the urban area. Also, if the average increase is 43.17 hectares, only 9 municipalities are above the average. Among them, Alcalá de Henares stands out with an approximate growth of 6 times the average, and San Martín de la Vega and Parla with a growth of more than two times the average. However, they are also above average growth Torres de la Alameda, Velilla de San Antonio, Ciempozuelos, Arganda del Rey, Cenicientos and Móstoles

As for land use discontinuous urban fabric:

Table 7. Difference in hectares between 2018 and 1990 of the discontinuous urban fabric land in the Municipalities of the Madrid Autonomous Community

Municipality	Difference 2018-1990	Municipality	Difference 2018-1990
Madrid	2122.43	Torrelodones	121.73
Pozuelo de Alarcón	931.15	Estremera	120.25
Getafe	849.23	Chapinería	115.09
Rozas de Madrid, Las	771.90	Valdemorillo	113.36
San Sebastián de los Reyes	547.55	Lozoyuela-Navas-Sieteiglesias	113.35
Torrejón de Ardoz	535.07	Camarma de Esteruelas	110.58
Alcorcón	508.10	Pedrezuela	110.51
Valdemoro	503.16	Chinchón	107.55
Villanueva de la Cajada	432.61	Torremocha de Jarama	107.37
Móstoles	424.72	Tres Cantos	107.08
Aranjuez	388.90	Talamanca de Jarama	105.28
Boalo, El	381.96	Quijorna	104.49
Leganés	380.98	Molinos, Los	100.34
Arganda del Rey	364.74	Alpedrete	99.98
San Martín de Valdeiglesias	359.49	San Martín de la Vega	99.84

continues on following page

Table 7. Continued

Municipality	Difference 2018-1990	Municipality	Difference 2018-1990
Boadilla del Monte	355.76	Álamo, El	97.33
Navalcarnero	350.34	Collado Mediano	91.68
Fuenlabrada	340.89	Brea de Tajo	87.20
Villaviciosa de Odón	331.45	Algete	85.09
Parla	326.00	Fresnedillas de la Oliva	81.17
Majadahonda	324.38	Valdeavero	80.26
Alcobendas	320.52	Coslada	76.22
Guadarrama	288.85	Colmenar del Arroyo	75.36
Moralzarzal	281.65	Fresno de Torote	66.12
Galapagar	267.44	Navalafuente	64.50
Colmenar Viejo	261.58	Manzanares el Real	62.25
Escorial, El	256.21	Rozas de Puerto Real	61.27
Villalbilla	241.04	Valdeolmos-Alalpardo	60.51
Villanueva del Pardillo	231.81	Ciempozuelos	56.89
Meco	229.55	Cercedilla	56.29
Cadalso de los Vidrios	205.03	Nuevo Baztán	56.20
Colmenarejo	195.49	Venturada	52.61
Becerril de la Sierra	191.46	Villarejo de Salvanis	52.16
Brunete	188.71	Pelayos de la Presa	51.73
Fuente el Saz de Jarama	187.17	Santa Mar	50.91
Robledo de Chavela	186.65	Ambite	48.02
Collado Villalba	177.39	Zarzalejo	45.34
Miraflores de la Sierra	172.13	Navalagamella	42.67
San Lorenzo de El Escorial	171.76	Corpa	38.11
San Agustín	169.07	Pozuelo del Rey	35.56
Griñón	161.77	Valverde de Alcalá	29.84
Valdetorres de Jarama	160.23	Lozoya	29.50
Mejorada del Campo	160.18	Batres	25.75
Pinto	159.62	ValdepiIlagos	22.57
Paracuellos de Jarama	159.16	Torres de la Alameda	16.92
Bustarviejo	158.10	Gargantilla del Lozoya y Pinilla de Buitrago	16.32
Cubas de la Sagra	151.41	Hoyo de Manzanares	9.20
Soto del Real	141.65	Valdemaqueda	7.32
Navacerrada	141.03	Villar del Olmo	2.39
San Fernando de Henares	138.17	Redue	-0.85
Fuentidueñas	132.03	Ribatejada	-3.90
Colmenar de Oreja	129.12	Patones	-8.90
Moraleja de Enmedio	126.63	Alcal	-17.29

continues on following page

Table 7. Continued

Municipality	Difference 2018-1990	Municipality	Difference 2018-1990
Villa del Prado	123.53	Aldea del Fresno	-52.59
Cabrera, La	122.51		

Source: authors

The variation of discontinuous urban fabric soil between 2018 and 1990 has been recorded in a greater number of municipalities (Table 7) than for the use of continuous urban fabric land. Because it has been registered in 109 municipalities, that is, more than half 62.64% of the municipalities varied the area occupied for discontinuous urban fabric. Among them almost the total, 104 experienced an increase in the surface and only 5 a decrease. The average growth rate is 204.73 hectares. Therefore, urban sprawl has been greater for the land use discontinuous urban fabric than for continuous urban fabric.

Above this average are 31 municipalities. In this regard, the municipality of Madrid stands out whose growth in this land use has been 10 times above the average. Likewise, the increase in this land use in the municipalities of Pozuelo de Alarcón and Getafe is noteworthy, with a growth of more than 4 times the average, Las Rozas de Madrid with a growth of more than 3 times the average and the municipalities of San Sebastián de los Reyes, Torrejón de Ardoz, Alcorcón, Valdemoro, Villanueva de la Cajada and Móstoles with an increase of more than 2 times the average.

SOLUTIONS AND RECOMMENDATIONS

The lowest urban expansion produced in some of the analyzed areas until 2018, among other factors, it could have been caused by the lack of adequate accessibility offered by transport infrastructures, and urban transport to the capital. If it were intended that an urban sprawl continues in the areas with less accessibility, it could be sought precisely an improvement in access to the city of Madrid, and, as a consequence, that transport infrastructures and public transport services in these more isolated areas of the capital are precisely increased. In this way, they could accommodate urban sprawl as it has happened in other areas with better transport infrastructure.

All municipalities should also have a sustainable spatial planning plan. In this way, they would be provided with adequate tools to achieve an optimal model of urban development. In the municipalities, where urban development has been lower in recent years analyzed, these tools should be available as soon as possible since once an inadequate urbanization process begins, the effects produced in the territory are practically irreversible.

FUTURE RESEARCH DIRECTIONS

This research provides information on the dynamics of urban sprawl produced in the Madrid Autonomous Community between 1990 and 2018. The thematic cartography obtained and the alphanumeric information extracted could serve as a reference or cartographic basis for the analysis of the urban sprawl of this Autonomous Community after the last year analyzed 2018.

CONCLUSION

There are three distinct strips according to the different land uses in the Madrid Autonomous Community that run obliquely. In this regard, in the northern strip predominate the uses of forest and semi-natural land areas, wetlands, and water bodies. Also, in the central strip the lands destined to artificial surfaces and agricultural areas prevail. And finally, in the third strip located to the south, there is a superiority of agricultural areas and forests and semi-natural areas.

In addition, the result obtained on the extension of urban land uses allows us to affirm that between 1990 and 2000 urban expansion is scarce. However, as the extension of urban centres increases, this effect is more intense in the part located to the northwest, north, and south of the municipality of Madrid. Equally, this urban expansion of the municipalities located south of the city of Madrid seems to be consolidated between 2000 and 2006. Consequently, in this area, a surrounding belt of great hegemony and urban development is formed. Subsequently, urban sprawl decreases in intensity year after year. Even if this trend continues, there could be a stagnation of urban expansion in areas where it had hitherto occurred with less energy.

Also, taking into account the population in the municipalities analyzed, it can be said that, in the most inhabited municipalities, in almost all of them except in Alcorcón between 2012 and 2018, urban growth has always been progressive and continuous. Thus, the enormous influence of the city of Madrid and the capacity of urban expansion of the municipalities most influenced by the capital is evident. In addition, hectares of the urban area has always been measured in the most inhabited municipalities, in all the years analyzed. On the contrary, in the case of the less inhabited municipalities, in the first years analyzed there were numerous municipalities without urban areas. However, later these same municipalities have been occupying land for urban use. As a result, the urbanization process in the latter municipalities has been less intense, because these municipalities were farther away from the capital.

ACKNOWLEDGMENT

This research was supported by the Consejería de Economía, Ciencia y Agenda Digital from Junta de Extremadura and by the European Regional Development Fund of the European Union to the Research Group of Environment and Spatial Planning [grant number RNM031]; and the National Funds Fundação para a Ciência e a Tecnologia, I.P. (Portuguese Foundation for Science and Technology) (VALORIZA—Research Centre for Endogenous Resource Valorization). [grant number UIDB/05064/2020]. Thanks also to Antonio López-Peláez Manoja for help with revising the English version of this text.

REFERENCES

Barnes, K. B., Morgan, J., Roberge, M. C., & Lowe, S. (2012). *Sprawl Development: Its patterns, consequences, and measurement*. Academic Press.

Burton, E., Williams, K., & Jenks, M. (1996). The compact city and urban sustainability: conflicts and complexities. In E. Burton, K. Williams, & M. Jenks (Eds.), *The compact city: a sustainable urban form* (pp. 231–247). Taylor & Francis Group.,

Carmona, F., & González, I. (2018). *Cartografía de la producción: dinámicas espaciales de localización de los tejidos industriales del área urbana de Madrid (1956-2014). Ciudad y formas urbanas: perspectivas transversales: Libro de resúmenes* [Production mapping: spatial dynamics of location of the industrial fabrics of the urban area of Madrid (1956-2014). City and urban forms: transversal perspectives: Book of summaries]. Academic Press.

Copernicus Europe's eyes on Earth. (2022). *CORINE Land Cover*. Retrieved 01/24/2022 from https://land.copernicus.eu/pan-european/corine-land-cover

Córdoba, R., & Morcillo, D. (2020). Marco territorial de la producción de espacio en la región funcional de Madrid [Territorial framework of the production of space in the functional region of Madrid]. *Ciudades: Revista del Instituto Universitario de Urbanística de la Universidad de Valladolid*, (23), 71–93. doi:10.24197/ciudades.23.2020.71-93

Daher, A. (2013). El sector inmobiliario y las crisis económicas [The real estate sector and economic crises]. *Eure (Santiago), 39*(118), 47-76.

Díaz-Pacheco, J., & García-Palomares, J. C. (2014). Urban Sprawl in the Mediterranean Urban Regions in Europe and the Crisis Effect on the Urban Land Development: Madrid as Study Case. *Urban Studies Research, 807381*, 1–13. Advance online publication. doi:10.1155/2014/807381

Domingo, D., Palka, G., & Hersperger, A. M. (2021). Effect of zoning plans on urban land-use change: A multi-scenario simulation for supporting sustainable urban growth. *Sustainable Cities and Society, 69*, 102833. doi:10.1016/j.scs.2021.102833

Dowall, D., & Clarke, G. (1996). *A framework for reforming urban land policies in developing countries*. The International Bank for Reconstruction and Development.

Dupuy, G. (1995). *Les territorios de l'automobile* [Automobile territories]. Anthropos-Economica.

European Environment Agency (EEA). (2020). *CORINE land cover*. Available at: www.eea.europa.eu/publications/COR0-landcover

Fernández, C., Cabezas, A., & Pinedo, P. (1998). *Urbanismo y arquitectura en el Madrid actual* [Urbanism and architecture in today's Madrid] (P. L. López Algora, Ed.). Comunidad de Madrid, Dirección General de Centros Docentes. http://www.madrid.org/bvirtual/BVCM000851.pdf

Fernández, R. (2006). *El tsunami urbanizador español y mundial: sobre sus causas y repercusiones devastadoras, y la necesidad de prepararse para el previsible estallido de la burbuja inmobiliaria* [The Spanish and global urbanizing tsunami: on its devastating causes and repercussions, and the need to prepare for the foreseeable bursting of the real estate bubble]. Virus Editorial. https://dspace-libros.metabiblioteca.com.co/jspui/bitstream/001/397/8/84-96044-74-2.pdf

Filion, P., Bunting, T., & Warriner, K. (1999). The Entrenchment of Urban Dispersion: Residential Preferences and Location Patterns in the Dispersed City. *Urban Studies (Edinburgh, Scotland), 36*(8), 1317–1347. doi:10.1080/0042098993015

Gaja i Diaz, F. (2013). Tras el tsunami inmobiliario. Salir del atolladero [After the real estate tsunami. Get out of the quagmire]. In Observatorio Metropolitano de Madrid (Ed.), *Paisajes devastados después del ciclo inmobiliario: impactos regionales y urbanos de la crisis* [Landscapes devastated after the real estate cycle: regional and urban impacts of the crisis] (pp. 313-354). Traficantes de sueños. https://www.traficantes.net/sites/default/files/pdfs/Paisajes%20Devastados-TdS.pdf

García-Palomares, J. C. (2010). Urban sprawl and travel to work: The case of the metropolitan area of Madrid. *Journal of Transport Geography*, *18*(2), 197–213. doi:10.1016/j.jtrangeo.2009.05.012

García-Palomares, J. C., & Gutiérrez, J. (2007). La ciudad dispersa: cambios recientes en los espacios residenciales de la Comunidad de Madrid [The dispersed city: recent changes in the residential spaces of the Community of Madrid]. *Anales de geografía de la Universidad Complutense, 27*(1), 45.

Giuliano, G., & Small, A. (1999). The determinants of growth of employment subcenters. *Journal of Transport Geography*, *7*(3), 189–201. doi:10.1016/S0966-6923(98)00043-X

Johnson, M. (2001). Environmental Impacts of Urban Sprawl: A Survey of the Literature and Proposed Research Agenda. *Environment and Planning A. Economy and Space*, *33*(4), 717–735. doi:10.1068/a3327

Naredo, J. M. (2014). El modelo inmobiliario español y sus consecuencias [The Spanish real estate model and its consequences]. *Boletín CF+S,* (44), 13-28. http://habitat.aq.upm.es/boletin/n44/ajnar.html

Plata, W., Gómez, M., & Bosque, J. (2008). Análisis de factores explicativos del crecimiento urbano en la Comunidad de Madrid a través de métodos estadísticos y sig [Analysis of explanatory factors of urban growth in the Community of Madrid through statistical methods and sig]. In *Tecnologías de la Información Geográfica para el desarrollo territorial: XIII Congreso Nacional de Tecnologías de la Información Geográfica* [Geographic Information Technologies for Territorial Development: XIII National Congress of Geographic Information Technologies]. Academic Press.

Romero, J., Brandis, D., & Melo, C. (2015). El giro neoliberal de las políticas para la ciudad en España. Balance a partir de los ejemplos de Madrid y Valencia [The neoliberal turn of policies for the city in Spain. Balance based on the examples of Madrid and Valencia]. *Boletín de la Asociación de Geógrafos Españoles*.

Solís, E. (2008). El horizonte urbano madrileño: Más allá de la región político-administrativa [The Madrilenian Urban Horizont: Beyond the Administrative Region]. *Anales de Geografía de la Universidad Complutense*, *28*(1), 133.

Solon, J. (2009). Spatial context of urbanization: Landscape pattern and changes between 1950 and 1990 in the Warsaw metropolitan area, Poland. *Landscape and Urban Planning*, *93*(3), 250–261. doi:10.1016/j.landurbplan.2009.07.012

Spanish Statistical Office. (2021). *Municipal Register*. Retrieved from https://www.ine.es/jaxiT3/Tabla.htm?t=2881

Squires, G. (2002). *Urban sprawl: Causes, consequences, & policy responses*. The Urban Institute.

Wang, X., Shi, R., & Zhou, Y. (2020). Dynamics of urban sprawl and sustainable development in China. *Socio-Economic Planning Sciences*, *70*, 100736. doi:10.1016/j.seps.2019.100736

Xu, Q., Yang, R., Dong, Y.-X., Liu, Y.-X., & Qiu, L.-R. (2016). The influence of rapid urbanization and land use changes on terrestrial carbon sources/sinks in Guangzhou, China. *Ecological Indicators, 70,* 304–316. doi:10.1016/j.ecolind.2016.05.052

Zhang, T. (2001). *Public participation in China's urban development.* Taylor & Francis Group.

ADDITIONAL READING

Akturk, E., & Guney, K. (2021). Vegetation Cover Change Analysis of Phytogeographic Regions of Turkey Based on CORINE Land Cover Datasets from 1990 to 2018. *Kastamonu University Journal of Forestry Faculty, 21*(2), 150–164. doi:10.17475/kastorman.1000406

Cieslak, I., Bilozor, A., Zrobek-Sokolnik, A., & Zagroba, M. (2020). The Use of Geographic Databases for Analyzing Changes in Land Cover-A Case Study of the Region of Warmia and Mazury in Poland. *ISPRS International Journal of Geo-Information, 9*(6), 358. Advance online publication. doi:10.3390/ijgi9060358

Cinar, I. (2015). Assessing the Correlation between Land Cover Conversion and Temporal Climate Change-A Pilot Study in Coastal Mediterranean City, Fethiye, Turkey. *Atmosphere, 6*(8), 1102–1118. doi:10.3390/atmos6081102

Grigorescu, I., Kucsicsa, G., Popovici, E.-A., Mitrica, B., Mocanu, I., & Dumitrascu, M. (2021). Modelling land use/cover change to assess future urban sprawl in Romania. *Geocarto International, 36*(7), 721–739. doi:10.1080/10106049.2019.1624981

Kaya, I. A., & Gorgun, E. K. (2020). Land use and land cover change monitoring in Bandirma (Turkey) using remote sensing and geographic information systems. *Environmental Monitoring and Assessment, 192*(7), 430. Advance onlinc publication. doi:10.100710661-020-08411-1 PMID:32535792

Maggiore, G., Semeraro, T., Aretano, R., De Bellis, L., & Luvisi, A. (2019). GIS Analysis of Land-Use Change in Threatened Landscapes by Xylella fastidiosa. *Sustainability, 11*(1), 253. Advance online publication. doi:10.3390u11010253

Stoica, I.-V., Virghileanu, M., Zamfir, D., Mihai, B.-A., & Savulescu, I. (2020). Comparative Assessment of the Built-Up Area Expansion Based on Corine Land Cover and Landsat Datasets: A Case Study of a Post-Socialist City. *Remote Sensing, 12*(13), 2137. Advance online publication. doi:10.3390/rs12132137

Ustaoglu, E., & Aydinoglu, A. C. (2019). Regional Variations of Land-Use Development and Land-Use/Cover Change Dynamics: A Case Study of Turkey. *Remote Sensing, 11*(7), 885. Advance online publication. doi:10.3390/rs11070885

Zomlot, Z., Verbeiren, B., Huysmans, M., & Batelaan, O. (2017). Trajectory analysis of land use and land cover maps to improve spatial-temporal patterns, and impact assessment on groundwater recharge. *Journal of Hydrology (Amsterdam), 554,* 558–569. doi:10.1016/j.jhydrol.2017.09.032

KEY TERMS AND DEFINITIONS

Autonomous Community: In Spain, it is an administrative-territorial division formed by one or more provinces. It has specific territorial limits and is endowed with legislative autonomy and executive powers, provided that it is not common with the rest of the Spanish State.

Geographic Information System: The system that uses georeferenced graphic and alphanumeric information stored in a spatial database, and to solve problems by analysing the spatial relationships of the elements that represent the Earth, using raster or vector models.

Land Use: Particular use made of a specific part of the land surface and which is essential for territorial planning.

Municipality: In Spain, it is the administrative-territorial division corresponding to the basic local entity in the territorial organization of the Spanish State governed by a town hall.

Population Density: An indicator that measures the number of people living in a given extension. The number of inhabitants living per square meter is usually related.

Remote Sensing: A technique for collecting information from the Earth's surface, by using sensors installed on space platforms, from the record of the electromagnetic interaction produced between the ground and the sensor, to interpret the Earth's surface.

Suburbanization: The action of urbanizing geographical spaces located on the periphery of the city in urban areas, or in the rural environment the rehabilitation of the old habitat of a village, construction of new houses on the outskirts of the villages, construction of houses and subdivisions of the houses on the outskirts of the village.

Urban Sprawl: Dispersed residential urban development that separates itself from other urban land uses, occupying rural land and heading towards the periphery of central urban areas in an uncoordinated manner. Usually through homes that tend to be single-family.

Chapter 5
Hydraulic Planning in Insular Urban Territories:
The Case of Madeira Island – Porto da Cruz

Sérgio António Neves Lousada
https://orcid.org/0000-0002-8429-2164
University of Madeira, Portugal

Leonardo Gonçalves
University of Madeira, Portugal

Erdem Kaya
Dumlupinar University, Turkey

ABSTRACT

This study aims to examine the flood propensity of the main watercourse of Porto da Cruz drainage basin and, if relevant, to propose two methodologies to alleviate the impacts (i.e., detention basin sizing and riverbed roughness coefficient adjustment). Geomorphological data were obtained from the watershed characterization process and used through the SIG ArcGIS software for the flood propensity assessment and then for the calculation of the expected peak flow rate for a return period of 100 years through the Gumbel Distribution. Subsequently, the drainage capacity of the river mouth was verified using the Manning-Strickler equation in order to establish whether the river mouth of the watershed has the capacity to drain the entire volume of rainwater in a severe flood event. In conclusion, the results demonstrate that the river mouth of the Porto da Cruz watershed does not have the capacity to drain the rain flow for the predetermined return period; thus, the detention basin was sized through the Dutch method and the simplified triangular hydrograph method.

DOI: 10.4018/978-1-6684-4548-8.ch005

INTRODUCTION

While people were civilizing, they took care to establish their settlements on the banks of rivers and streams. For this reason, floods and overflows have been one of the most destructive natural disasters on human settlements (Wetherald & Manabe, 2007). Especially after the industrial revolution, the increase in temperature and the climate change seen as a result of it have also been the trigger of extreme natural events. In parallel with this, floods and overflows gradually cause great material and moral losses on societies (Trenberth & Dai & Rasmussen & Parsons, 2007).

Today, flood and overflow events are not limited to rivers and streams, but they have also started to be effective in densely populated metropolises (Barichivichi & Gloor & Peylin & Brienen & Schöngart & Espinoza & Pattnayak, 2018). The reason for this is that the rain waters cannot drain fast enough in the city centers. Due to the high rate of concretization, the connection of the soil with water is cut off and the soil loses its ability to absorb water to a great extent (Gonçalves & Lousada, 2020). Especially the settlements established around the flood areas are frequently faced with floods today. In addition, mountainous regions can be exposed to destructive floods because they are suitable for high-speed flows due to gravity (Gonçalves & Rodrigues & Curitiba & Torres & Lousada, 2020).

Today, as a result of the increase in flood disasters due to extreme natural events, communication and technological interventions have become necessary (Gonçalves & Lousada & Lis, 2020). Traditional methods are based on changing the direction of the flow from the source of the flood (Tucci & Porto & Barros, 1995). Although this method works in the source area, it causes further exacerbation of floods in the downstream areas. For this reason, it has become essential to find a solution to the problem with more technological and effective interventions without directing the flow from one region to another (Lousada & Gonçalves & Atmaca, 2022).

Taking all of this into consideration, this study aims to perform a hydrological analysis of the parish of Porto da Cruz, estimating its peak flowrate for a recurring period of 100 years, and establish a comparison with its watershed´s river mouth drainage capacity. Based on the premise that the streams river mouth hydraulic features are insufficient to drain the expected peak flow rate, it was designed a detention basin to further control the downstream flowrate and avoiding the need to change the stream cross section. This structural measure was also chosen as it results in considerably reduced urban effects and can be complemented with small changes of the streambed and walls roughness coefficient, thus increasing the drainage capacity of the river mouth without affecting its cross section.

MATERIALS AND METHODS

Area of Study

The area of study - i.e. the Porto da Cruz watershed - is located on the southern slope of Madeira Island, latitude 32º 46’ N and longitude 16º 49’ W, as can be observed in Figure 1 (Fernandes, 2009; Prada & Gaspar & Sequeira & Nunes, 2005). This watershed belongs to the homonymous parish, Porto da Cruz, and corresponds to the precipitation catchment area that supplies the main stream of the parish.

Figure 1. The Porto da Cruz watershed
(Source: Authors by ESRI ArcGIS, 2021)

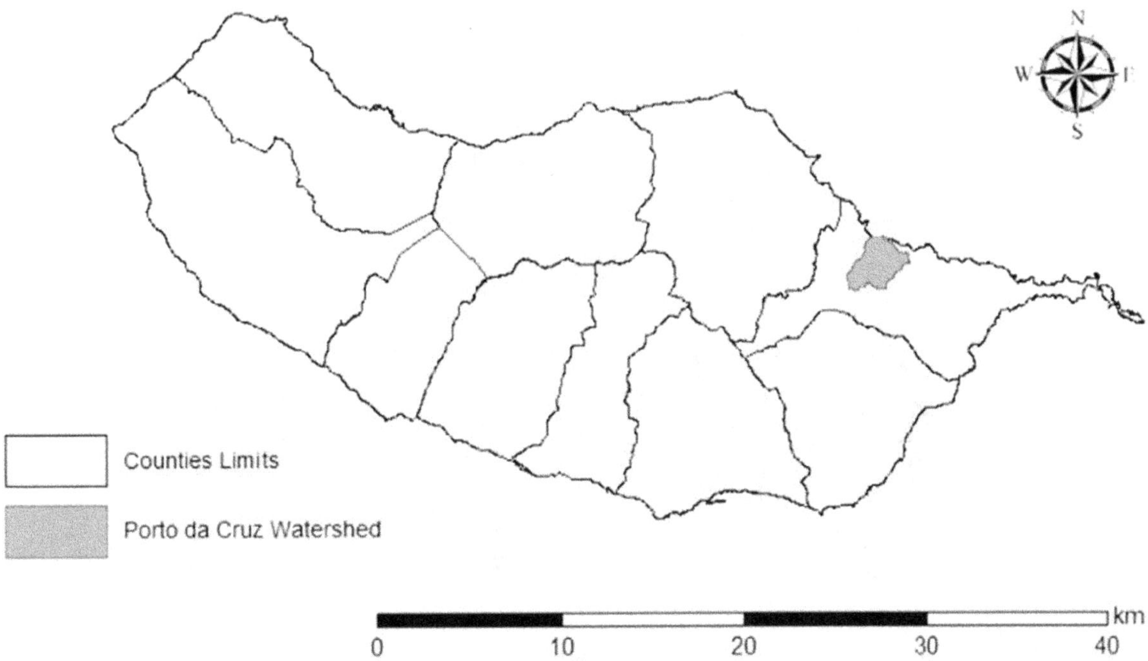

As well as Funchal, the main municipality of the island, the Porto da Cruz watershed suffers from severe flooding problems, as observed in 2010 and 2013, when the region was impacted with significant losses, both in material and human aspects. Because it is located in a considerably urbanized area, the Porto da Cruz watershed has a significant rate of soil sealing from the buildings and pavements present (Moura & Silva & Gonçalves & Lousada, 2020; Silva & Lousada & Moura & Gonçalves, 2020). In addition, as can be observed in Figure 2, the river mouth of the Porto da Cruz watershed has the presence of vegetation and sedimentation, which significantly reduces the drainage capacity of the channel.

The state of conservation of the stream is practically the same throughout its length within the urban perimeter, which can be easily verified *in situ*. The excess of sedimentation and presence of vegetation is explained by the significantly reduced slope, which reduces the drainage velocity and the capacity to drag sediments with larger grain size.

Schematic of the Methodology

The methodology adopted can be synthesized in 6 stages, as shown in Figure 3.

The methodology adopted started from a literature review, with the objective of gathering all the necessary information for the hydrological and morphometric characterization of Porto da Cruz basin. Therefore, through the aforementioned review, the methodologies recommended by several authors in the field were selected, as a way to obtain a more reliable analysis of the flood propensity in the basin under study. Finally, the various steps of the adopted methodology presented in Figure 3 are described below.

Figure 2. State of Conservation of Porto da Cruz main watercourse river mouth

Figure 3. Organogram of the adopted methodology

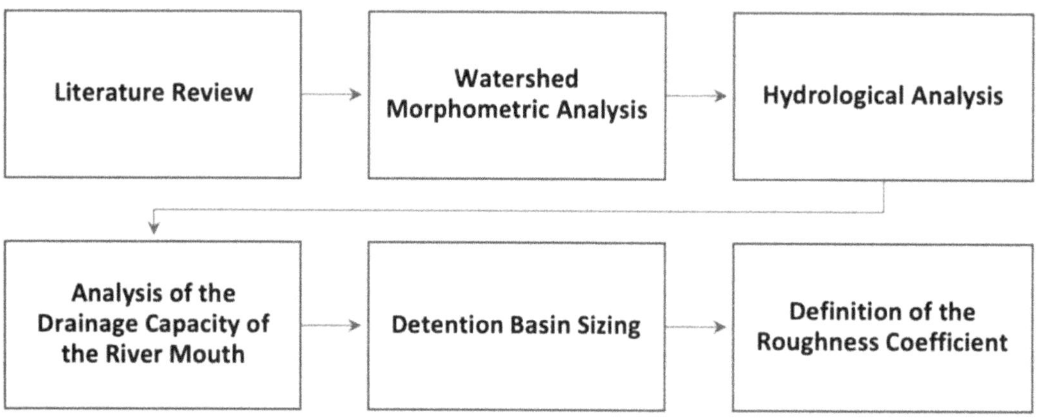

Morphometric Characterization of the Watershed

The main parameters used for the morphometric characterization of a watershed are (Tucci, 1993; Gonçalves & Lousada, 2020; Gonçalves & Rodrigues & Curitiba & Torres & Lousada, 2020; Lousada & Gonçalves & Velarde, 2020; Chow, 1988; Gonçalves, 2020; Lousada & Camacho, 2018; Christofoletti, 1980):

- **Gravelius Index – K_C:** establishes the relationship between the actual perimeter of the basin under study and the perimeter of a perfectly circular basin, yet with the same area, thus establishing how close the geometric shape of the watershed is to a perfect circle (Lousada & Gonçalves &

Velarde, 2020; Lousada & Camacho, 2018). It can be calculated using Equation 1. It is a dimensionless parameter, and the closer it is to "1", the more rounded the watershed is and the greater its propensity to flood (Lousada & Camacho, 2018).

$$K_C = P / 2 \times \sqrt{\pi \times A} \tag{1}$$

with:

P = Perimeter of the watershed, km;
A = Watershed area, km^2.

- **Elongation Factor – K_L:** establishes the relationship between the watershed under study with an equivalent rectangle of the same area. It can be calculated using Equation 2. This parameter is dimensionless and, if it is higher than "2", it is an elongated watershed (Lousada & Camacho, 2018).

$$K_L = \frac{L_E}{l_E} = \frac{\dfrac{K_C \times \sqrt{A}}{1.128} \times \left|1 + \sqrt{1 - \left(\dfrac{1.128}{K_C}\right)^2}\right|}{\dfrac{K_C \times \sqrt{A}}{1.128} \times \left|1 - \sqrt{1 - \left(\dfrac{1.128}{K_C}\right)^2}\right|} \tag{2}$$

with:

L_E = Equivalent length, km;
l_E = Equivalent width, km;
K_C = Gravelius Index, dimensionless;
A = Watershed area, km^2.

- **Shape Factor – K_F:** establishes the relationship between the average width and length of the watershed under analysis. It can be calculated using Equation 3. This parameter is dimensionless and also indicates the degree of elongation of the watershed; however, the lower the value obtained the greater the elongation and lower the propensity to flood. Values equal to the unit "1" indicate a square basin.

$$K_F = A / L_B^2 \tag{3}$$

with:

A = Watershed area, km^2;
L_B = Length of the watershed, km.

The length of the watershed can be determined by measuring the distance between the mouth of the stream and the furthest point. However, it is noted that the length of the watershed does not necessarily correspond to the length of its main watercourse, since the latter generally has a larger dimension due to its sinuosity. Based on the MDE file (whose spatial resolution is 5 meters) provided by LREC-RAM (the Regional Civil Engineering Laboratory of the Autonomous Region of Madeira), it became possible to perform the morphological characterization of the Porto da Cruz watershed and its respective main watercourse. The morphological data collected in this analysis were used in the empirical equations of several authors, as a way to avoid single method restrictions.

A priori, a morphometric analysis must have its origin in the establishment of a hierarchy of the watercourses - e.g. Strahler or Shreve - according to their order or magnitude, respectively (Christofoletti, 1980). The aforementioned classifications can be obtained by performing a hydrological analysis of the DEM file, obtaining the rasters "flow accumulation" and "flow direction", coming from the "flow order" tool (Gonçalves, 2020). Furthermore, it is evidenced that Strahler hierarchy is deeply associated with the branching or bifurcation ratio of a watershed, in which each degree of branching or bifurcation can be calculated using the Equation 4 (Tucci, 1993; Chow, 1988; Tucci & Porto & Barros, 1995; Gonçalves, 2020; Lousada & Camacho, 2018; Christofoletti, 1980; Strahler, 1964).

$$R_B = \frac{N_i}{N_{i+1}} \quad (4)$$

with:

N_i = Number of watercourses classified as "i"; dimensionless;
N_{i+1} = Number of watercourses classified as "i+1", dimensionless.

This is a dimensionless coefficient, obtained by the ratio of the number of watercourses of a given order by the number of watercourses of the immediately higher order. The average bifurcation value can be calculated based on Equation 5.

$$\overline{R_B} = \sqrt[i-1]{\prod_{i=1}^{i-1} \frac{N_i}{N_{i+1}}} = \sqrt[i-1]{N_1} \quad (5)$$

with:

N_i = Number of watercourses classified as "i"; dimensionless;
N_{i+1} = Number of watercourses classified as "i+1", dimensionless;
N_1 = Number of first-order watercourses.

Like the previous parameters, the average bifurcation ratio is a dimensionless parameter, since it only represents an arithmetic mean of the bifurcation ratios. Moreover, one of the fundamental parameters for a good morphometric characterization of a watershed refers to the concentration time, which establishes the time required for the entire watershed area to contribute to the process of rainfall drainage to the river mouth (Tucci, 1993; Chow, 1988; Christofoletti, 1980; Strahler, 1964).

Since the equations for determining the concentration time have an empirical character, each methodology will indicate different values for the same parameter, where it is prudent to establish the arithmetic mean, as a way to avoid extreme values. For the present study the arithmetic mean was used for the values obtained by Kirpich (Equation 6), Témez (Equation 7) and Giandotti´s (Equation 8) (Gonçalves, 2020).

$$t_C = 57 \times (L^3/(H_{MAX} - H_{MIN}))^{0.385} \tag{6}$$

with:

t_C = Time of concentration, minutes;
L = Length of the main watercourse, km;
H_{MAX} = Maximum height of the main watercourse, m;
H_{MIN} = Minimum height of the main watercourse l, m.

$$t_C = \left(\frac{L}{i^{0.25}}\right)^{0.76} \tag{7}$$

with:

t_C = Time of concentration, hours;
L = Length of the main watercourse, km;
i = Slope of the main watercourse, m/m.

$$t_C = \frac{\left(4+\sqrt{A}\right)+\left(1.5\times L\right)}{0.8\times\sqrt{H_M}} \tag{8}$$

with:

t_C = Time of concentration, hours;
A = Watershed area, km^2;
L = Length of the main watercourse, km;
H_M = Average height of the watershed, m.

Precipitation Analysis

The hydrological study of precipitation consisted of a probabilistic analysis of extreme events of short duration and high intensity that occurred in the Porto da Cruz watershed. The data used in this analysis were obtained through public sources such as the National Water Resources Information System (SNIRH), which provide precipitation data recorded automatically at precipitation stations. Regarding the probabilistic methodology adopted, the Gumbel Distribution was chosen, since it showed a better fit with the data obtained and the expected projection for watersheds on the island of Madeira (Camacho, 2015). Therefore, the maximum annual daily precipitation can be calculated based on Equation 9.

$$P_{EST} = P_M + S' \times K_T \tag{9}$$

with:

P_{EST} = Estimated annual maximum daily precipitation, mm;
P_M = Average annual precipitation, mm;
S' = Sample standard deviation, mm;

K_T = Frequency Factor, dimensionless.

with:

$$S' = \left(\frac{\Sigma (X_i - X_M)^2}{n'} \right)^{0.5} \tag{10}$$

with:

X_i = Sample value, mm;
X_M = Sample mean, mm;
n' = Number of samples.

$$K_T = -\frac{6^{0.5}}{\pi} \times \left\{ 0.577216 + \ln\left(\ln\left(\frac{T_R}{T_R - 1} \right) \right) \right\} \tag{11}$$

with:

T_R = Return period, years.

After determining the daily precipitation for an extreme phenomenon, the precipitation intensity with a specific duration can be obtained using Equation 12.

$$I = \frac{P_{EST} \times k}{t_C} \tag{12}$$

with:

I = Precipitation intensity, mm/h;
P_{EST} = Estimated annual maximum daily precipitation, mm;
t_C = Time of concentration, hours;
k = Time distribution coefficient, dimensionless.

with:

$$k = 0.181 \times \ln(t_C) + 0.4368 \tag{13}$$

with:

t_C = Time of concentration, hours.

The time distribution coefficient presents itself as a primary parameter, since the annual maximum daily precipitation is only valid for events lasting 24 hours. Therefore, since the duration of precipitation is equal to the time of concentration of the watershed, using the total amount of daily precipitation in the hydrologic analysis would cause oversizing of hydraulic structures (Vieira & Barreto & Figueira & Lousada & Prada, 2016; Gonçalves, 2020).

Drainage Capacity of the River Mouth and Peak Flow rate

The drainage capacity of the river mouth was calculated using the Manning-Strickler equation (Equation 14), and then compared to the expected flow for an extreme event with a recurrence time of 100 years. To calculate the expected flow, empirical methodologies already widespread at the global level were used, namely: Forti (Equation 16); Rational (Equation 17); Giandotti (Equation 18), and Mockus (Equation 19).

$$Q_M = \left(\frac{1}{n}\right) \times A_M \times R^{\frac{2}{3}} \times \sqrt{i} \tag{14}$$

with:

Q_M = Drainage capacity of the river mouth, m^3/s.
A_M = Area of the river mouth cross-section, m^2;
R = Hydraulic radius, m;
i = Average slope of the river mouth region, m/m;
n = Roughness coefficient of the riverbed and walls, $m^{-1/3}$ s, Table A1.

with:

$$R = \frac{B + 2 \times h}{A_M} \tag{15}$$

with:

B = Width of the river mouth runoff section, m;
h = Height of the river mouth runoff section, m;
A_M = Area of the river mouth cross-section, m^2.

Both the width and height of the stream in the region of the mouth were obtained through bibliographies that have conducted previous studies in the region (Gonçalves, 2020), with first parameter being confirmed through the georeferencing process.

$$Q_{Forti} = A \times \left(b \times \frac{500}{125 + A}\right) + c \tag{16}$$

with:

Q_{Forti} = Peak flow rate by Forti, m^3/s;
A = Watershed area, km^2;
b = 2.35 for maximum daily precipitation below 200 mm and 3.25 for values above 200 mm;
c = 0.5 for maximum daily precipitation below 200 mm and 1 for values above 200 mm.

$$Q_{\text{Rational}} = \frac{C \times I \times A}{3.6} \tag{17}$$

with:

Q_{Rational} = Peak flow rate by the rational methodology, m^3/s;
C = Surface runoff coefficient, Table A2;
I = Precipitation intensity, mm/h;
A = Watershed area, km^2.

$$Q_{Giandotti} = \frac{\lambda \times A \times P_{MAX}}{t_C} \tag{18}$$

with:

$Q_{\text{Giandotti}}$ = Peak flow rate by Giandotti, m^3/s;
λ = Reduction coefficient, Table A3;
A = Watershed area, km^2;
P_{MAX} = Precipitation height for a duration equal to the concentration time, mm;
t_C = Concentration time, hours.

$$Q_{\text{Mockus}} = \frac{2.08 \times A \times P_{\text{EST}} \times C}{\sqrt{t_C} + 0.6 \times t_C} \tag{19}$$

with:

Q_{Mockus} = Peak flow rate by Mockus, m^3/s;
A = Watershed area, km^2;
P_{EST} = Estimated precipitation, cm;
C = Surface runoff coefficient, Table A2;
t_C = Concentration time, hours.

When sizing hydraulic works, one of the main design criteria is to establish a Fill Rate value lower than 85%, as a way to determine a significant margin of error and thus ensure the safety of the population and their property (Gonçalves, 2020; Porto & Filho & Tucci & Bidone, 1993). Therefore, for the control of the flow that will follow to the river mouth, it becomes necessary to implement runoff regulation mechanisms, such as spillways.

As mentioned above, the Fill Rate value must be calculated using Equation 20 and, in case the run-off capacity of the river mouth is insufficient to drain the rain flow in the watershed and guarantee the established safety margin, it becomes necessary to size mitigation mechanisms, such as detention basins.

$$FR = \frac{Q_P}{Q_M} \times 100 \tag{20}$$

with:

FR = Fill Rate, %;

Q_P = Peak flow rate of each methodology, m^3/s;

Q_M = Drainage capacity of the river mouth, m^3/s.

The Fill Rate parameter, in short, refers to the ability of a drainage section to drain a given flow. In this sense, if the Fill Rate value is greater than 100%, it means that the section is no longer able to drain the entire volume of water without overflowing (Gonçalves, 2020).

Detention Basin Sizing

As previously mentioned, if the runoff capacity of the river mouth is insufficient to drain all the rain and drained water superficially to the mouth, it becomes necessary to size a spillway to regulate the flow that will reach the mouth, keeping the downstream flow below the expected limit. For the present study, a Cipolleti type of spillway was chosen, as this model has characteristics that reduce turbulence in the regions of contact with water and facilitate the runoff (Gonçalves & Lousada, 2020; Gonçalves & Rodrigues & Curitiba & Torres & Lousada, 2020; Vieira & Barreto & Figueira & Lousada & Prada, 2016). The sizing of this spillway can be done using Equation 21.

After the definition and regularization of the flow that will be drained to the river mouth, it becomes possible to estimate the volume of water that will be retained along the detention basin. For that, two methodologies were used, named Dutch Method (Equation 22), and Simplified Triangular Hydrograph – STH (Equation 23).

$$Q_S = 1.86 \times L_{SD} \times H_D^{1.5} \quad (21)$$

with:

Q_S = Flow drained by spillway, m^3/s;

L_{SD} = Width of the sill, m^3/s;

H_D = Height of the waterline above the sill, m.

$$V_A = (Q_P - Q_S) \times t_C \times 3600 \quad (22)$$

$$V_A = \frac{(Q_P - Q_S) \times \left(2 \times t_C - 2 \times \left[Q_S / \{Q_P / t_C\}\right]\right)}{2} \quad (23)$$

with:

V_A = Storage Volume, m^3;

Q_P = Peak flow rate of each methodology, m^3/s;

Q_S = Flow drained by the spillway, m^3/s;

t_C = Concentration time, hours.

Note that Equation 23 was formulated based on the geometric analysis of the STH (Figure A1), considering an event with a total duration of at least twice the concentration time of the watershed under study. This consideration took into account that the last rainwater particle to arrive at the river mouth would occur at the last instant of precipitation and in the farthest region, which suggests that it would need a value equal to the time of concentration to be summarily considered in the amount drained by the mouth (Gonçalves, 2020).

Regarding the methodologies chosen, it is due to the fact that the Dutch Method does not contemplate the delay and damping of the precipitation hydrograph, causing a certain oversizing of the structure (David & Carvalho, 2008), as shown in Figure 4, where qs: runoff capacity of the spillway; tc: concentration time; tmax: maximum precipitation duration (base); td: time delay until the beginning of water accumulation in the detention basin; Ha,MAX: maximum storage capacity; i(tMAX): precipitation intensity corresponding to the maximum duration.

Figure 4. (A) Dutch Method; (B) STH Method
(Source: David & Carvalho, 2008)

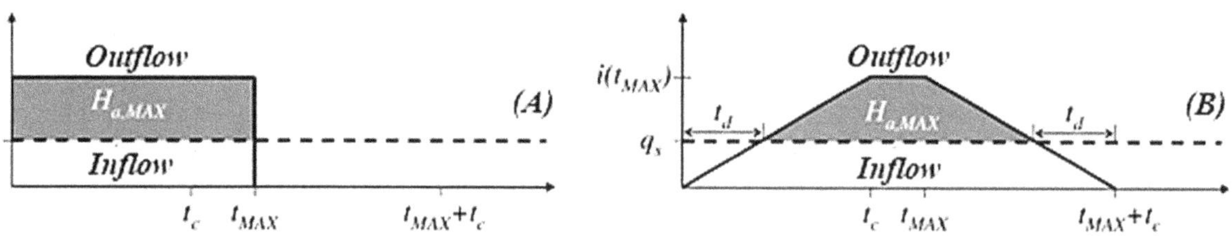

Therefore, it is verified that in the Dutch Method, storage begins simultaneously with precipitation, which does not correspond to reality, since storage will only begin when the flow drained downstream is greater than the spillway's runoff capacity.

Modification of the Roughness Coefficient_

In addition, a structural mitigation measure taken into consideration was the modification of the roughness coefficient of the walls and riverbed of the watercourse, thus avoiding the reduction of drainage capacity due to friction. This methodology consists of changing the value of the parameter "n" in the Manning-Strickler equation in order to improve the flow of a given watercourse by considering another material for the coverage of the stream wall and riverbed (Gonçalves, 2020).

RESULTS

The results presented here correspond to the data obtained by applying the formulas described above. Therefore, to evaluate the morphometric characteristics of the main watercourse of Porto da Cruz, an individual analysis of each of the parameters listed in Table 1, was undertaken, correlating them with reference values suggested in several bibliographies.

Table 1. Parameters calculated or extracted from ArcGIS

Parameter	Unit of Measurement	Value
Area	km^2	5.658
Perimeter	Km	13.930
Length of Main Watercourse	Km	4.189
Maximum Height of Main Watercourse	M	661.990
Minimum Height of Main Watercourse	M	0.000
Average Concentration Time	hours	0.933
Gravelius Coefficient of Compactness	dimensionless	2.487
Elongation Factor	dimensionless	17.388
Shape Factor	dimensionless	0.723
Number of Watercourses	units	100.000
Average Bifurcation Ratio	dimensionless	4.273
Strahler Classification	dimensionless	4.000

The first parameter to be analyzed concerns the watershed area, which has a primary role for the analysis of the volume of water drained to the river mouth. This parameter can be classified as: Very Large> 20 km^2; Large> 10 km^2; Medium> 1 km^2 and Small <1 km^2 (Beck & Bruijnzeel & Van Dijk & McVicar & Scatena & Schellekens, 2013). In this sense, as can be seen in the previous table, the watershed under study has a "Medium" classification, indicating a higher propensity to flood compared to smaller watersheds. However, it is noted that the aforementioned reference values are arbitrary and may differ according to the type of analysis to be performed (Beck & Bruijnzeel & Van Dijk & McVicar & Scatena & Schellekens, 2013), as well as the propensity to flood.

As shown in Figure 5, the Porto da Cruz watershed has borders with significantly higher altitudes than the central region, which denotes a steep slope that tends to supply the main watercourse quickly and thus increase the volume of water present in the stream that, consequently, will be redirected to the river mouth.

Regarding the drainage system of the watershed under study, Figure 6, the high number of watercourses indicates a high drainage capacity, i.e., the basin has many low and medium-order watercourses that will supply the main watercourse. Moreover, this index represents the hydrographic behavior of a given area, where the fundamental aspect is its propensity to generate new watercourses. Therefore, basins with high hydric density have a tendency to present a greater amount of ephemeral channels, precisely due to the ability to generate new watercourses (Gonçalves, 2020; Christofoletti, 1980).

The analysis of precipitation was based on data provided by the National Information System on Water Resources (SNIRH) (SNIRH, 2021), through sample data from sixteen years, presented in Table A4 and Figure A2. Thus, through the probabilistic processing of Gumbel Distribution the values presented in Table 2 were obtained.

Figure 5. Hypsometric map – DEM file
(Source: Authors by ESRI ArcGIS, 2021)

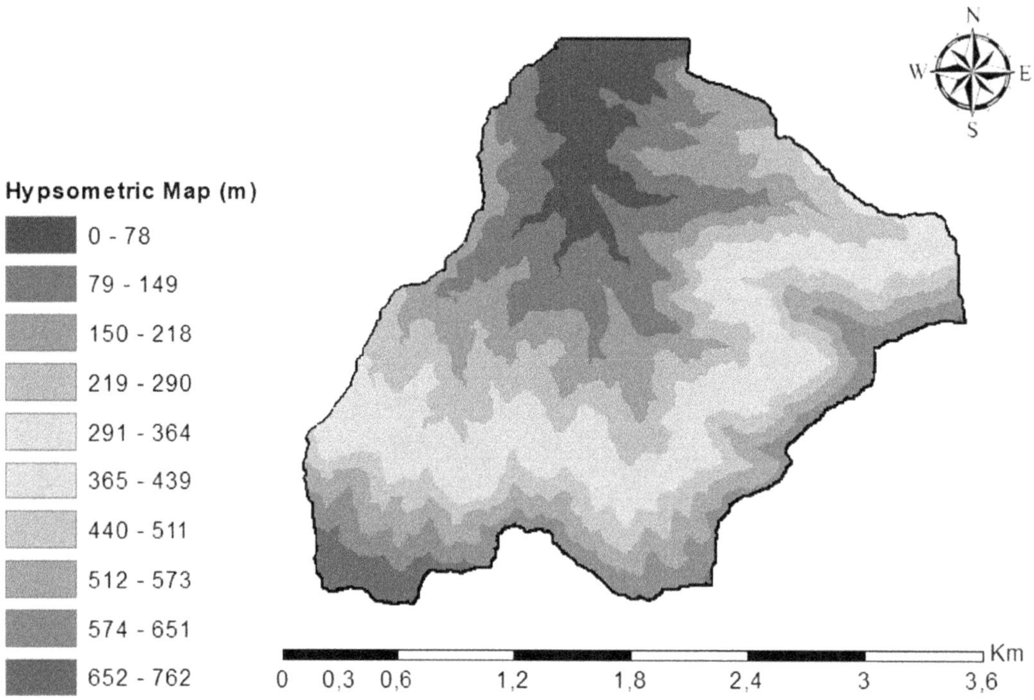

Figure 6. Strahler classification
(Source: Authors by ESRI ArcGIS, 2021)

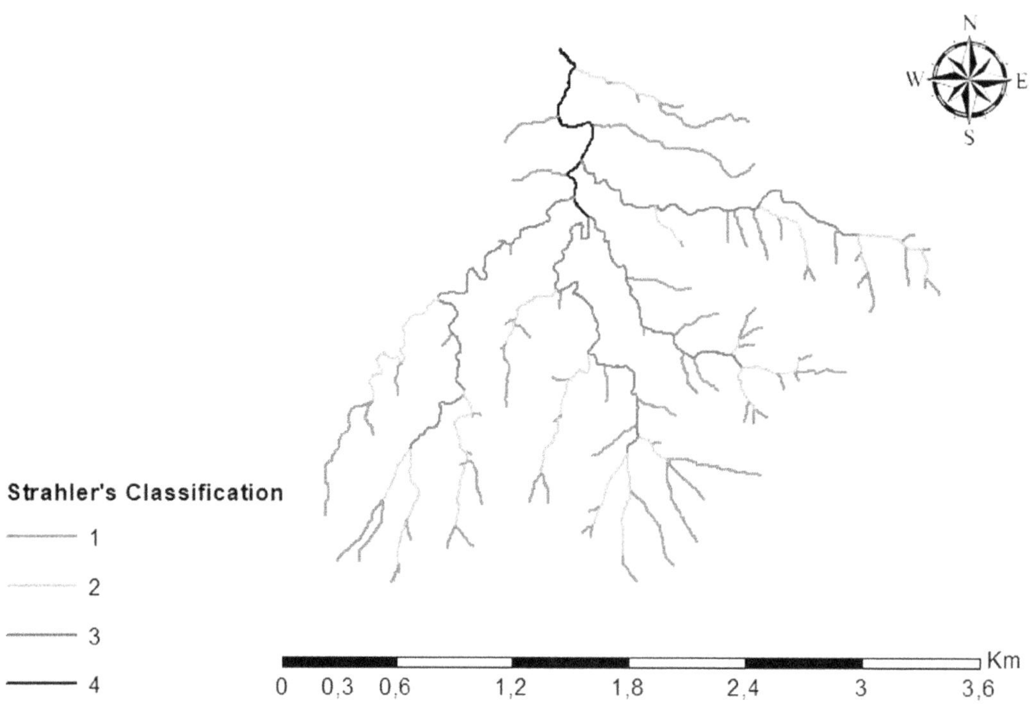

Table 2. Precipitation parameters

Parameter	Symbol	Unit of Measurement	Value
Average Annual Precipitation	P_M	mm	112.025
Standard Deviation	S'	mm	53.844
Frequency Factor	K_T	dimensionless	3.1367
Time Distribution Coefficient	k	dimensionless	0.424
Annual Maximum Daily Precipitation	P_{EST}	mm	112.025
Precipitation Intensity	I	mm/h	127.736

After determining the precipitation intensity estimated for a recurrence time of 100 years, the calculation of the peak flow rates was made, Table 3, through the formulas and methodologies indicated in the previous chapter. The surface drainage coefficient used in the Rational Method corresponds to the value of 0.500, Table 4, since the region under study is a peripheral region and with the presence of commercial buildings. This value corresponds, basically, to the percentage of water that tends to drain superficially, i.e., 50% of the total precipitation.

Table 3. Peak flow rate

Methodology	Flow (m^3/s)
Forti	76.027
Rational	100.202
Giandotti	267.483
Mockus	108.324

Table 4. Surface drainage coefficient adopted

Urban Areas		
Land Occupation		Surface Drainage Coefficient
Commercial Area	City Center	0.700 – 0.950
	Peripheral Areas	**0.500** – 0.700

(Source: Chow, 1964)

The value of the reduction coefficient (λ) used in the calculation of the flow by Giandotti's methodology is shown in Table 5.

Table 5. Adopted Giandotti's reduction coefficient

Area (km^2)	λ	Equivalent "C"
< 300	0.346	1.250

(Source: Gonçalves, 2016)

Regarding the analysis of the drainage capacity of the river mouth, the Manning-Strickler equation was used to verify the need to implement a detention basin, where the values obtained are summarized in Table 6. However, it is necessary to emphasize that the stream walls and bed have different roughness coefficients. Therefore, the drainage capacity of the river mouth was calculated by means of the weighted mean of the respective coefficients, where the stone and mortar walls are in good condition (n=0.020) and the riverbed consists of a stony and vegetated surface in poor condition (n=0.040). Another important factor to note is the very low slope in the river mouth region, which tends to decrease the water flow velocity and, consequently, the drainage capacity of the section. As a way to simulate a critical situation, it was opted to take into account a slope of 0.01 m/m in the reference section.

Table 6. Assessment of the need for Detention Basin implementation

Parameter	*Unit of Measurement*	*Value*
Width of the River Mouth	m	10.000
Height of the River Mouth	m	4.000
Drainage Capacity of the River Mouth	m^3/s	218.946
Fill Rate - Forti (pre-regularization)	%	35
Fill Rate - Rational (pre-regularization)	%	46
Fill Rate - Giandotti (pre-regularization)	%	122
Fill Rate - Mockus (pre-regularization)	%	49

As shown in Table 6, the Fill Rate is higher than the established limit of 85% for the Rational, Giandotti and Mockus methods, which denotes the need to implement mitigation and flow control measures for the river mouth region. Therefore, a detention basin was sized with the flows found in the aforementioned methodologies, considering the urban and spatial limitations coming from already existing infrastructures near the stream.

Since the sizing of the detention basin depends on the flow that exceeds the limit established for the river mouth, first a Cipolletti trapezoid spillway was sized for the regularization and control of the flow that will drain downstream. The characteristics of the spillway are presented in Table 7.

Table 7. Application of the Cipolletti Spillway

Parameter	**Unit of Measurement**	**Value**
Width of the Spillway	m	8.000
Height of the Spillway Sill	m	4.000
Spillway Outflow	m^3/s	119.040
Fill Rate – Giandotti (post-regularization)	%	54

A posteriori, the detention basins were sized using the Dutch Method and the STH, which are simplified methodologies that neglect several factors and, thus, can present an oversizing of the hydraulic structure. Additionally, the height and width of the detention basin were fixed with the same values of the existing cross-section, as a way to reduce the environmental and urban impacts from the implementation works of this measure. Therefore, the only geometric variable of the detention basin will be its length, which is limited by the total length of the main watercourse.

After making the calculations with the aforementioned methodologies, the values shown in Table 8 were found.

Table 8. Detention Basin sizing

Parameter	Unit of Measurement	Value
Width	m	10.000
Height	m	4.000
Length – Dutch Method (Giandotti)	m	12464.759
Length – STH Method (Giandotti)	m	6917.472

Finally, the modification of the roughness coefficient was used as an alternative measure to mitigate the impacts of the flood, while maintaining the characteristics of the riverbed vegetation. In this context, the values shown in Table 9 correspond, more precisely, to the improvement of the condition of conservation of the river bed, as a way to reduce the loss of drainage capacity caused by excessive friction between the fluid and the covering material.

Table 9. Modification of the roughness coefficient

Parameter	Unit of Measurement	Value
Wall Roughness Coefficient - Modified	$m^{-1/3}$	0.011
Riverbed Roughness Coefficient - Modified	$m^{-1/3}$	0.030
Drainage Capacity of the River Mouth - Modified	m^3/s	316.004
Fill Rate – Giandotti (post-modification)	%	84.64

In short, the modified roughness coefficients of the walls correspond to the surface with concrete finishing in very good condition, while the riverbed remains with the stony and vegetated characteristic; yet, in good condition. The values used for these coefficients are summarized in Table 10.

Table 10. Adopted roughness coefficient

Channel Typology	Very Good	Good	Regular	Bad
Channel with stony and vegetated slope	0.025	**0.030**	0.035	0.040
Surface with concrete finishing	**0.011**	0.012	0.013	0.015

(Source: Gonçalves, 2016)

DISCUSSION

Since the primary objective of this research was to verify the need to implement simplified measures to mitigate the impacts of floods in the Porto da Cruz watershed, the use of the detention basin proved to be effective in controlling the flow at the river mouth, which can be characterized as a structural measure (Lousada & Castanho, 2021). Initially, the Fill Rate values were 122% for the Giandotti methodology, respectively, passing to 54% after the implementation of the measure. Therefore, it is evidenced that the chosen mitigation measure allows the river mouth to work below the 85% threshold, as indicated earlier. Furthermore, this study corroborates the flood risk analysis prepared by the Regional Directorate for Territorial Ordering and Environment (DROTA), Table 11, indicating an acceptable accuracy for the present study.

Table 11. Watersheds with high flood risk

Municipality	Watershed
Machico	Porto da Cruz

(Source: DROTA, 2017)

As the proposal of this research aims to cause the least possible impact on the existing waterway and its surroundings, since the presence of natural elements and values in cities is nowadays a fundamental condition for the environmental recovery of the urban territory (Castanho & Fernández & Loures & Pozo, 2017). Moreover, natural and urban systems are coexistent components in the same space, and their integrated management is a primary requirement of regional space and a condition for achieving the goal of sustainability in territories and cities (Fadigas, 2015; Loures, 2011), because urban growth or development can occur in a disorganized way, creating urban voids (Castanho & Lousada & Gómez & Escorcio &Cabezas & Pozo-Fernández & Loures, 2019).

Thus, it was decided not to change the dimensions of the cross-section of the streams, both in width and height, with the length being the only dimensional variable. Based on this assumption, the Dutch Method presented considerable oversizing, as the total length of the detention basin found is greater than the total length of the main waterway, which denotes the need to change one more of the dimensions - i.e. height or width. In this context, despite the efficiency in flow regularization, the Dutch Method is not applicable when considering the urban conditions previously imposed.

The same problem was found for the STH Method, because despite considering the damping of the flood hydrograph, the length of the main watercourse is also insufficient for the implementation of this measure.

With regard to the change in the roughness coefficient of the stream, it was decided to remain with the stony and vegetated characteristics of the riverbed, only improving the conditions of the state of conservation. This option was made because the total removal of stones, sediments and vegetation from the riverbed would be a costly, time consuming and frequent process. For the walls, maintenance should not be frequent, since wear by abrasion would occur exclusively in an alluvial that tends to drain a high volume of water and large granular sediments.

Despite being a simple measure, the modification of the roughness coefficient of the stream presented considerable effectiveness, allowing the river mouth to work below the pre-established filling limit. Furthermore, it becomes necessary to highlight that both methodologies can be implemented together, in order to reduce the length required for the detention basin and, thus, optimizing the sizing performed.

It should be noted that the methodologies used are simplified in nature, i.e., they do not consider local peculiarities. Therefore, this causes the results to have a very high safety margin and, consequently, causes the oversizing of hydraulic structures.

CONCLUSION

The results obtained in this study denote that the Porto da Cruz watershed is susceptible to floods during an extreme precipitation event, which is reinforced by the Flood Risk Report prepared by DROTA. This occurs because the stream bed has characteristics that are very unfavorable to surface runoff, as it has a surface with a significant presence of stones and vegetation. These characteristics of the riverbed reduce the velocity of the water flow and, consequently, reduce the drainage capacity, especially in areas with very low slope, such as the river mouth. Therefore, as one of the methodologies – i.e. Giandotti – presented a Fill Rate higher than the aforementioned limit, there was a need to implement mitigating measures.

Regarding the proposed simplified solutions, the Dutch Method and Simplified Triangular Hydrograph Method did not present an applicability coherent with the pre-established urban premises, because it suggests very high lengths for the detention basin, higher than the length of the main watercourse.

However, the change in the roughness coefficient also presented satisfactory and effective results for the mitigation of flood impacts, because it is a relatively simple measure to be proceeded and still meets the demand of the watershed under study.

Due to the impossibility of considering all the aspects that make up a more complete and effective analysis in this study, other analyses can be carried out in order to complement or optimize the results presented here, such as the analysis of the drainage capacity of the urban hydraulic system implemented, in order to reduce the storage volume of the detention basins; analysis of sediment deposition according to the entrainment velocity present in the main watercourse (Yu & Wu & Sui & Ni & Whitcombe, 2020); verification of the deterioration of the artificial canal walls by abrasion, and the analysis of the maximum time to proceed with maintenance (silting and desilting processes); influence on the degradation of the artificial water channels according to the quality of the water discharged by its tributaries (Shrestha & Wang, 2020; Li & Li & Shi, 2020); analysis of the urban growth perspective of the studied municipalities and its influence on the flow increase; budgeting for the implementation of the mitigation measures of this study; analysis of the influence of the tide level on the drainage in artificial water channels, direct relationship with the risk of downstream flooding; characterizing the influence of artificial water channels in territorial planning processes - i.e. adaptation to rural watersheds.

The obtained findings corroborate with the ideas and conclusions put forward on similar studies which used case study analysis and simulations as drivers for scientific development (Vargues & Loures, 2008; Nunes & Ramos-Miras & Lopez-Pineiro & Loures & Gil & Coelho & Loures, 2014).

Funding: This study was conducted under the project "Projeto Turismo: caraterização, impacto e sustentabilidade do turismo da Madeira" (M1420-01-0145-FEDER-000007), co-financed by the Madeira Program 14-20, Portugal 2020 and the European Union, through the European Regional Development Fund, which belongs to the Research Group CITUR - Madeira - Centre for Tourism Research, Development and Innovation.

REFERENCES

Barichivichi, J., Gloor, E., Peylin, P., Brienen, R., Schöngart, J., Espinoza, J., & Pattnayak, K. (2018). Recent intensification of Amazon flooding extremes driven by strengthened Walker circulation. *Science Advances*, *4*(9), 8785. doi:10.1126ciadv.aat8785 PMID:30255149

Beck, H. E., Bruijnzeel, L. A., Van Dijk, A. I. J. M., McVicar, T. R., Scatena, F. N., & Schellekens, J. (2013). The impact of forest regeneration on streamflow in 12 mesoscale humid tropical catchments. *Hydrology and Earth System Sciences*, *17*(7), 2613–2635. doi:10.5194/hess-17-2613-2013

Camacho, R. F. (2015). *Caractererização, Simulação (à escala) e Modelação do Escoamento em Canais Artificiais. Aplicação a caso de estudo.* Dissertação para obtenção do grau de Mestre em Engenharia Civil pela Universidade da Madeira, Funchal, Portugal.

Castanho, R., Lousada, S., Gómez, J. M. N., Escorcio, P., Cabezas, J., Pozo-Fernández, L., & Loures, L. (2019). *Dynamics of the Land Use Changes and the Associated Barriers and Opportunities for Sustainble. Development on Peripheral and Insular Territories.* Intechopen. doi:10.5772/intechopen.80827

Castanho, R. A. C., Fernández, J. C., Loures, L., & Pozo, L. F. (2017). Evolución del procedimiento de planeamiento urbano en la península ibérica y sus huellas en el paisaje urbano. Retos de futuro. *Revista Científica Monfragüe Desarrollo Resiliente*, *8*(2).

Chow, V. (1964). *Handbook of Applied Hydrology*. McGraw-Hill.

Chow, V. T. (1988). Applied Hydrology. McGraw-Hill.

Christofoletti, A. (1980). *Geomorfologia*. Edgard Blücher.

David, L. M. & Carvalho, R. F. (2008). *Bacias de retenção para controlo de cheias: Reflexão sobre os métodos de dimensionamento.* 13º Encontro Nacional de Saneamento Básico.

DROTA. (2017). *Identificação das Zonas Críticas a Cartografar Considerando os Impactos da Ocorrência de Inundações (Com Risco Potencial Significativo).* Direção Regional de Ordenamento do Território e Ambiente.

Fadigas, L. (2015). *Urbanismo e Território - As Políticas Públicas*. Editorial Estampa.

Fernandes, M. J. P. (2009). *Riscos no Concelho da Ribeira Brava: Movimentos de vertente cheias rápidas e inundações*. Dissertação para obtenção do grau de Mestre em Dinâmicas Sociais e Riscos Naturais pela Universidade de Coimbra, Coimbra, Portugal.

Gonçalves, J. (2016). *Caracterização do coeficiente de rugosidade e seu efeito no escoamento em Canais Naturais. Simulação e Modelação (à escala) no laboratório de Hidráulica. Aplicação às ribeiras do Funchal.* Dissertação para obtenção do grau de Mestre em Engenharia Civil pela Universidade da Madeira, Funchal, Portugal.

Gonçalves, L. B. (2020). *Análise teórico-prática do risco de cheias no Arquipélago da Madeira – Caso de estudo dos concelhos do Funchal, Machico, Ribeira Brava e São Vicente.* Dissertação para obtenção do grau de Mestre em Engenharia Civil pela Universidade da Madeira, Funchal, Portugal.

Gonçalves, L. B., & Lousada, S. A. N. (2020). Análise Probabilística de Cheias e o Uso de Bacias de Detenção como Medida Mitigadora: Aplicação à Bacia de Santa Luzia. *Revista Científica Monfragüe Desarrollo Resiliente, 13.*

Gonçalves, L. B., Lousada, S. A. N., & Lis, M. A. (2020). Utilização da Bacia de Detenção Para o Controlo e Regularização de Caudais a Jusante, sua Sustentabilidade. In *Espacios y Sociedades em Transformación* (1st ed., Vol. 6, pp. 505–518). Thomson Reuters Aranzadi.

Gonçalves, L. B., Rodrigues, W. T., Curitiba, A. S., Torres, H., & Lousada, S. A. N. (2020). *Utilização de Bacias de Detenção Para Mitigação de Cheias em Áreas Urbanas, Espacios y Sociedades em Transfo, Rmación* (1st ed., Vol. 6). Thomson Reuters Aranzadi.

Li, Z., Li, J. J., & Shi, X. P. (2020). A Two-Stage Multisite and Multivariate Weather Generator. *Journal of Environmental Informatics, 35*(2), 148–159.

Loures, L. (2011). *Planning and Design in Post-industrial Landscapes: East Bank Arade River – Lagoa, Case Study* [Ph.D Thesis]. University of Algarve, Portugal.

Lousada, S., & Castanho, R. A. (2021). GIS-based Assessment of Morphological and Hydrological Parameters of Ribeira dos Socorridos and Ribeira do Vigário Basins, Madeira Island, Portugal. *Current World Environment, 16*(2), 408–426. doi:10.12944/CWE.16.2.08

Lousada, S., Gonçalves, L., & Atmaca, A. (2022). Hydraulic Planning in Insular Urban Territories: The Case of Madeira Island—São Vicente. *Water (Basel), 2022*(14), 112. doi:10.3390/w14010112

Lousada, S. A. N., & Camacho, R. (2018). *Hidrologia, recursos hídricos e ambiente - Aulas Teóricas* (Vol. 1). Universidade da Madeira.

Lousada, S. A. N., Gonçalves, L. B., & Velarde, J. G. (2020). *Controlo e regularização de caudais excedentes por meio de bacia de detenção: simulação para bacia hidrográfica de João Gomes, Funchal. Espacios y Sociedades em Transformación* (1st ed., Vol. 6). Thomson Reuters Aranzadi.

Moura, A., Silva, A., Gonçalves, L., & Lousada, S. (2020). Numerical modelling of the flow rate in artificial water channels: Application to Ribeira Brava's stream. *Revista Brasileira de Planejamento e Desenvolvimento, Curitiba, 9*(1), 39–59. doi:10.3895/rbpd.v9n1.10974

Nunes, J. R., Ramos-Miras, J., Lopez-Pineiro, A., Loures, L., Gil, C., Coelho, J., & Loures, A. (2014). Concentrations of available heavy metals in Mediterranean agricultural soils and their relation with some soil selected properties: A case study in typical Mediterranean soils. *Sustainability (Switzerland), 6*(12), 9124–9138. doi:10.3390u6129124

Porto, R. L., Filho, K. Z., Tucci, C. E. M., & Bidone, F. (1993). *Drenagem urbana. Hidrologia: Ciência e Aplicação. 2ª Edição*. Editora da Universidade UFRGS.

Prada, S., Gaspar, A., Sequeira, M. M., & Nunes, A. (2005). Disponibilidades Hídricas na Ilha da Madeira. In C. de Lanzarote & C. I. de Aguas de Lanzarote (Eds.), *AQUAMAC – Técnicas e métodos para a gestão sustentável da água na Macaronésia. Disponibilidades Hídricas na Ilha da Madeira, Publisher: Instituto Tecnológico das Canarias* (pp. 261–294). Instituto Tecnológico de Canarias.

Shrestha, N. K., & Wang, J. (2020). Water Quality Management of a Cold Climate Region Watershed in Changing Climate. *Journal of Environmental Informatics*, *35*(1), 56–80.

Silva, A. R. F. & Lousada, S. A. N. & Moura, A. D. S. & Gonçalves, L. B. (2020). Modelação numérica do escoamento em canais artificiais: aplicação à ribeira da Ribeira Brava. *Revista Científica Monfragüe Desarrollo Resiliente, 13*.

SNIRH. (2021). *Sistema Nacional de Informação de Recursos Hídricos*. Disponível em: https://snirh.apambiente.pt/index.php?idMain=2&idItem=1&objCover=920123704&objSite=920685506

Strahler, A. N. (1964). Quantitative geomorphology of drainage basins and channel networks. McGraw-Hill.

Trenberth, K. E., Dai, A., Rasmussen, R. M., & Parsons, D. B. (2003). *The Changing Character of Precipitation*. American Meteorological Society. doi:10.1175/BAMS-84-9-1205

Tucci, C. E. M. (1993). *Controle de Enchentes. Hidrologia: Ciência e Aplicação. 2ª Edição*. Editora da Universidade UFRGS.

Tucci, C. E. M., Porto, R. L., & Barros, M. T. (1995). *Drenagem urbana. Universidade Federal do Rio Grande do Sul*. UFRGS.

Vargues, P., & Loures, L. (2008). Using Geographic Information Systems in Visual and Aesthetic Analysis: The case study of a golf course in Algarve. *WSEAS Transactions on Environment and Development*, *4*(9), 774–783.

Vieira, I. L. S., Barreto, V., Figueira, C., Lousada, S., & Prada, S. (2016). The use of detention basins to reduce flash flood hazard in small and step volcanic watersheds – a simulation from Madeira Island. *Journal of Flood Risk Management*. Advance online publication. doi:10.1111/jfr3.12285

Wetherald, R. T., & Manabe, S. (2007). Climate Change. IPCC.

Yu, B. Y., Wu, P., Sui, J., Ni, J., & Whitcombe, T. (2020). Variation of Runoff and Sediment Transport in the Huai River–A Case Study. *Journal of Environmental Informatics*, *35*(2), 138–147. doi:10.3808/jei.202000429

KEY TERMS AND DEFINITIONS

Hydraulics: Branch of science concerned with the practical applications of fluids, primarily liquids, in motion. It is related to fluid mechanics, which in large part provides its theoretical foundation. Hydraulics deals with such matters as the flow of liquids in pipes, rivers, and channels and their confinement by dams and tanks. Some of its principles apply also to gases, usually in cases in which variations in density are relatively small. Consequently, the scope of hydraulics extends to such mechanical devices as fans and gas turbines and to pneumatic control systems.

Hydrology: A science dealing with the properties, distribution, and circulation of water on and below the earth's surface and in the atmosphere.

Insular Territories: Independent territory that is composed of an island or an archipelago.

Spatial Analysis: Is the process of extracting or creating new information about a set of geographic features to perform routine examination, assessment, evaluation, analysis, or modeling of data in a geographic area based on pre-established and computerized criteria and standards.

Territorial Management: Mean the decision-making processes of economical and institutional social actors of a particular space in the process of appropriation and the use of territories.

Urban Planning: Is a technical and political process concerned with the control of the use of land and design of the urban environment, including transportation networks, to guide and ensure the orderly development of settlements and communities. It concerns itself with research and analysis, strategic thinking, architecture, urban design, public consultation, policy recommendations, implementation, and management.

Chapter 6
Extreme Precipitation Events in Chile:
Latitudinal and Altitudinal Variations

Javier Lozano-Parra
Universidad Autónoma de Madrid, Spain

Jacinto Garrido Velarde
University of Extremadura, Spain

Ignacio Aguirre
https://orcid.org/0000-0002-7722-9677
University of Saskatchewan, Canada

ABSTRACT

Extreme precipitation has not only detrimental effects on ecosystems and social and economic sectors, but it is a natural hazard that can trigger floods or soil erosion. This study tries to analyze the extreme rainfalls on different geomorphological units and geographical regions of Chile. For this, data from 87 meteorological stations latitudinally and altitudinally distributed and covering a long period (1980–2018) were used. Results showed that precipitation concentration displays an exponential curve where 30% of the rainiest days were concentrated in only 10% of days with precipitation, proving high irregularity. The decisive weight on annual precipitation falls on a few rainy days with very high rainfall amounts. For return periods > 100 years, extreme events of daily precipitation could reach 109 mm and 305 mm in Northern and Southern Andes Mountains, respectively, while in Northern and Southern Central Depression, their values could be 70 mm and 170 mm, respectively.

DOI: 10.4018/978-1-6684-4548-8.ch006

INTRODUCTION

Concern about extreme weather events has earned growing attention in the last decades because of their natural and socioeconomic impacts (*IPCC,* 2018). Scientific studies commonly accord on pattern change of some meteorological phenomena, for example, it is widely recognized that global temperature is increasing, while some extreme meteorological events are more difficult to predict owing to they are complex and vary regionally (*López-Moreno, Vicente-Serrano et al.,* 2010).

The study of extreme daily rainfalls is a subject of great interest due to hydrological issues resulting from the high daily amounts and the irregular temporal distribution in large areas, highlighting their role in the hydrological cycle. This type of rainfalls can come from high precipitation rates in short time intervals or large daily accumulations on successive days that end up saturating the soils (*Dingman,* 2015). However, despite their importance as a natural hazard, heavy daily rainfalls have frequently been less tackled than other weather issues, such as temperature anomalies, especially in South America (*Martin-Vide,* 2004; *Valenzuela and Garreaud,* 2019).

In regions with very great latitudinal and altitudinal gradients as Chile, the rainfall variability creates a large range of weather conditions in short distances: from the wettest and driest areas on the southern and on the northern, respectively, to wetter and drier areas on higher and lower elevations, respectively. The study of the extreme daily concentrations of rainfall and the definition of their return periods could be very useful in this country to minimize the vulnerability of society to extreme weather events and reduce the impact that they may cause (*Meehl, Karl et al.,* 2000). However, although current studies have increasingly used bigger spatiotemporal-scale databases of precipitation, their analysis have often focused either on multi-station averages and their spatiotemporal anomalies, or on climate modelling, with less attention on extreme events (*Garreaud, López et al.,* 2013; *Sippel, Mitchell et al.,* 2015). Therefore, an analysis of extreme rainfall events and their latitudinal and altitudinal variations in Chile is still needed.

In the study of extreme rainfalls, some challenging topics should especially be considered before proceeding to the analysis. On one hand, extreme rainfall events are, by definition, rare and stochastics events, so due to its both cyclic and chaotic variability, constant observations over a long time period are required (*Martin-Vide,* 2004). On the other hand, in terms of statistical structure, the extreme daily rainfalls do not exhibit a Gaussian distribution but rather a Generalized Extreme Value (GEV) distribution, which can be fitted through a distribution of extreme events such as those from annual maximum precipitations (*Zeder and Fischer,* 2020). Finally, the most extreme precipitation events that trigger the largest natural and socioeconomic impacts can show divergent behaviors over small geographic domains, so defining their frequency of exceedance by the return period can suppose a very useful tool to avoid these impacts. Nevertheless, short databases can cause bias in the estimation of return period, proving the dependence of this statistical tool on the size of used databases (*Zwiers, Zhang et al.,* 2011; *Easterling, Kunkel et al.,* 2016).

Due to the importance of precipitation for the maintenance and development of natural ecosystems as well as societies and economies in Chile, this work aims to define the extreme rainfall events in this country in different latitudinal and altitudinal regions, using spatially distributed long databases. Thus, the objectives of this study are the following: I) To analyze variations of annual precipitation on different altitudinal and latitudinal regions of Chile; II) to define the rainfall concentration of extreme events regarding rainy days; III) to determine the probability of non-exceedance and return periods of extreme events of precipitation.

STUDY AREA

Chile's surface is greater than 755000 km^2 and, from north to south, reaches a greater length than 4200 km (Figure 1). It has a population close to 18 million inhabitants, mainly concentrated in the central area of the country, between the Valparaiso and the Metropolitan regions.

The Chilean geology is especially conditioned by its location next to the western edge of the South American tectonic plate, which has allowed it to interact with the subducting plates located to the west, resulting in very heterogeneous and tangled geological features. Thus, its geomorphology can be considered extraordinarily varied and complex. The current reliefs have their origin in the erosion processes of material in higher areas and its subsequent accumulation and sedimentation in valley bottom and depressions. Likewise, the glacier, volcanic or fluvial processes have finally outlined the current reliefs throughout the country. The predominant reliefs are the Andes Mountain, the Coastal Mountain, the Intermediate Depression, and the Transverse Valleys.

The climate is conditioned by latitudinal and altitudinal gradients, and by the maritime influence of the Pacific Ocean. Thus, the most important climatic variations are produced by the influence of latitude and elevation, ranging almost all climatic types except for a few, such as the tropical rainy (Figure 1).

The soil types are very varied due to their different factors which influence on their formation, such as geomorphology, latitude, vegetation and climate variations. The main types of soils are Cambisols, Andosols, Regosols, and Leptosols.

The Chilean vegetation is also conditioned by latitudinal and altitudinal gradients. Consequently, it presents a great diversity of biomes characterized by strong endemism (it is estimated that there are approximately more than 6,300 endemic species). Despite this, the plant diversity is less than those found in neighboring countries such as Brazil.

MATERIAL AND METHODS

Geomorphology and Geographical Regions of Chile

In order to carry out an altitudinal and latitudinal analysis of precipitation in Chile, its geomorphology was considered. The spatial distribution of the geomorphological units was obtained from the geomorphologic classification for Chile developed by *Paskoff* (1996), who initially distinguished 62 different relief units. To simply them, they were grouped into 8 units following the classification of *IGM* (2018), of which only two were selected: Andes Mountain and Central Depression (Figure 2). It was justified because they depict a latitudinal distribution, show altitudinal difference, and cover a significant area of Chile. Finally, they were distributed from north to south in 2 geographical regions (Figure 2), according to the *BNC* (2016). The northern geographical region (GR1) includes desert plains and pre-altiplanic mountain ranges and corresponds to the administrative zones of Arica and Parinacota, Tarapacá, Antofagasta and part of Atacama. The southern geographical region (GR2) includes central lacustrine zones and glacio-volcanic plains. It extends approximately from the Araucanía to the Los Lagos Region.

Figure 1. Chile in the South America region and its main climates according to the Köppen climate classification
(based on Rubel and Kottek (2010))

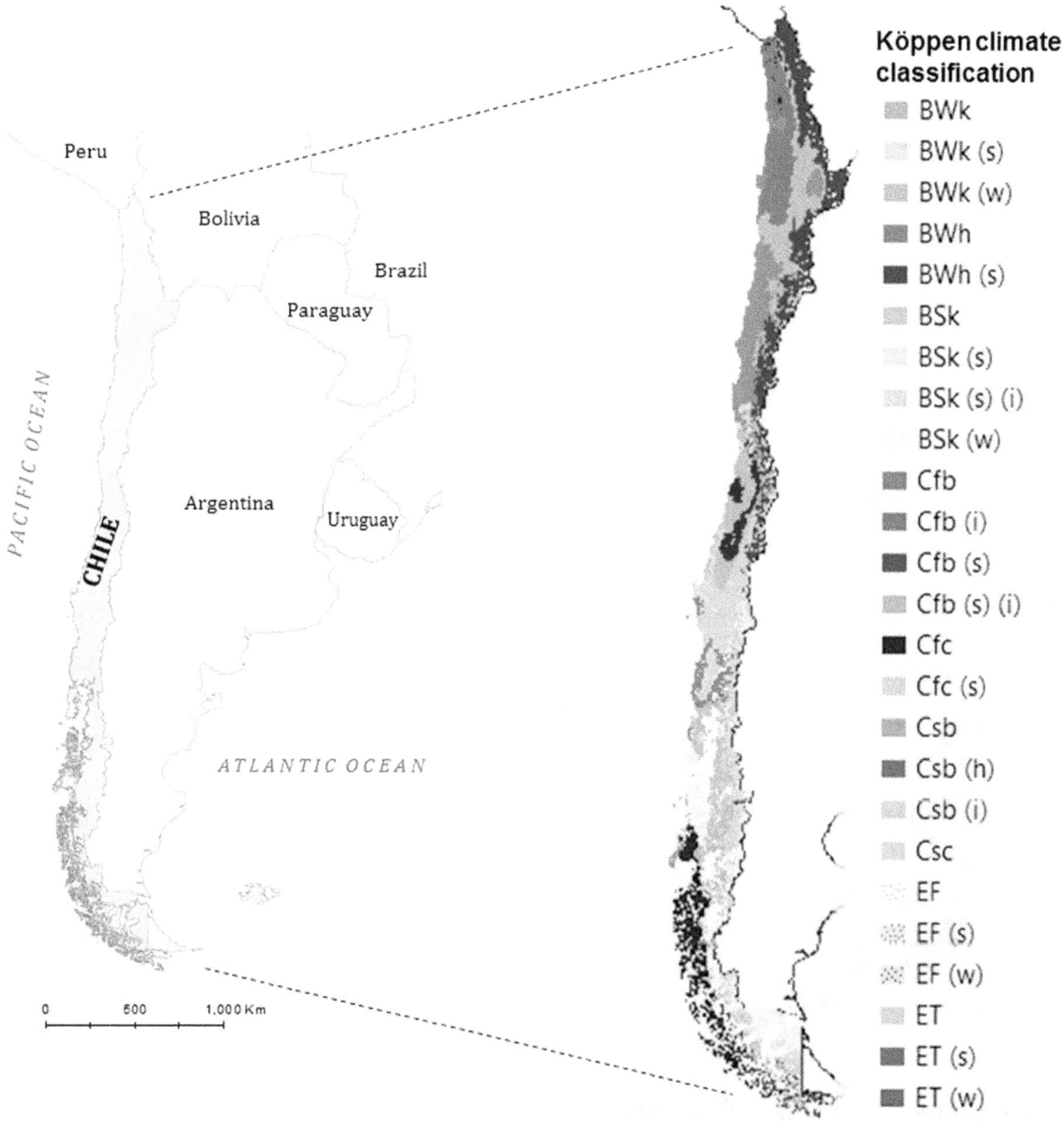

Figure 2. Geomorphological units and geographical regions (GR) of Chile, according to IGM (2018) and BNC (2016), respectively
SD = Standard deviation.

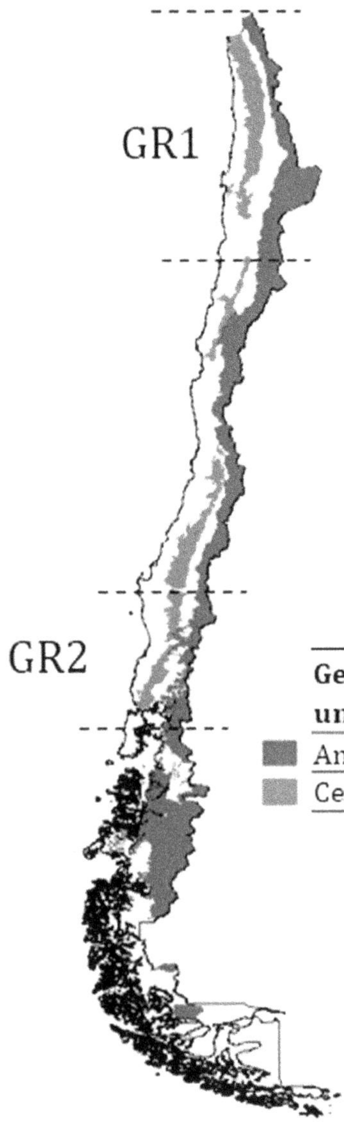

Geomorphological units	Surface (km^2)	Elevation (m)	
		Mean	SD
Andes Mountains	227000	2225	1542
Central Depression	88000	907	777

Precipitation and Rainy Days Analysis

Precipitation data were obtained from the *Center for Climate and Resilience Research*. This institution includes meteorological stations spread throughout Chile and belonging to two official Chilean organisms, the *Dirección Meteorológica de Chile* and the *Dirección General de Aguas*. Data were obtained at a daily timescale from 1980 to 2018. The chosen geomorphological units and geographical regions were used to select meteorological stations located on their zones. Only the meteorological stations with more than 75% of the complete database were used. Thus, from 816 meteorological stations, a total of 87 were selected and distributed as is shown in Table 1:

Table 1. Meteorological stations distributed by geomorphological units and geographical regions

Geomorphological units	Geographical Regions	
	GR1	GR2
Andes Mountain	36	19
Central Depression	13	19

The quantitative analysis of precipitation allows to characterize the rainfall of a territory and evaluate it in a relative way, in a framework defined by the set of daily contributions. The temporal distribution of annual precipitation was analyzed using all the meteorological stations belonging to each geomorphological unit and each geographical region. For each defined group of meteorological stations, daily precipitation was classified by quartiles (25%, 50%, 75% and 100%) and the annual precipitation by quartile was calculated, which allowed to describe their temporal irregularity and categorize them from the lightest to the heaviest ones.

Relations between rainy days and daily events with maximum precipitation was also described by the cumulative frequencies. The used procedure is based on calculating the curves formed by the relative cumulative rainfall and the total cumulative rainfall (both on the Y axes) regarding the cumulative rainy days (on the X axis) (*Martin-Vide,* 2004; *Lozano-Parra, Schnabel et al.,* 2015), and can be expressed as follow:

$$CF = \frac{n_i + n_{i-1}}{N}$$

where, CF is the cumulative frequency; n_i is the variable in the day i; n_{i-1} is the variable in the day before to i; and N is the sample size. Note that to obtain the relative CF in percentage, the quotient must be multiplied by one hundred, while for the absolute CF it should not be multiplied.

To define these curves were used two variables, the rainy days and the extreme precipitation of these rainy days, i.e., the maximum value of rainfall observed among all the meteorological stations belonging to each geomorphological unit and each geographical region. The analyzed period spanned 14702 days by each meteorological station.

Extreme Rainfalls and Return Periods

Extreme rainfall analysis was carried out by the Gumbel distribution, which is a GEV probability distribution used to model extreme values from a sample of independent, identically distributed random variables, as the size of the sample increases. It is known in hydrology because lets to analyze monthly and annual maximum values of daily rainfall and lets to predict the probability of extreme rainfall will occur (*Maity,* 2018). Thus, the probability of non-exceedance of a precipitation (*x*) by Gumbel distribution can be defined as follows:

$$F(x) = e^{-e^{-\left(\frac{x-\beta}{\alpha}\right)}}$$

where, $F(x)$ is the probability that a precipitation lower than x will occur, and α and β are scale and location parameters, respectively. These parameters can be obtained by $\alpha = \frac{\sqrt{6} * S_x}{\pi}$ and $\beta = \bar{x} - 0.5772 \cdot \alpha$; being 0.5772 the Euler–Mascheroni constant, and $\bar{x}$ and S_x the mean and standard deviation of the sample, respectively.

The return period of a hydrologic event can be related to probability of exceedance of that hydrologic event. The probability that a precipitation (P) is exceeded ($Pr\ (P < x)$) will be given by the inverse of the probability of non-exceedance:

$$Pr(P < x) = 1 - F(x)$$

The return period (Tr), in years, can then be defined as follows:

$$T_r = \frac{1}{1 - F(x)}$$

RESULTS AND DISCUSSION

Temporal Distribution of Annual Precipitation

The annual precipitation by quartiles showed a clear spatial variability and temporal irregularity among the studies regions, except for Andes Mountain in southern Chile where rainfall distribution was more homogeneous in time (Figure 3). The temporal irregularity was more pronounced in the northern regions (regardless the geomorphological unit) relative to the southern ones, which is prove by a larger alternation of dry-wet years and by the lower values of quartiles. It was obvious that the southern regions were rainier than the northern ones, however, is striking that the Central Depression showed a clear decreasing trend of their rainfalls from beginning of the current century to the last studied year.

The geomorphological and regional difference here observed could be explained because of altitudinal and latitudinal issues (*Hornberger, Wiberg et al.,* 2014). On one hand, there is a relationship between altitude and precipitation, being the higher elevations traditionally rainier than the lower ones (*Qing, Zhu-Guo et al.,* 2011). On the other hand, the northern areas are drier than the southern ones owing to a combination of factors, such as dynamics of atmospheric circulation, ocean currents, and landmass distribution (*Garreaud, Vuille et al.,* 2009). A geomorphological and geographical effect on torrential precipitation in DPR Korea was also found by *Om, Ren et al.* (2018), who observed that rainiest areas were those located on the windward side of the southwest or southeast wind during the summer monsoon season.

Figure 3. Annual precipitation by quartiles in each geomorphological unit and each geographical region

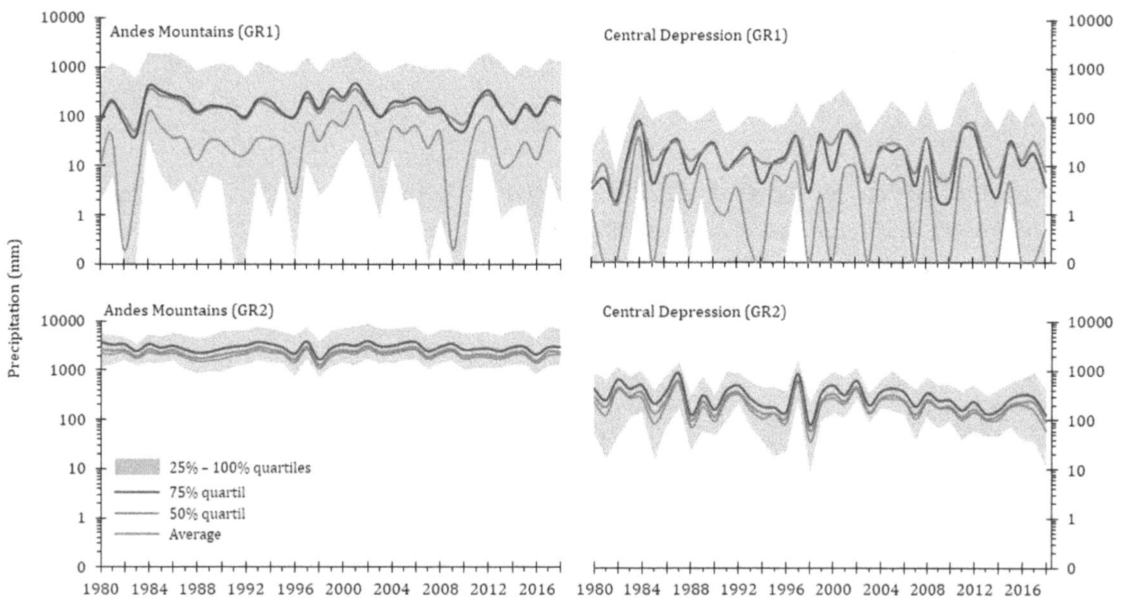

Rainy Days and Daily Events with Maximum Precipitations

Relations between rainy days and daily events with maximum precipitation can be described by a pair of curves reflecting, on one hand, the relationship between the cumulative of maximum daily precipitation and the cumulative rainy days (both in %) and, on the other hand, the relationship between the cumulative of maximum daily precipitation (mm) and the cumulative rainy days (%) (Figure 4). The first relationship results in an exponential curve where, in all situations, 50% of days with precipitation produced less than 15% of the rainiest days, 75% of days with precipitation generated less than 40% of the rainiest days, and 90% of days with precipitation represented less than 68% of the rainiest days. Thus, 30% of the rainiest days were concentrated in only 10% of days with precipitation, which proves the high irregularity of the precipitation and the climate importance of this fact, due to the decisive weight on annual precipitation that shows the appearance of a few days with very high rainfall amounts. The second relationship results in a logistic curve (Figure 4), where if Andes Mountain in the south (GR2) is excluded, 50% of days with precipitation produced less than 7 mm of rainfall, 75% of days with precipitation generated less than 15 mm of rainfall, and 90% of days with precipitation resulted in less than 33 mm of rainfall. Considering that events with more precipitation were 98.5 mm and 274 mm for the northern and the southern of Andes Mountain, respectively, and 189 mm and 155 mm for the northern and the southern of Central Depression, respectively, the rainfall distribution showed a clearly very high concentration on very few precipitation events.

Despite this, the events of maximum daily precipitation had clear regional differences. In the Andes Mountain, the rainiest days produced more rainfall amount than in the Central Depression. However, in this last zone, the extreme rainfalls were more concentrated, since only 5% of days with precipitation produced more than 30% of the rainiest days. Quite similar results were observed by *Schnabel* (1998) in an analysis of a 17-year belonging to the meteorological station of Caceres (Spain), with Mediterranean climate. This study found an exponential curve that defined the daily rainfall throughout the year. Furthermore, it was reported that 80.8% of the daily amounts were less than 10 mm. On the other hand, *Martín-Vide and Gómez* (1999) determined that the percentage of days with rainfall less than 10 mm can account for 95% in Mediterranean coastal areas.

Figure 4. Relationship between the relative accumulated of daily precipitation (%) and the relative rainy days (%) (blue line), and relationship between the daily precipitation (mm) and the relative rainy days (%) (red line), for each geomorphological unit and each geographical region over the period 1980-2018

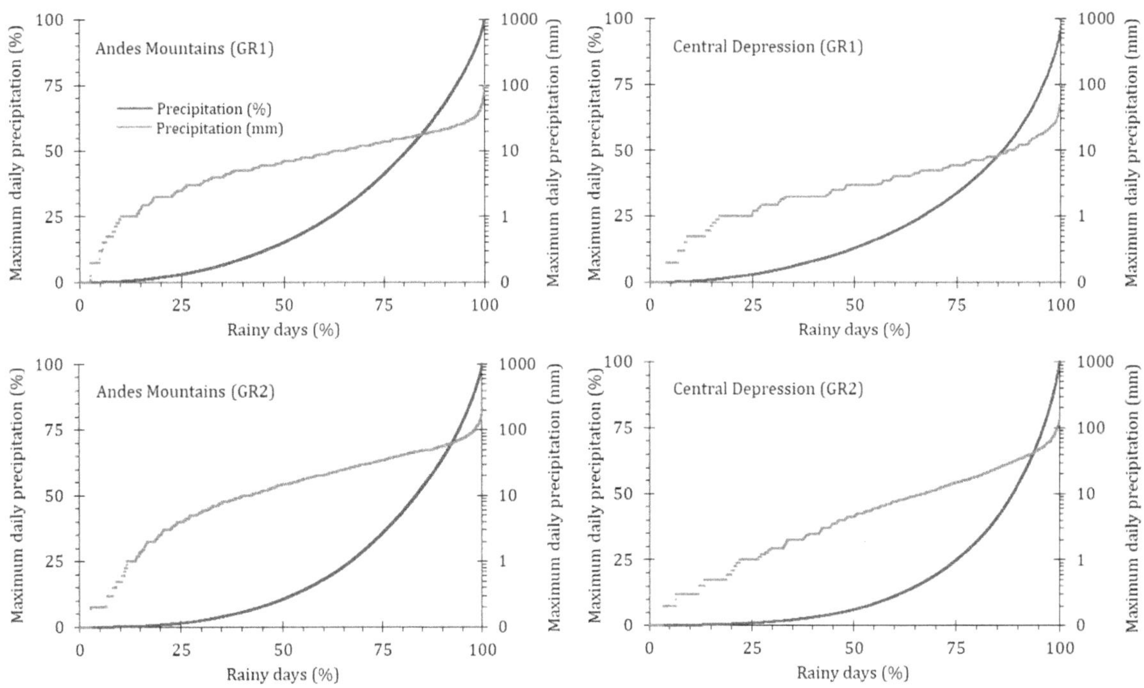

Extreme Rainfalls and Return Periods

The relationships between maximum daily precipitation and probability of non-exceedance and return periods are depicted in Figure 5 and Figure 6, respectively. The results show spatially variable differences in extreme events of daily precipitation, with larger rainfalls towards southern areas than northern ones. The lowest values of extreme precipitation were observed in the Central Depression of northern Chile, where the probability of an event lower than 30 *mm day*$^{-1}$ was 80% (Figure 5), while return periods for rainfall events lower than this one were shorter than 4 years (Figure 6). Conversely, the highest values of extreme precipitation were found in the Andes Mountain of southern Chile, where the probability of occurrence for rainfall events lower than 200 *mm day*$^{-1}$ was 80%, while period returns for these events were shorter than 5.5 years.

For small return periods (<20 years), extreme events of daily precipitation could reach 85 mm and 248 mm in northern and southern of Andes Mountain, respectively, while in northern and southern of Central Depression their values could be 50 mm and 132 mm, respectively. For longer return periods

(> 100 years), extreme events of daily precipitation could reach 109 mm and 305 mm in northern and southern of Andes Mountain, respectively, while in northern and southern of Central Depression their values could be 70 mm and 170 mm, respectively (Figure 6).

These results showed a clear positive bias in total rainfall by extreme events, being more frequent those with lower precipitation. However, Central Depression produced most of the annual total precipitation in a few very rainy days, which could affect areas where the population is concentrated and could impact on possible floods and other geomorphological processes. The frequencies of extreme rainfall events will likely increase under the future projections of climate (*IPCC,* 2014). The projected rise could have implications on different issues such as infrastructure, hydrological cycle and water resources, soil erosion, and the energy sector in Chile (*Valenzuela and Garreaud,* 2019).

Figure 5. Probability of non-exceedance of maximum daily precipitation in each geomorphological unit and each geographical region (GR)

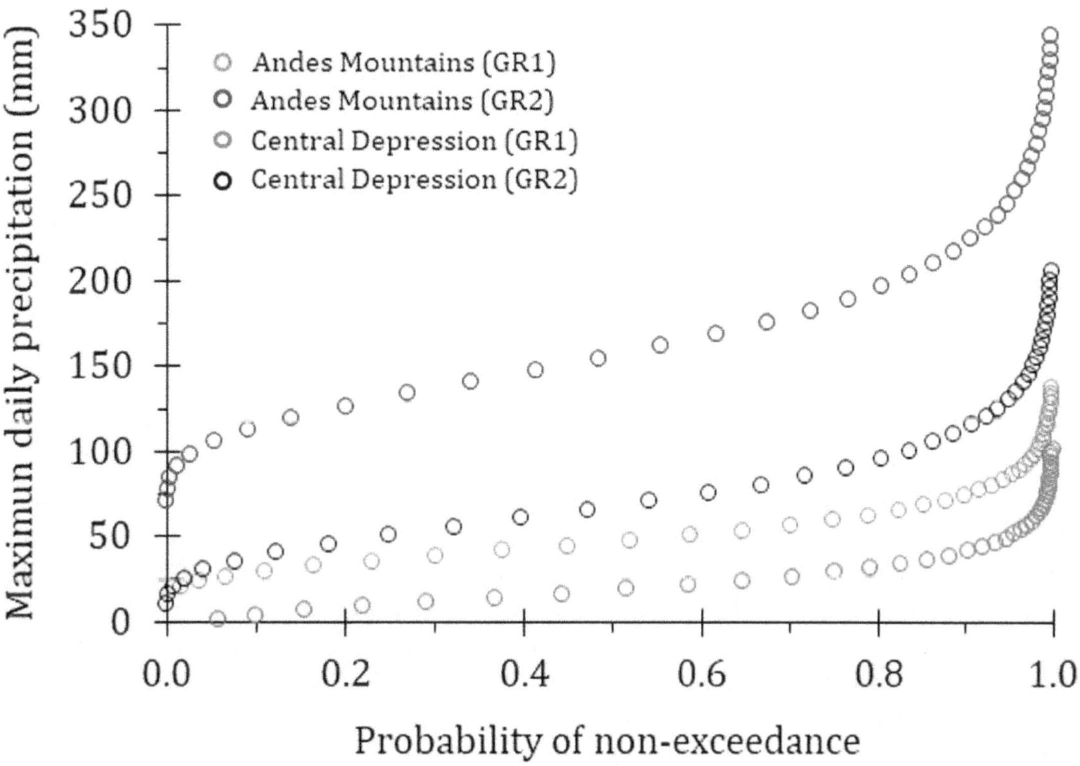

Figure 6. Return periods for maximum daily precipitation in each geomorphological unit and each geographical region (GR)

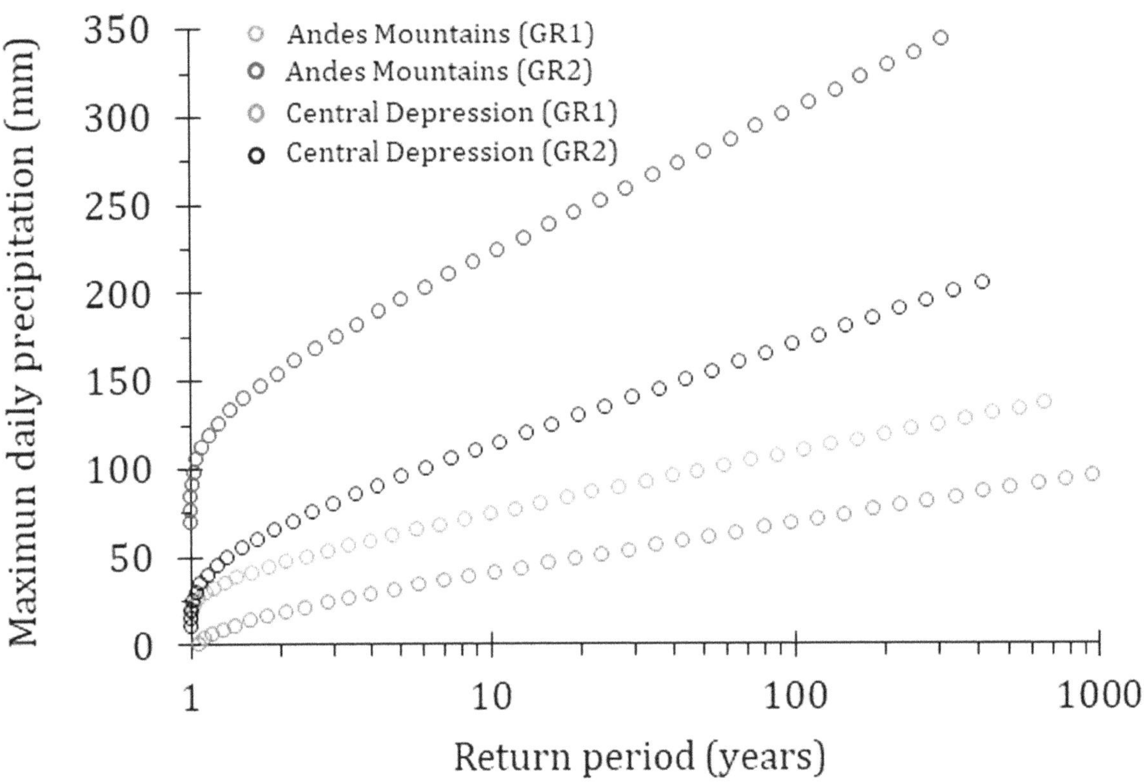

CONCLUSION

This study has analyzed the extreme rainfalls on different geomorphological units and geographical regions of Chile. Results have showed a clear positive bias in total rainfall by extreme events, being more frequent those with lower precipitation, nevertheless, most of the annual total precipitation occurred in a few very rainy days. Northern Chile registered both more extreme events and more rainfall by event than southern Chile. For return periods > 100 years, extreme events of daily precipitation could reach 109 mm and 305 mm in northern and southern of Andes Mountain, respectively, while in northern and southern of Central Depression their values could be 70 mm and 170 mm, respectively.

Greater increase in extreme events of precipitation could have an interest not merely climatic, but also environmental and socioeconomic, for example, on infrastructure, human demand, or natural environment. The frequency of precipitation extremes is projected to rise more prominently in northern and central Chile through the mid and end of the 21st century under the worst representative concentration pathways. Next studies could be addressed to define the daily extreme precipitation in all Chile and to determine their return periods.

ACKNOWLEDGMENT

This study was supported by the project SI3/PJI/2021-00398 financed by the *Comunidad Autónoma de Madrid* and *Universidad Autónoma de Madrid* by means of the *Programa de Estímulo a la Investigación de Jóvenes Doctores para la realización de Proyectos de I+D.*

REFERENCES

BNC. (2016). *Geomorfología de Chile*. Biblioteca Nacional de Chile.

Dingman, S. L. (2015). *Physical Hydrology* (3rd ed.). Waveland Press, Inc.

Easterling, D., Kunkel, K., Wehner, M. F., & Sun, L. (2016). Detection and attribution of climate extremes in the observed record. *Weather and Climate Extremes, 11*, 17–27. doi:10.1016/j.wace.2016.01.001

Garreaud, R., López, P., Minvielle, M., & Rojas, M. (2013). Large-Scale Control on the Patagonian Climate. *Journal of Climate, 26*(1), 215–230. doi:10.1175/JCLI-D-12-00001.1

Garreaud, R. D., Vuille, M., Compagnucci, R., & Marengo, J. (2009). Present-day south american climate. *Palaeogeography, Palaeoclimatology, Palaeoecology, 281*(3), 180–195. doi:10.1016/j.palaeo.2007.10.032

Hornberger, G., & Wiberg, P. (2014). *Elements of physical hydrology*. Johns Hopkins University Press.

IGM. (2018). *Mapa Geomorfológico de Chile*. Instituto Geográfico Militar.

IPCC. (2014). *Climate Change 2014: Synthesis Report. Contribution of Working Groups I, II and III to the Fifth Assessment Report of the Intergovernmental Panel on Climate Change*. IPCC.

IPCC. (2018). Summary for Policymakers. In *Global Warming of 1.5°C. An IPCC Special Report on the impacts of global warming of 1.5°C above pre-industrial levels and related global greenhouse gas emission pathways, in the context of strengthening the global response to the threat of climate change, sustainable development, and efforts to eradicate poverty*. World Meteorological Organization.

López-Moreno, J. I., Vicente-Serrano, S. M., Angulo-Martínez, M., Beguería, S., & Kenawy, A. (2010). Trends in daily precipitation on the northeastern Iberian Peninsula, 1955–2006. *International Journal of Climatology, 30*(7), 1026–1041. doi:10.1002/joc.1945

Lozano-Parra, J., Schnabel, S., & Ceballos-Barbancho, A. (2015). The role of vegetation covers on soil wetting processes at rainfall event scale in scattered tree woodland of Mediterranean climate. *Journal of Hydrology (Amsterdam), 529*, 951–961. doi:10.1016/j.jhydrol.2015.09.018

Maity, R. (2018). *Statistical Methods in Hydrology and Hydroclimatology*. Springer. doi:10.1007/978-981-10-8779-0

Martin-Vide, J. (2004). Spatial distribution of a daily precipitation concentration index in peninsular Spain. *International Journal of Climatology, 24*(8), 959–971. doi:10.1002/joc.1030

Martín-Vide, J., & Gómez, L. (1999). Regionalization of peninsular Spain based on the length of dry spells. *International Journal of Climatology*, *19*(5), 537–555. doi:10.1002/(SICI)1097-0088(199904)19:5<537::AID-JOC371>3.0.CO;2-X

Meehl, G., Karl, T., Easterling, D. R., Changnon, S., Pielke, R. Jr, Changnon, D., Evans, J., Groisman, P. Y., Knutson, T. R., Kunkel, K. E., Mearns, L. O., Parmesan, C., Pulwarty, R., Root, T., Sylves, R. T., Whetton, P., & Zwiers, F. (2000). An Introduction to Trends in Extreme Weather and Climate Events: Observations, Socioeconomic Impacts, Terrestrial Ecological Impacts, and Model Projections. *Bulletin of the American Meteorological Society*, *81*(3), 413–416. doi:10.1175/1520-0477(2000)081<0413:AITTIE>2.3.CO;2

Om, K.-C., Ren, G., Li, S., & Kang-Chol, O. (2018). Climatological characteristics and long-term variation of rainy season and torrential rain over DPR Korea. *Weather and Climate Extremes*, *22*, 48–58. doi:10.1016/j.wace.2018.09.003

Paskoff, R. (1996). Atlas de las formas de relieve de Chile. Instituto Geográfico Militar de Chile.

Qing, Y., Zhu-Guo, M., & Liang, C. (2011). A Preliminary Analysis of the Relationship between Precipitation Variation Trends and Altitude in China. *Atmospheric and Oceanic Science Letters*, *4*(1), 41–46. doi:10.1080/16742834.2011.11446899

Rubel, F., & Kottek, M. (2010). Observed and projected climate shifts 1901–2100 depicted by world maps of the Koppen-Geiger climate classification. *Meteorologische Zeitschrift (Berlin)*, *19*(2), 135–141. doi:10.1127/0941-2948/2010/0430

Schnabel, S. (1998). La precipitación como factor en los procesos hidrológicos y erosivos. Análisis de datos de Cáceres capital. *Revija za Geografijo*, *X*, 137–152.

Sippel, S., Mitchell, D., Black, M. T., Dittus, A. J., Harrington, L., Schaller, N., & Otto, F. E. L. (2015). Combining large model ensembles with extreme value statistics to improve attribution statements of rare events. *Weather and Climate Extremes*, *9*, 25–35. doi:10.1016/j.wace.2015.06.004

Valenzuela, R., & Garreaud, R. (2019). Extreme Daily Rainfall in Central-Southern Chile and Its Relationship with Low-Level Horizontal Water Vapor Fluxes. *Journal of Hydrometeorology*, *20*(9), 1829–1850. doi:10.1175/JHM-D-19-0036.1

Zeder, J., & Fischer, E. (2020). Observed extreme precipitation trends and scaling in Central Europe. *Weather and Climate Extremes*, *29*, 100266. doi:10.1016/j.wace.2020.100266

Zwiers, F., Zhang, X., & Feng, Y. (2011). Anthropogenic Influence on Long Return Period Daily Temperature Extremes at Regional Scales. *Journal of Climate*, *24*(3), 881–892. doi:10.1175/2010JCLI3908.1

Chapter 7
Reservoir Time-Series Filling From Remote Sensing Data in the Central Valley, Chile

Ignacio Aguirre
https://orcid.org/0000-0002-7722-9677
University of Saskatchewan, Canada

Javier Lozano-Parra
Universidad Autónoma de Madrid, Spain

Jacinto Garrido Velarde
University of Extremadura, Spain

ABSTRACT

Reservoirs play a fundamental role in the hydrological planning of the central valley of Chile as they provide water for human and animal consumption, energy generation, and crop irrigation, especially during the summer season. In agriculture, reservoirs represent a significant source to keep the food security standard for more than half of the population of the country. The water management plans need complete records of their volume to calculate rules of operation or future scenarios; however, currently, these time series include gaps that do not allow better analysis, which increases uncertainty. To address this, the authors test a methodology to assess Sentinel 2 imagery through normalized difference water index (NDWI). The results correctly represent the temporality and seasonality of reservoir dynamics; however, the magnitude of the changes is not well represented when the reservoir is delivering water. This research allows more data-based planning of water resources in the central zone, contributing to better decision-making and more efficient water management.

DOI: 10.4018/978-1-6684-4548-8.ch007

INTRODUCTION

Reservoirs play a fundamental role in watershed management since they allow water to be delivered during dry periods, both for human consumption and for agricultural use, increasing water security, allowing the generation of hydroelectric power, and the possibility of mitigating the effects of floods. (Deng et al., 2020; Scola et al., 2014). They also alter the dynamics of river sedimentation, the concentration of certain pollutants, and can alter the regime of the basins (Kondolf et al., 2014). In Chile, there are three categories of reservoirs according to their volume and the height of their wall. Small reservoirs (category A) with a wall between 5 and 15 m and a capacity between 5 x 104 m3 and 1.5 x 106 m3, medium-sized reservoirs (category B) with a wall between 15 and 30 m and a capacity between 1.5 x 106 m3 and 6.0 x107 m3 and large reservoirs (category C) with a wall greater than 30m, and a capacity equal to or greater than 6.0 x107 m3 (Ministry of Public Works - Chile, 2013).

The 26 large reservoirs currently built have a combined volume of 12,961 million cubic meters, including reservoirs of 22 million cubic meters such as Conchi or Tutuvén, to larger ones such as Colbún (1,544 mm3), and Laja (5,582 mm3).) (General Water Directorate, 2021a). These have played a fundamental role in preventing major consequences of the drought in Chile (Garreaud et al., 2017), maintaining food security, and irrigation security in the main valleys of the country (Cordova, 2017; DGA, 2013). The records of these reservoirs include both the outgoing flow from the wall, and the registered volume based on the elevation, and allow knowing the resources delivered and their timing; however, these time series are not complete due to multiple problems: (a) measurement instrument failures, (b) problems associated with floods or other risks, (c) measurement planning or administration problems (Tencaliec et al. al., 2015).

An alternative to reconstructing these time series is through satellite images (Klein et a., 2017; Yao et al., 2019). This technique has several advantages, given its low cost, its global coverage, its availability since the 1980s, its independence from other hydrological stations, and the multiple associated validation methodologies (Arvor et al., 2016).

The use of satellite images in this field embodies several challenges. In the first place, the need for a complete database of reservoirs that includes basic data such as coordinates or geographic information, date of construction, and maximum volume. Globally, there is the GOODD project with more than 38,000 reservoirs (Mulligan et al., 2020), GRAND that includes more than 7,320 with an area greater than 0.1 km3 (Lehner et al., 2011), Global WaterPack (Klein et al. ., 2017) or the AQUASTAT database (FAO, 2021). In Chile, the official database is developed by different offices of the Ministry of Public Works, publishing monthly the volume of the main bodies of water (Direccion General de Aguas, 2021a).

Secondly, the selection of images, since there are daily information sources such as MODIS but with a large cell size of 250 to 500m (Friedl et al., 2015) has been used as support in the generation of time series of reservoir volume (Deng et al., 2020; Zhang et al., 2014), while others with bi-weekly review with better spatial resolution such as Landsat (30m) or Sentinel (10 to 60m). These have been used individually by Van Den Hoek et al., (2019) or in combination with Schwatke et al. (2019) or Militino et al., (2020). Additionally, it should be noted that not all images can be used due to sensor problems such as Landsat 7 (Hossain et al., 2015; Jabar et al., 2014) or cloud cover (Foga et al., 2017). ; Zhao & Gao, 2018)

Thirdly, the selection of spectral indices to occupy, where the resolution and availability of the considered bands must be taken into account. The NDWI index developed by McFeeters (1996) has been used on several occasions both in the generation of reservoir time series (Condeça et al., 2022; Militino et al., 2020) and in their delimitation (Jones et al. ., 2017; Meng et al., 2019). Other indices that have also been considered are the Modified Normalized Difference Water Index (Xu, 2006) and the NDWI developed (Gao, 1996).

Fourth, the methodology to relate the data obtained from the images, usually the area covered by water, with the data observed as volume or outgoing flow from the reservoir wall. Condeça et al., (2022) used the area/volume relationships established by the authority in charge of reservoir management, while Van Den Hoek et al., (2019) used altimetric information from GLAS/ICESat. In this study, we compare both series in a dimensionless way to avoid the uncertainty of the area/volume conversion methods.

Fifth, the problems associated with the reproducibility of the results, especially in the context of continuous monitoring to mitigate droughts or provide reliable information for agricultural and hydrological planning. Militino et al., (2020) developed a package in the R statistical software to be able to automate the process, while others have developed their analyzes in Google Earth Engine (Gorelick et al., 2017) that provides a high analysis capacity such as Condeça et al., (2022) or Schwatke et al., (2019) or the present study.

Following these methodologies, Schwatke et al. (2019) obtained information from 32 globally distributed reservoirs using Landsat and Sentinel 2 images and four spectral indices. One of the places considered was Ray Roberts, Lake (Texas, USA) where an R2 of 0.934 was obtained, comparing the observed area and the area obtained with satellite images. In Poço da Cruz (Brazil) an R2 = 0.92 was obtained and then a data filling post-process was generated, obtaining R2 = 0.981. Deng et al. (2020) used a similar methodology in Yellow River (China) with 149 reservoirs, using Landsat images and the support of MODIS images.

Militino et al., (2020) combined Landsat 8 and Sentinel 2 images to generate a time series of the Itoiz reservoir in Spain. They obtained a mean absolute error (MAE) of 0.66 globally, with the satellite MAE being 1.04m for Landsat and 0.53 for Sentinel. In eastern Brazil, Van Den Hoek et al. (2019) studied 13 reservoirs during the years 2003 to 2014, obtaining time series with good results (R2 = 0.88, RMSE = 0.89).

An alternative to reconstruct these time series is through satellite images (Klein et a., 2017; Yao et al., 2019). This technique has several advantages, given its low cost, its global coverage, its availability since the 1980s, its independence from other hydrological stations, and the multiple associated validation methodologies (Arvor et al., 2016).

There are several applications of satellite image methodologies to obtain time series of reservoirs, such as Schwatke et al. (2019) who obtained information from 32 globally distributed reservoirs using Landsat and Sentinel 2 images and four spectral indices. One of the places considered was Ray Roberts, Lake (Texas, USA) where an R2 of 0.934 was obtained, comparing the observed area and the area obtained with satellite images. In Poço da Cruz (Brazil) an R2 = 0.92 was obtained and then a data filling post-process was generated, obtaining R2 = 0.981. Deng et al. (2020) used a similar methodology in Yellow River (China) with 149 reservoirs, using Landsat images and the support of MODIS images.

In eastern Brazil, Van Den Hoek et al. (2019) developed a more complex methodology to study 13 reservoirs from 2003 to 2014. First, they processed Landsat images to quantify the area covered by water, then generated volume versus area curves from digital elevation models, and then obtained time series with good results (R2 = 0.88, RMSE = 0.89).

In Chile there is no study to address the problems of reservoir registrations; Therefore, this article studies how satellite images can be used as a proxy to complete the records of reservoirs in Chile, capturing their dynamics both in terms of magnitude (proportion of filling or emptying), as well as temporality and seasonality (when it is filled and empty). This is a first approximation to evaluate the methodology in the central zone, which presents several peculiarities such as the aridity of the terrain, the irregular shape of the reservoirs, and the recent lack of rainfall (Garreaud et al., 2017).

For this, the reservoirs of the Choapa basin are selected, a series of Sentinel 2 images reclassified at 10m are obtained, the NDWI index is calculated and they are reclassified according to a threshold value, adding the area and comparing it with the volume of the reservoir of normalized form. Normalization prevents observed problems from being the result of unit conversion or transfer from area to volume via a polynomial equation.

The results of this research will allow a better characterization of satellite images as a proxy to fill a series of reservoirs in Chile, suggesting new indices or sources. These data will serve as input in projects of strategic basin management plans, studies of irrigation associations, and food safety, considering that they will improve their quality by having information without missing records.

STUDY AREA

The analysis was carried out in the Choapa basin located in the central zone of Chile. Its average rainfall is 293 mm/year, with an aridity index of 3.5 (Alvarez-garreton et al., 2018). It has the El Bato reservoir built in 2010 and the Corrales reservoir built-in 2014, described in detail in Table 1, both were built for irrigation.

Table 1. Information on the El Bato and Corrales Reservoirs

	Embalse El Bato	Embalse Corrales
Año de Construcción	2010	2004
Volumen	26 millones de m^3	50 millones de m^3
Área de inundación	117 ha	270 ha
Coordenadas Muro	31.556°S, 70.873°W	31.911°S, 70.915°W
Altitud Muro	853 msnm	870 msnm
Cuenca	Choapa	Choapa
Subcuenca	Rio Illapel	Rio Choapa Medio
Rio	Rio Illapel	Estero Camisas
Uso Principal	Riego	Riego
Promedio histórico	13 millones de m^3	37 millones de m^3
Promedio (2016 - 2020)	18 millones de m^3	31 millones de m^3

Source. Own elaboration (CNR (National Irrigation Commission), 2004; DGA (General Water Directorate), 2021a; DOH (Hydraulic Works Directorate), 2017)

METHODOLOGY

The methodology is made up of four parts: (a) digitalization of the reservoirs, (b) selection of images, (c) application of the Normalized Differential Water Index (NDWI) and reclassification based on a threshold value to obtain the area; and (d) data post-processing.

Digitization

All reservoirs described in the study area are digitized as polygons. The coordinates of each one were obtained from the spatial database of the General Directorate of Waters (DGA, 2021). This information is in point format, so it was digitized in Google Earth Pro using high-resolution images to obtain the maximum surface area of each body of water.

Image Evaluation

The images selected in this study correspond to the Sentinel 2 mission, which includes both satellites 2A launched in June 2015, and 2B launched in July 2016 (ESA, 2015). The mission comprises 12 spectral bands, with a resolution of 10 to 60 meters for different bands (ESA, 2015). In this study, 6 described in Table 1 have been selected, which will be reclassified to 10 meters. Each of the images considered must have a cloud coverage of less than 40%.

The images used are corrected to the top of the atmosphere (TOA) previously generated in Google Earth Engine (Gorelick et al., 2017) (https://developers.google.com/earth-engine/datasets/catalog/COPERNICUS_S2). This makes it possible to avoid distortions in the conversion process to radiance (ESA, 2015).

Table 2. Description, band number, wavelength and resolution of the selected Sentinel 2 Bands

Description	Band Number	Wavelength	Resolution
Blue	2	496.6nm(S2A)/492.1nm(S2B)	10m
Green	3	560nm(S2A)/559nm(S2B)	10m
Red	4	664.5nm(S2A)/665nm(S2B)	10m
NIR	8	835.1nm(S2A)/833nm(S2B)	10m
SWIR 1	11	1613.7nm(S2A)/1610.4nm(S2B)	20m
SWIR 2	12	2202.4nm(S2A)/2185.7nm(S2B)	20m

Source. ESA, 2015.

Application of the Index

From the compilation of images generated in the previous step, the Normalized Differential Water Index (NDWI) is applied to each image in Google Earth Engine (Gorelick et al., 2017). This index was developed by McFeeters (1996) and allows the identification of water bodies from the difference between the green and near-infrared bands. It is defined by Equation 1:

$$NDWI = \frac{Green - NIR}{Green + NIR} \tag{1}$$

The results are constrained between -1 and 1, where theoretically from zero there is the presence of water. Once the index has been applied to each image, its result is reclassified into two categories, above a threshold t or below it. All cells that meet the condition of being over t are added to obtain an area.

Threshold Definition

To determine the threshold t, one starts with an initial value of 0.05 following Deng et al., (2020) and contrasts the normalized area result versus the normalized reservoir volume result using R2 (Equation 2) and root-mean-square deviation (RMSE, Equation 3):

$$R^2 = \frac{\left(\sum_{i=1}^{n}(O_i - \bar{O})(S_i - \bar{S})\right)^2}{\sum_{i=1}^{n}(O_i - \bar{O})^2 \sum_{i=1}^{n}(S_i - \bar{S})^2} \tag{2}$$

$$RMSE = \left(\frac{1}{n}\sum_{i=1}^{n}(S_i - \bar{S})^2\right)^{\frac{1}{2}} \tag{3}$$

Where O_i is the observed normalized value of reservoir volume, $\bar{O}$ is its average, S_i is the normalized value of area obtained through satellite images, and S is its average. The Python implementation was developed by Roberts et al. (2018). This value is iterated until optimal performance is obtained in both metrics. Due to the conditions of the terrain, a threshold can be generated for each reservoir.

Post Process

Finally, a post-process of the time series is generated, normalizing them according to Equation 4:

$$NV = \frac{V - m}{M - m} \tag{4}$$

Where NV is the normalized value, V is the value, m is the minimum value of the series, M is the maximum value, the maximum capacity in the case of reservoirs, and the maximum area in the case of images. Both the observed series obtained from the General Directorate of Water and the series obtained by satellite images are normalized, which allows immediate comparison of the two series.

RESULTS

El Bato Reservoir

The El Bato reservoir was digitized, obtaining a polygon of 117 ha in Google Earth Pro, and then its topology was corrected in ArcMap 10.4. From this, a compilation of Sentinel 2 images with less than 40% cloud cover was generated and the NDWI index was applied to each of them. With this, it was possible to select the threshold value (Table 3) that corresponds to 0.15, because an optimal value of both objective functions is achieved.

Table 3. Selection of NDWI threshold of El Bato Reservoir

	El Bato Reservoir	
Threshold	R^2	RMSE
0.05	0.9397	0.1883
0.1	0.9390	0.1845
0.15	0.9388	0.1824
0.2	0.8964	0.1809
0.25	0.4864	0.2177

Source. Own elaboration

The results of the post-processing are presented in Figure 1, which includes three sub-figures. The upper one presents the observed volume time series and the area obtained by satellite images, both normalized, where it is observed that the period between 2016 and the end of 2018 where the reservoir is full is captured in a good way; however, the period in which it begins to empty, as a result of the drought (Garreaud et al., 2017) and its management, is overestimated in magnitude, although the temporality and seasonality of the changes are well represented.

The sub-figure located in the lower-left zone presents the monthly average, where it is observed that the methodology used tends to overestimate every month, especially during August. The scatter plot to its right shows the same situation, an overestimation with the selected methodology; although the period where the reservoir is at its maximum capacity (1.0, 1.0) is well represented on the 1:1 line.

Corrales Reservoir

The Corrales reservoir was digitized, generating a polygon of 270 ha in Google Earth Pro, and its topology was also corrected in ArcMap 10.4. A compilation of Sentinel 2 images was generated in Google Earth Engine using the JavaScript implementation. The index was applied to each image in the set. With this, a threshold of 0.15 (Table 4) was determined since it presented an optimal value in both metrics.

The upper sub-figure of Figure 2 presents the monthly time series, where a good fit is observed in the 2016-2018 period, and then an overestimation during the remaining period; however, the temporality of the changes is well represented. For example, when the reservoir is at a minimum level during 2020, the same signal is observed in the satellite images, and also the period of filling at the end of 2020 until 2021.

Figure 1. Normalized volume of El Bato Reservoir and area obtained by Remote Sensing (RS). Plot of monthly time series, monthly average, and scatter plot.
Source. Own elaboration

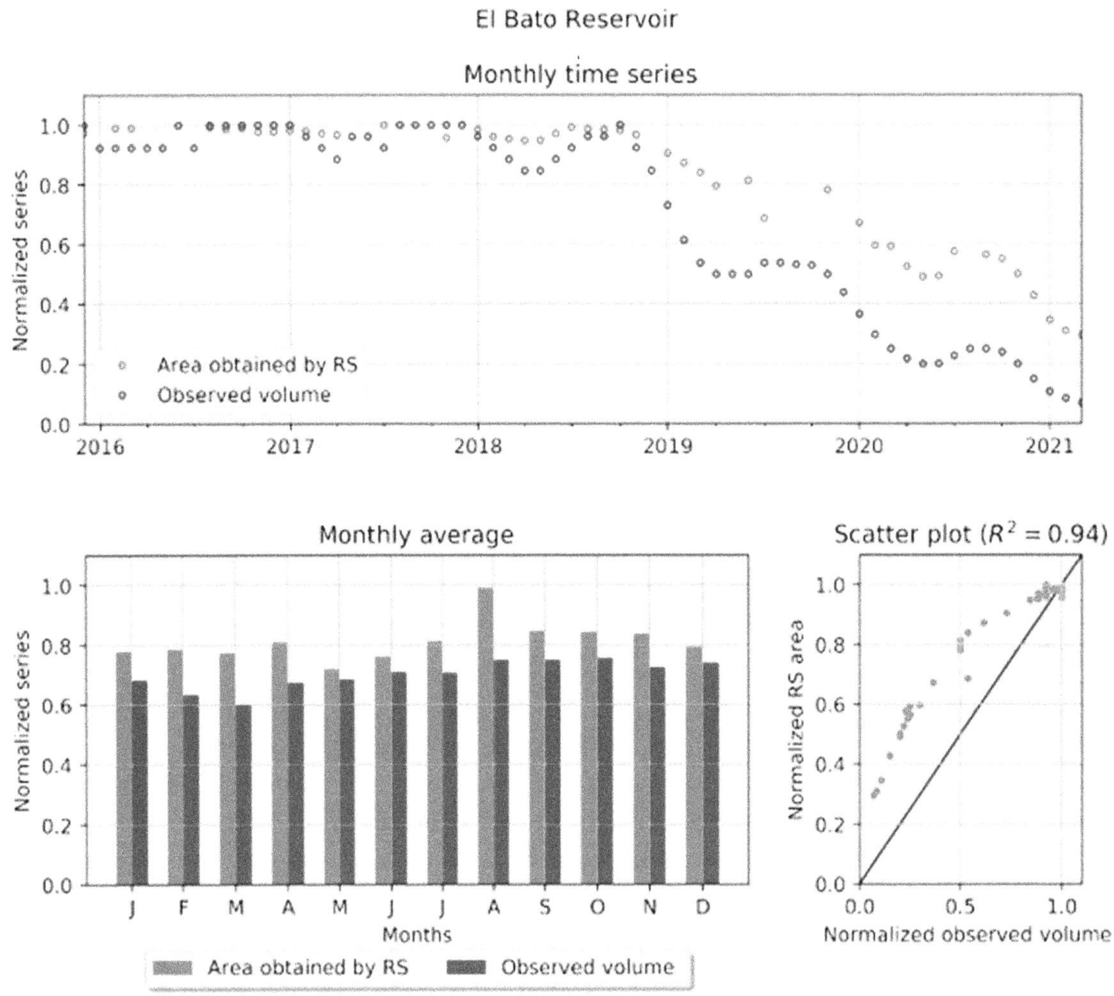

Table 4. Selection of NDWI threshold of Corrales Reservoir

	Corrales Reservoir	
Threshold	**R^2**	**RMSE**
0.05	0.1292	0.4735
0.1	0.9759	0.1257
0.15	0.9768	0.1222
0.2	0.9767	0.1185
0.25	0.9767	0.1141

Source. Own elaboration

The left sub-figure shows the monthly average, where the overestimation is observed as in El Bato, however, during May and June the value is underestimated. The scatterplot shows a better fit, especially the period when the reservoir is at full capacity, placing its results on line 1:1

Figure 2. Normalized volume of Corrales Reservoir and area obtained by Remote Sensing (RS). Plot of monthly time series, monthly average, and scatter plot.
Source. Own elaboration

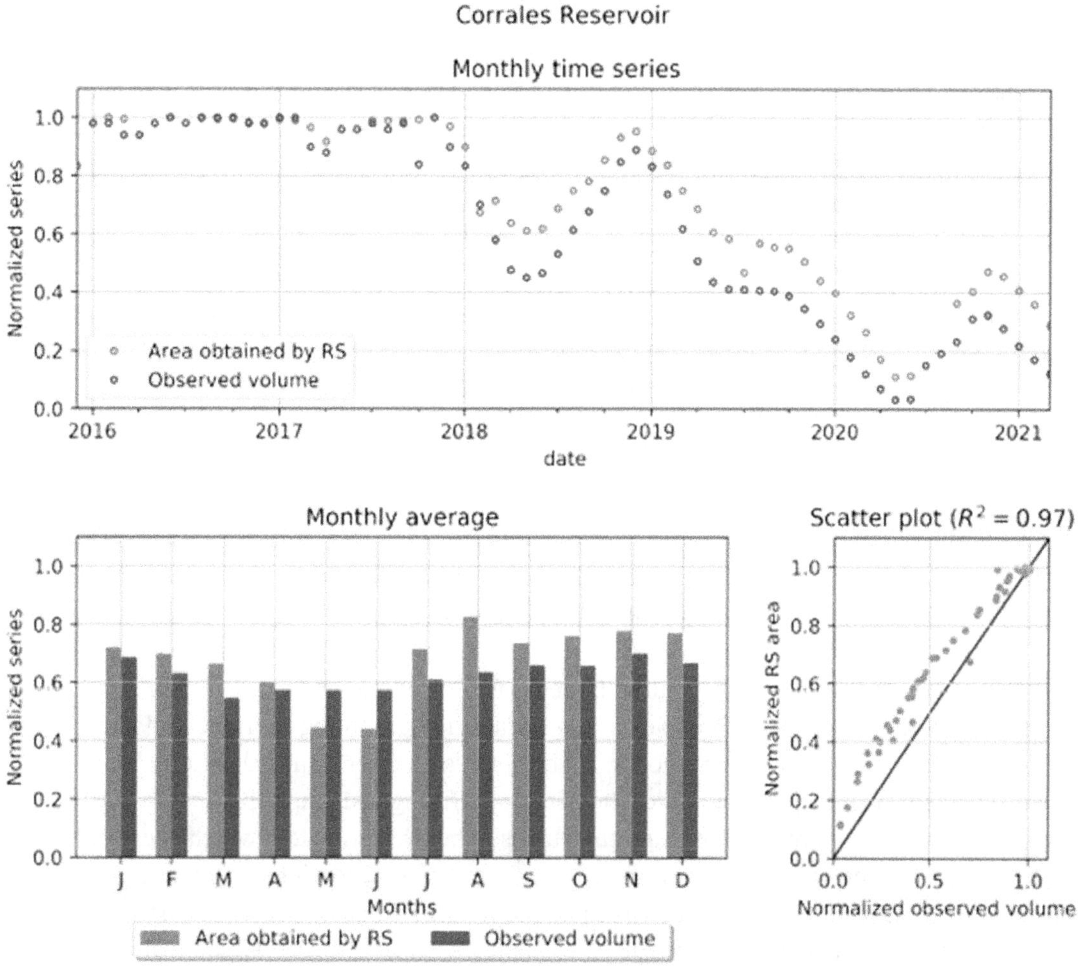

DISCUSSION

Following Van Den Hoek et al. (2019), several elements can be analyzed to evaluate the results. First of all, it should be noted that both in Bato and in Corrales the interannual seasonality of the cycles of filling and emptying of the reservoirs is correctly represented, a central aspect in several studies such as Deng et al., (2020) or Schwatke et al., (2019). Secondly, the precision in the magnitude is more complex to analyze, given that on the one hand, a good level of representation is reached in the period when the reservoirs are full; however, as they begin to empty, the values begin to be overestimated. In Bato, this overestimation is considered close to 25%, while in Corrales it is 15%.

This has different problems since simplified equations could not be used to estimate the volume from the area such as those used by Rodrigues et al., (2012), since these require a good representation of the magnitude, and in this case, it is not achieved. Schwatke et al., (2019) address this using various indices such as Modified Normalized Difference Water Index (MNDWI), NewWater Index (NWI), Automated Water Extraction Index for Non-Shadow Areas, and Tasseled Cap for Wetness. These can provide new information by using the longwave radiation band, which is the modified version of the NDWI index developed by McFeeters (1996) used in this study. Jones et al. (2017) use both NDWI and two modified versions with long wavebands.

The analysis of the monthly variation allows us to observe that the seasonality at the monthly level is well represented, and at the same time that the differences in magnitude are not focused on a month but are distributed uniformly, which means that the bias product of the images or the index is applied homogeneously to the entire series.

The data obtained from satellite images were post-processed at a monthly level, as were the data from the reservoirs, given that Sentinel 2 captures information every 16 days, which prevents obtaining a view with a greater temporal resolution of the information, given that they are being averaged two snapshots in a month, versus a monthly average generated from daily observed data. The data from the MODIS sensor will allow solving this problem of temporal resolution, although losing the spatial resolution, which could affect magnitude problems more significantly.

CONCLUSION

This study evaluated how satellite images can help us complete the missing records of reservoirs in central Chile, allowing us as an initial step to observe trends and seasonality. To do this, the El Bato reservoir and the Corrales reservoir were analyzed with the NDWI index on the Google Earth Engine platform using Sentinel 2 images. Each of the images was reclassified according to a threshold value and then this response was compared with the records. of the reservoirs, after standardization.

The results show that the temporality of the periods of filling and emptying of the reservoirs can be observed in a good way; however, the magnitude tends to be overestimated, which is critical in periods when the resources accumulated in the reservoir begin to be used. Reviewing other approaches in the literature, other spectral indices that take advantage of longwave bands can be used (Deng et al., 2020; Jones et al., 2017; Schwatke et al., 2019), or other images such as MODIS (Klein et al., 2021). Solving these challenges will allow volume time series to be generated accurately.

This work was a first approximation in the filling of time series of reservoirs with this methodology in the country, and with this methodology, it will be possible to evaluate other reservoirs in the same, to have a diagnosis and then generate a database of estimated volume to from satellite images, which will complement the input data in the basin management plans and other planning instruments.

Future Lines of Research

These results open several avenues for new research. In the first place, the generation of time series filled with satellite images for all the reservoirs in the country considering the three categories mentioned in Decree 50. This will allow having a complete monthly database of the area covered by the reservoirs.

Second, is the generation of a library to generate this analysis in an automated way, considering both the choice of images, the exclusion of images with cloudiness, the construction of the index, and the normalization of the results. The package generated by Militino et al. (2020) in R, and the routine provided by Condeça et al. (2022) in GEE-Javascript.

ACKNOWLEDGMENT

Thanks to Copernicus for making the Sentinel images freely available, and to the Google Earth Engine developer team for the tool.

Disclosure Statement: No potential conflict of interest was reported by the authors.

REFERENCES

Alvarez-garreton, C., Mendoza, P. A., Boisier, J. P., Addor, N., Galleguillos, M., Zambrano-bigiarini, M., ... Garreaud, R. (2018). The CAMELS-CL dataset : Catchment attributes and meteorology for large sample studies – Chile dataset. *Hydrol. Earth Syst*, *22*(11), 5817–5846. doi:10.5194/hess-22-5817-2018

Arvor, D., Daher, F., Corpetti, T., Laslier, M., & Dubreuil, V. (2016). Monitoring of artificial water reservoirs in the Southern Brazilian Amazon with remote sensing data. *Remote Sensing for Agriculture, Ecosystems, and Hydrology XVIII*, *9998*, 999816. doi:10.1117/12.2241905

CNR (Comision Nacional de Riego). (2004). *Diagnóstico base de los embalses El Bato y Corrales, IV Región*. Retrieved from http://bibliotecadigital.ciren.cl/handle/123456789/9601

Condeça, J., Nascimento, J., & Barreiras, N. (2022). Monitoring the Storage Volume of Water Reservoirs Using Google Earth Engine. *Water Resources Research*, *58*(3). Advance online publication. doi:10.1029/2021WR030026

Cordova, J. (2017). *Estudio del concepto de seguridad de riego, su funciones de beneficio agrícola, caso embalse las palmas, v region, Chile*. Universidad de Chile. Departamento de ingenería civil.

Deng, X., Song, C., Liu, K., Ke, L., Zhang, W., Ma, R., … Wu, Q. (2020). Remote sensing estimation of catchment-scale reservoir water impoundment in the upper Yellow River and implications for river discharge alteration. *Journal of Hydrology*, *585*(December), 124791. doi:10.1016/j.jhydrol.2020.124791

DGA. (2013). *Estrategia nacional de recursos hídricos 2012-2025*. Retrieved from https://www.mop.cl/Documents/ENRH_2013_OK.pdf

DGA (Direccion General de Aguas). (2021a). Información pluviométrica,fluviométrica, estado de embalses y aguas subterráneas. DGA.

DGA (Direccion General de Aguas). (2021b). *Mapoteca Digital*. Retrieved from https://dga.mop.gob.cl/estudiospublicaciones/mapoteca/Paginas/default.aspx

DOH (Direccion Obras Hidraulicas). (2017). *Actualización modelo de operación embalse el bato*. DOH.

ESA. (2015). Sentinel-2 User Handbook. Sentinel 2. doi:10.1021/ie51400a018

FAO. (2021). *Geo-referenced Database on Dams*. Retrieved from AQUASTAT - FAO's Global Information System on Water and Agriculture website: https://www.fao.org/aquastat/en/overview/

Foga, S., Scaramuzza, P. L., Guo, S., Zhu, Z., Dilley, R. D. Jr, Beckmann, T., Schmidt, G. L., Dwyer, J. L., Joseph Hughes, M., & Laue, B. (2017). Cloud detection algorithm comparison and validation for operational Landsat data products. *Remote Sensing of Environment*, *194*, 379–390. doi:10.1016/j.rse.2017.03.026

Friedl, M., Gray, J., & Sulla-Menashe, D. (2015). *MCD12Q2 MODIS/Terra+Aqua Land Cover Dynamics Yearly L3 Global 500m SIN Grid V006*. Academic Press.

Gao, B. (1996). NDWI—A normalized difference water index for remote sensing of vegetation liquid water from space. *Remote Sensing of Environment*, *58*(3), 257–266. doi:10.1016/S0034-4257(96)00067-3

Garreaud, R. D., Alvarez-Garreton, C., Barichivich, J., Pablo Boisier, J., Christie, D., Galleguillos, M., ... Zambrano-Bigiarini, M. (2017). The 2010-2015 megadrought in central Chile: Impacts on regional hydroclimate and vegetation. *Hydrology and Earth System Sciences*, *21*(12), 6307–6327. Advance online publication. doi:10.5194/hess-21-6307-2017

Gorelick, N., Hancher, M., Dixon, M., Ilyushchenko, S., Thau, D., & Moore, R. (2017). Google Earth Engine: Planetary-scale geospatial analysis for everyone. *Remote Sensing of Environment*, *202*, 18–27. doi:10.1016/j.rse.2017.06.031

Hossain, M. S., Bujang, J. S., Zakaria, M. H., & Hashim, M. (2015). Assessment of Landsat 7 Scan Line Corrector-off data gap-filling methods for seagrass distribution mapping. *International Journal of Remote Sensing*, *36*(4), 1188–1215. doi:10.1080/01431161.2015.1007257

Jabar, A. S. A., Sulong, G., & George, L. E. (2014). Survey on gap filling algorithms in Landsat 7 ETM+ images. *Journal of Theoretical and Applied Information Technology*, *63*(1), 136–146.

Jones, S. K., Fremier, A. K., DeClerck, F. A., Smedley, D., Pieck, A. O., & Mulligan, M. (2017). Big data and multiple methods for mapping small reservoirs: Comparing accuracies for applications in agricultural landscapes. *Remote Sensing*, *9*(12), 1307. Advance online publication. doi:10.3390/rs9121307

Klein, I., Gessner, U., Dietz, A. J., & Kuenzer, C. (2017). Global WaterPack – A 250 m resolution dataset revealing the daily dynamics of global inland water bodies. *Remote Sensing of Environment*, *198*, 345–362. doi:10.1016/j.rse.2017.06.045

Klein, I., Mayr, S., Gessner, U., Hirner, A., & Kuenzer, C. (2021). Water and hydropower reservoirs: High temporal resolution time series derived from MODIS data to characterize seasonality and variability. *Remote Sensing of Environment, 253*(November), 112207. doi:10.1016/j.rse.2020.112207

Kondolf, G. M., Gao, Y., Annandale, G. W., Morris, G. L., Jiang, E., Zhang, J., Cao, Y., Carling, P., Fu, K., Guo, Q., Hotchkiss, R., Peteuil, C., Sumi, T., Wang, H.-W., Wang, Z., Wei, Z., Wu, B., Wu, C., & Yang, C. T. (2014). Sustainable sediment management in reservoirs and regulated rivers: Experiences from five continents. *Earth's Future*, *2*(5), 256–280. doi:10.1002/2013EF000184

Lehner, B., Liermann, C. R., Revenga, C., Vörösmarty, C., Fekete, B., Crouzet, P., Döll, P., Endejan, M., Frenken, K., Magome, J., Nilsson, C., Robertson, J. C., Rödel, R., Sindorf, N., & Wisser, D. (2011). High-resolution mapping of the world's reservoirs and dams for sustainable river-flow management. *Frontiers in Ecology and the Environment*, *9*(9), 494–502. doi:10.1890/100125

McFeeters, S. (1996). The use of the Normalized Difference Water Index (NDWI) in the delineation of open water features. *International Journal of Remote Sensing*, *17*(7), 1425–1432. doi:10.1080/01431169608948714

Meng, L., Zhang, Z., Zhang, W., Ye, J., Wu, C., Chen, D., & Song, C. (2019). An Automatic Extraction Method for Lakes and Reservoirs Using Satellite Images. *IEEE Access: Practical Innovations, Open Solutions*, *7*, 62443–62456. doi:10.1109/ACCESS.2019.2916148

Militino, A. F., Montesino-SanMartin, M., Pérez-Goya, U., & Ugarte, M. D. (2020). Using RGISTools to estimate water levels in reservoirs and lakes. *Remote Sensing*, *12*(12), 1934. Advance online publication. doi:10.3390/rs12121934

Ministerio de Obras Publicas - Chile. (2013). *Decreto 50*. Author.

Mulligan, M., van Soesbergen, A., & Sáenz, L. (2020). GOODD, a global dataset of more than 38,000 georeferenced dams. *Scientific Data*, *7*(1), 1–8. doi:10.103841597-020-0362-5 PMID:31964896

Roberts, W., Williams, G. P., Jackson, E., Nelson, E. J., & Ames, D. P. (2018). Hydrostats: A Python package for characterizing errors between observed and predicted time series. *Hydrology*, *5*(4), 66. Advance online publication. doi:10.3390/hydrology5040066

Rodrigues, L. N., Sano, E. E., Steenhuis, T. S., & Passo, D. P. (2012). Estimation of small reservoir storage capacities with remote sensing in the Brazilian Savannah region. *Water Resources Management*, *26*(4), 873–882. doi:10.100711269-011-9941-8

Schwatke, C., Scherer, D., & Dettmering, D. (2019). Automated extraction of consistent time-variable water surfaces of lakes and reservoirs based on Landsat and Sentinel-2. In Remote Sensing (Vol. 11). doi:10.3390/rs11091010

Scola, L. A., Takahashi, R. H. C., & Cerqueira, S. A. A. G. (2014). Multipurpose Water Reservoir Management: An Evolutionary Multiobjective Optimization Approach. *Mathematical Problems in Engineering, 2014*(2). doi:10.1155/2014/638259

Tencaliec, P., Favre, A., Prieur, C., & Mathevet, T. (2015). Reconstruction of missing daily streamflow data using dynamic regression models. *Water Resources Research*, *51*(12), 9447–9463. doi:10.1002/2015WR017399

Van Den Hoek, J., Getirana, A., Jung, H. C., Okeowo, M. A., & Lee, H. (2019). Monitoring reservoir drought dynamics with landsat and radar/lidar altimetry time series in persistently cloudy eastern Brazil. *Remote Sensing*, *11*(7), 827. Advance online publication. doi:10.3390/rs11070827

Xu, H. (2006). Modification of normalised difference water index (NDWI) to enhance open water features in remotely sensed imagery. *International Journal of Remote Sensing*, *27*(14), 3025–3033. doi:10.1080/01431160600589179

Yao, F., Wang, J., Wang, C., & Crétaux, J. F. (2019). Constructing long-term high-frequency time series of global lake and reservoir areas using Landsat imagery. *Remote Sensing of Environment*, *232*(July), 111210. doi:10.1016/j.rse.2019.111210

Zhang, S., Gao, H., & Naz, B. (2014). Monitoring reservoir storage in South Asia from multisatellite remote sensing. *Water Resources Research*, *50*(11), 5375–5377. doi:10.1002/2014WR015829

Zhao, G., & Gao, H. (2018). Automatic Correction of Contaminated Images for Assessment of Reservoir Surface Area Dynamics. *Geophysical Research Letters*, *45*(12), 6092–6099. doi:10.1029/2018GL078343 PMID:35095126

Chapter 8
The Sustainability Level of Agricultural Food Markets:
Selected Indicators From Mediterranean Countries

Seda Yildirim
https://orcid.org/0000-0003-4367-6652
Tekirdag Namık Kemal University, Turkey

Durmus Cagri Yildirim
Tekirdag Namık Kemal University, Turkey

Fatma Erdogan
Tekirdag Namık Kemal University, Turkey

ABSTRACT

Food crisis and hunger issues will challenge many countries due to the effects of climate change and global warming. When considering the truth that the Mediterranean Basin is the region that has been challenged by climate change and invasive alien species much more than others in the world, food insecurity is the most critical issue. As achieving food security, Mediterranean countries should keep the balance between the population and food sources in the related region. Accordingly, determining indicators and variables guiding policy makers to develop sustainable strategies for food security will be important cases for researchers. This study aims to review the sustainability level of the agricultural food market among Mediterranean countries. It is thought to give a holistic view for agricultural food markets in the Mediterranean Basin. This study focuses on examining some selected indicators for agricultural food markets among Mediterranean countries by showing recent descriptive statistics for agricultural products as crops and cereals.

DOI: 10.4018/978-1-6684-4548-8.ch008

INTRODUCTION

The humanity has accepted that natural resources have been declining rapidly and also there has been a huge pollution among natural environment including seas, oceans and lands (Yıldırım et.al., 2018). Due to the effect of rising climate change and global warming, scientist warns economies to adapt sustainable policies (NASA Global Climate Change). Recent indicators also support the assumption that climate change should be stopped in the short term to save the next generation (World Meteorological Organization, 2021). In this point, 2030 Sustainable Development Goals (SDGs) guides economies to follow sustainable policies for critical titles such as food, water, energy, climate change etc. (United Nations Department of Economic and Social Affairs, n.d.). In 2015, the 2030 Agenda for Sustainable Development was adopted by all United Nations Member States, resulting in a global guide for the realization of 17 basic sustainable development goals. The 2030 SDGs contain very comprehensive sub-targets in order to eliminate other deprivations and problems, especially poverty, and to accelerate the fight against climate change. For many years, United Nations has been continuing its important initiatives for the fight against climate change and sustainability. For example, Agenda 21 was adopted in June 1992. The Millennium Declaration was adopted in September 2000. At this point, it can be said that the first sustainable development goals emerged. 8 key targets have been accepted as Millennium Development Goals (MDGs) to be achieved by 2015. Paris Agreement on Climate Change and 2030 Sustainable Development Goals, which took place in 2015, are important initiatives (United Nations Department of Economic and Social Affairs, n.d.). The acceptance of the SDGs as a global project necessitates a coordinated implementation of sustainable policies. The 2030 SDGs have a broader framework, being more comprehensive than the millennium goals and emphasizing global cooperation (Weber, 2017). Sustainable development goals (SDGs) have an important place in the success of food security. Sustainable development goals are not essentially authoritative or strictly binding laws, but serve as an adaptive guide that guides countries in the global framework. There may be some differences between priority areas for each country or economy. Therefore, each economy may have targets that can be prioritized in sustainability targets according to its own priority (Echendu, 2022).

Although food insecurity and water stress are old problems in the world, the last decade has pointed out that food insecurity (Yıldırım and Yıldırım, 2021) and water stress (Esen et.al., 2020) have challenged developed and developing countries. Especially, the Covid-19 pandemic showed that food insecurity was a real problem for each economy in the world. In fact, there was no any good signal for achieving food security or ending hunger until 2030 before the Covid-19 pandemic. Based on the report as "The State of Food Security and Nutrition in the World 2021", it was seen that the hunger increased by 2020. For example, it was determined that there was higher numbers of people with hunger in 2020 rather than 2019 (FAO, IFAD, UNICEF, WFP and WHO, 2021). While FAO continues to work for sustainable food and agriculture, it also emphasizes the importance of the sustainable food relationship with sustainable development goals. According to FAO, there are 5 basic principles. In these principles, there are 20 basic actions related to sustainable development goals (FAO, 2018). These actions serve as a guide for countries to link the many sectors of agriculture and rural development with a broader development program covering poverty eradication, job creation, national growth, urban renewal and natural resource richness. The basic principles given for sustainable food and agriculture principles defined by FAO are shown in Figure 1 (FAO, 2018):

Figure 1. Five principles for achieving sustainable food and agriculture
Source: adapted from FAO, 2018 and created by the authors

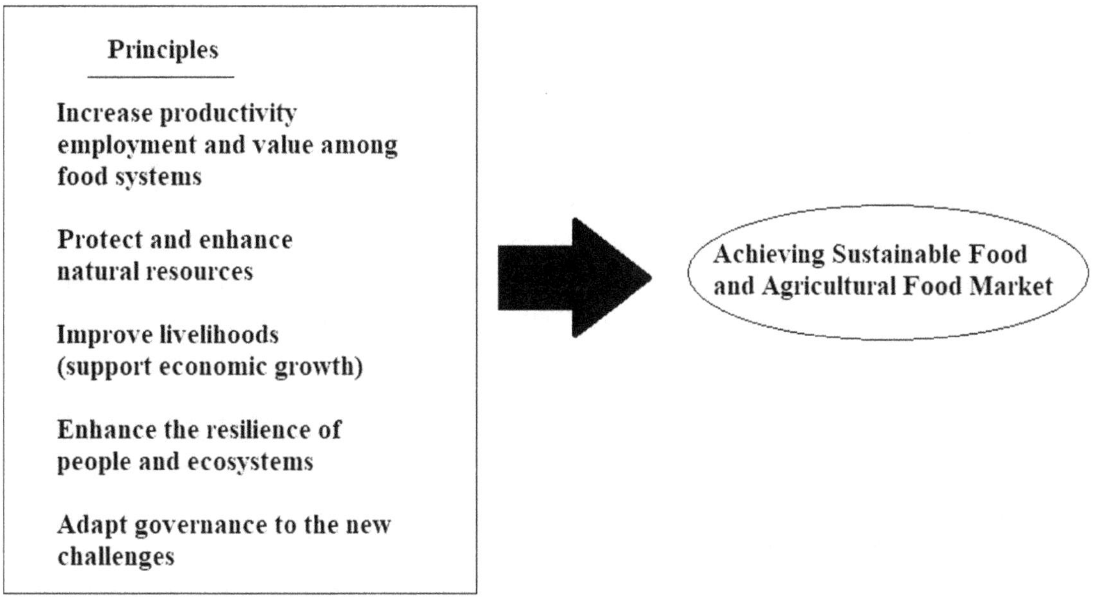

Climate change appears to have a significant impact on food availability, physical and economic access and use of food, as well as food stability. In this respect, it can be said that these effects of climate change will create significant limitations in terms of supply and access in the global food system (U.S. Department of Agriculture, n.d.). Achieving sustainable food and agriculture is an important issue both for developed and developing countries since the last decades. So it is urgent that sustainable agriculture policies should be adapted for global food security in the long term. As it has been known for a long time, climate change is still rising and it is changing the native biodiversity and current weather conditions. Especially, the Mediterranean Basin has been challenged with the climate change and it is under attack of invasive alien species (Yıldırım and Kaplan, 2022). The Mediterranean Basin seems a lucky region as having several food sources from sea and lands. But, climate change and water stress threaten food security in this region for many countries. Many studies have discussed the effect of climate change, invasion of or human activities on biodiversity and food sources in the Mediterranean Basin. This study aims to show current profile for agriculture food market for the Mediterranean Basin. It is purposed to give recent news, reports and statistics for agriculture production and sustainable agriculture food and food security in this region. This study will provide less literature but more statistical information for agricultural food market for Mediterranean countries. Based on FAOSTAT, the study will give recent numbers for agriculture food production in Mediterranean countries. In this study, statistics for agriculture food market includes crops and cereal products. So, livestock statistics is not included here.

AGRICULTURAL FOOD MARKET: MEDITERRANEAN BASIN

Urbanization is a rising trend all over the world. Therefore, populations began to concentrate in cities. Today, 55% of the world's population lives in cities and it is estimated that the population density in cities will reach 70 percent before 2050. When rural areas are compared with urban areas, it is seen that urban areas have important developmental features such as more social and economic development and more workforce opportunities. However, urban areas also intensify poverty. The challenges of urban life highlight poverty, hunger and other deprivations more. Sustainable development goals also reveal the burden of urbanization in terms of food insecurity (Vilar-Compte et.al., 2021).

Food security is closely related to nutrition security and health, but food safety and nutrition safety do not mean the same thing. Nutritional security can be thought of as the ability to reach the optimal amount of nutrients needed by the human body. Food security, on the other hand, is a broader concept that includes nutrition security. At this point, in order to have food security, first of all, there must be food security in the households. From a global perspective, the ability of countries to maintain economic and healthy food supplies brings global food security policies to the agenda. The 2030 Sustainable Development Goals placed a special emphasis on food security (Pérez-Escamilla, 2017). Food security is an important issue among the Mediterranean Basin. In the Mediterranean Basin, food production from the land and from the sea is under danger due to rising climate change and environmental pollution. Over warming weather threatens agricultural food production in Mediterranean countries and also climate change effects seafood security in the Mediterranean Sea (Mrabet et.al., 2020). In general, the Mediterranean Basin includes these countries: "Albania, Algeria, Bosnia and Herzegovina, Croatia, Cyprus, Egypt, France, Greece, Israel, Italy, Lebanon, Libya, Malta, Monaco, Montenegro, Morocco, Slovenia, Spain, Syria, Tunisia, and Turkey (Barcelona Convention - Mediterranean 2017 Quality Status Report, n.d.)". Table 1 shows the population of Mediterranean countries by 2018.

Table 1. Annual population-2018 (1000 persons)

Area	Element	Value
Albania	Rural population	1164.388
Albania	Urban population	1769.975
Algeria	Rural population	11498.038
Algeria	Urban population	30510.016
Bosnia and Herzegovina	Rural population	1813.259
Bosnia and Herzegovina	Urban population	1690.295
Bulgaria	Rural population	1758.643
Bulgaria	Urban population	5278.205
Croatia	Rural population	1793.077
Croatia	Urban population	2371.706
Cyprus	Rural population	394.652
Cyprus	Urban population	794.433
France	Rural population	12756.906
France	Urban population	52476.365
Greece	Rural population	2333.39
Greece	Urban population	8808.771
Israel	Rural population	640.922
Israel	Urban population	7811.919

Table 1. Continued

Area	Element	Value
Italy	Rural population	17527.87
Italy	Urban population	41763.099
Lebanon	Rural population	695.098
Lebanon	Urban population	5398.411
Malta	Rural population	23.28
Malta	Urban population	408.809
Montenegro	Rural population	208.816
Montenegro	Urban population	420.403
Slovenia	Rural population	946.127
Slovenia	Urban population	1135.133
Spain	Rural population	9130.38
Spain	Urban population	37267.072
Sudan	Rural population	27131.309
Sudan	Urban population	14380.217
Syrian Arab Republic	Rural population	8381.17
Syrian Arab Republic	Urban population	9903.237
Tunisia	Rural population	3620.748
Tunisia	Urban population	8038.426
Turkey	Rural population	20362.183
Turkey	Urban population	61554.688

Source: FAOSTAT, (n.d.a).

It covers the Mediterranean, which covers an area of 2.54 million square kilometers. While the widest place is 800 km; The longest sea is 3,800 km long. It is a region bordered by Europe to the North, Africa to the South, and Asia to the East. Different sources can classify Mediterranean countries differently. The Mediterranean is essentially surrounded by 16 countries. Six of these countries (Spain, France, Monaco, Italy, Greece and Malta) are in Europe, five (Egypt, Libya, Tunisia, Algeria and Morocco) are in Africa, four (Syria, Lebanon, Israel and Cyprus) and one one (turkey) is in both Europe and Asia. populous Mediterranean island with 502,650 people living within its territory. (WorldAtlas, n.d.). In the Mediterranean, about one-third of the population is concentrated in coastal areas. Meanwhile, approximately 55% of the total population lives in coastal hydrological basins. In the southern region of the Mediterranean, 65% of the population is located (European Environment Agency (EEA), 2015). It can be said that the Mediterranean Basin has several countries with different population size. Accordingly, it can be said that the food demand will be different for these countries, too.

Descriptive Statistics for Agricultural Food Market in the Mediterranean Countries

This part aims to show a brief framework for the current status of agriculture food production and food market for the Mediterranean countries. Table 2 shows production of agricultural production in the Mediterranean countries by 2020. There are descriptive statistics for "area harvested (ha) and production (tonnes)" for countries.

Table 2. Production of crops products in the Mediterranean countries (2020)

Area	Element	Item	Unit	Value	Area	Element	Item	Unit	Value
Albania	Area harvested	Fruit, citrus nes	ha	1	Croatia	Area harvested	Cereals nes	ha	1040
Albania	Production	Fruit, citrus nes	tonnes	16	Croatia	Production	Cereals nes	tonnes	1900
Albania	Area harvested	Fruit, fresh nes	ha	125	Croatia	Area harvested	Fruit, citrus nes	ha	0
Albania	Production	Fruit, fresh nes	tonnes	3345	Croatia	Production	Fruit, citrus nes	tonnes	0
Albania	Area harvested	Fruit, stone nes	ha		Croatia	Area harvested	Maize	ha	288400
Albania	Production	Fruit, stone nes	tonnes		Croatia	Production	Maize	tonnes	2430600
Albania	Area harvested	Maize	ha	56727	Croatia	Area harvested	Olives	ha	20280
Albania	Production	Maize	tonnes	399125	Croatia	Production	Olives	tonnes	33230
Albania	Area harvested	Olives	ha	48102	Croatia	Area harvested	Rice, paddy	ha	0
Albania	Production	Olives	tonnes	131971	Croatia	Production	Rice, paddy	tonnes	0
Albania	Area harvested	Rice, paddy	ha		Croatia	Production	Rice, paddy (rice milled equivalent)	tonnes	0
Albania	Production	Rice, paddy	tonnes		Croatia	Area harvested	Vegetables, fresh nes	ha	890
Albania	Production	Rice, paddy (rice milled equivalent)	tonnes		Croatia	Production	Vegetables, fresh nes	tonnes	14040
Albania	Area harvested	Vegetables, fresh nes	ha	1650	Croatia	Area harvested	Wheat	ha	147840
Albania	Production	Vegetables, fresh nes	tonnes	28706	Croatia	Production	Wheat	tonnes	867530
Albania	Area harvested	Vegetables, leguminous nes	ha	22	Cyprus	Area harvested	Fruit, pome nes	ha	20
Albania	Production	Vegetables, leguminous nes	tonnes	110	Cyprus	Production	Fruit, pome nes	tonnes	60
Albania	Area harvested	Wheat	ha	53946	Cyprus	Area harvested	Fruit, stone nes	ha	30
Albania	Production	Wheat	tonnes	233430	Cyprus	Production	Fruit, stone nes	tonnes	320
Algeria	Area harvested	Fruit, citrus nes	ha	468	Cyprus	Area harvested	Fruit, tropical fresh nes	ha	360
Algeria	Production	Fruit, citrus nes	tonnes	569	Cyprus	Production	Fruit, tropical fresh nes	tonnes	1640
Algeria	Area harvested	Fruit, fresh nes	ha	12969	Cyprus	Area harvested	Olives	ha	11110
Algeria	Production	Fruit, fresh nes	tonnes	133801	Cyprus	Production	Olives	tonnes	22100
Algeria	Area harvested	Fruit, stone nes	ha	2125	Cyprus	Area harvested	Rice, paddy	ha	0
Algeria	Production	Fruit, stone nes	tonnes	26534	Cyprus	Production	Rice, paddy	tonnes	0
Algeria	Area harvested	Fruit, tropical fresh nes	ha		Cyprus	Production	Rice, paddy (rice milled equivalent)	tonnes	0

Table 2. Continued

Area	Element	Item	Unit	Value	Area	Element	Item	Unit	Value
Algeria	Production	Fruit, tropical fresh nes	tonnes		Cyprus	Area harvested	Vegetables, fresh nes	ha	20
Algeria	Area harvested	Maize	ha	530	Cyprus	Production	Vegetables, fresh nes	tonnes	340
Algeria	Production	Maize	tonnes	3221	Cyprus	Area harvested	Vegetables, leguminous nes	ha	150
Algeria	Area harvested	Olives	ha	438828	Cyprus	Production	Vegetables, leguminous nes	tonnes	2480
Algeria	Production	Olives	tonnes	1079508	Cyprus	Area harvested	Wheat	ha	12500
Algeria	Area harvested	Rice, paddy	ha	168	Cyprus	Production	Wheat	tonnes	32270
Algeria	Production	Rice, paddy	tonnes	300	Egypt	Area harvested	Fruit, citrus nes	ha	390
Algeria	Production	Rice, paddy (rice milled equivalent)	tonnes	200	Egypt	Production	Fruit, citrus nes	tonnes	5482
Algeria	Area harvested	Vegetables, fresh nes	ha	56757	Egypt	Area harvested	Fruit, fresh nes	ha	23815
Algeria	Production	Vegetables, fresh nes	tonnes	1458965	Egypt	Production	Fruit, fresh nes	tonnes	491076
Algeria	Area harvested	Vegetables, leguminous nes	ha	33125	Egypt	Area harvested	Fruit, stone nes	ha	142
Algeria	Production	Vegetables, leguminous nes	tonnes	287235	Egypt	Production	Fruit, stone nes	tonnes	1081
Algeria	Area harvested	Wheat	ha	1848083	Egypt	Area harvested	Fruit, tropical fresh nes	ha	3409
Algeria	Production	Wheat	tonnes	3106754	Egypt	Production	Fruit, tropical fresh nes	tonnes	26745
Bosnia and Herzegovina	Area harvested	Cereals nes	ha		Egypt	Area harvested	Maize	ha	1458881
Bosnia and Herzegovina	Production	Cereals nes	tonnes		Egypt	Production	Maize	tonnes	7500000
Bosnia and Herzegovina	Area harvested	Fruit, fresh nes	ha	147	Egypt	Area harvested	Olives	ha	100826
Bosnia and Herzegovina	Production	Fruit, fresh nes	tonnes	395	Egypt	Production	Olives	tonnes	932927
Bosnia and Herzegovina	Area harvested	Maize	ha	200544	Egypt	Area harvested	Rice, paddy	ha	554205
Bosnia and Herzegovina	Production	Maize	tonnes	1426356	Egypt	Production	Rice, paddy	tonnes	4893507
Bosnia and Herzegovina	Area harvested	Olives	ha	570	Egypt	Production	Rice, paddy (rice milled equivalent)	tonnes	3263969
Bosnia and Herzegovina	Production	Olives	tonnes	985	Egypt	Area harvested	Vegetables, fresh nes	ha	123134
Bosnia and Herzegovina	Area harvested	Vegetables, fresh nes	ha	110101	Egypt	Production	Vegetables, fresh nes	tonnes	723416
Bosnia and Herzegovina	Production	Vegetables, fresh nes	tonnes	519204	Egypt	Area harvested	Vegetables, leguminous nes	ha	3688
Bosnia and Herzegovina	Area harvested	Wheat	ha	69914	Egypt	Production	Vegetables, leguminous nes	tonnes	52469
Bosnia and Herzegovina	Production	Wheat	tonnes	321804	Egypt	Area harvested	Wheat	ha	1370235
Bulgaria	Area harvested	Cereals nes	ha	5660	Egypt	Production	Wheat	tonnes	9000000
Bulgaria	Production	Cereals nes	tonnes	9670	France	Area harvested	Cereals nes	ha	100520
Bulgaria	Area harvested	Fruit, citrus nes	ha	0	France	Production	Cereals nes	tonnes	310090

Table 2. Continued

Area	Element	Item	Unit	Value	Area	Element	Item	Unit	Value
Bulgaria	Production	Fruit, citrus nes	tonnes	0	France	Area harvested	Fruit, tropical fresh nes	ha	6070
Bulgaria	Area harvested	Fruit, pome nes	ha	50	France	Production	Fruit, tropical fresh nes	tonnes	49090
Bulgaria	Production	Fruit, pome nes	tonnes	140	France	Area harvested	Maize	ha	1691130
Bulgaria	Area harvested	Fruit, stone nes	ha	0	France	Production	Maize	tonnes	13419140
Bulgaria	Production	Fruit, stone nes	tonnes	0	France	Area harvested	Olives	ha	17620
Bulgaria	Area harvested	Fruit, tropical fresh nes	ha	0	France	Production	Olives	tonnes	30240
Bulgaria	Production	Fruit, tropical fresh nes	tonnes	0	France	Area harvested	Rice, paddy	ha	14810
Bulgaria	Area harvested	Maize	ha	581530	France	Production	Rice, paddy	tonnes	76320
Bulgaria	Production	Maize	tonnes	3014090	France	Production	Rice, paddy (rice milled equivalent)	tonnes	50905
Bulgaria	Area harvested	Olives	ha	0	France	Area harvested	Vegetables, fresh nes	ha	3760
Bulgaria	Production	Olives	tonnes	0	France	Production	Vegetables, fresh nes	tonnes	35520
Bulgaria	Area harvested	Rice, paddy	ha	12350	France	Area harvested	Vegetables, leguminous nes	ha	7720
Bulgaria	Production	Rice, paddy	tonnes	65810	France	Production	Vegetables, leguminous nes	tonnes	34190
Bulgaria	Production	Rice, paddy (rice milled equivalent)	tonnes	43895	France	Area harvested	Wheat	ha	4512420
Bulgaria	Area harvested	Vegetables, fresh nes	ha	920	France	Production	Wheat	tonnes	30144110
Bulgaria	Production	Vegetables, fresh nes	tonnes	7210	Greece	Area harvested	Cereals nes	ha	270
Bulgaria	Area harvested	Wheat	ha	1200180	Greece	Production	Cereals nes	tonnes	810
Bulgaria	Production	Wheat	tonnes	4847940	Greece	Area harvested	Fruit, citrus nes	ha	1000
Israel	Area harvested	Fruit, citrus nes	ha	3385	Greece	Production	Fruit, citrus nes	tonnes	22280
Israel	Production	Fruit, citrus nes	tonnes	10078	Greece	Area harvested	Fruit, pome nes	ha	0
Israel	Area harvested	Fruit, fresh nes	ha	2865	Greece	Production	Fruit, pome nes	tonnes	0
Israel	Production	Fruit, fresh nes	tonnes	79217	Greece	Area harvested	Fruit, stone nes	ha	860
Israel	Area harvested	Fruit, stone nes	ha	0	Greece	Production	Fruit, stone nes	tonnes	10550
Israel	Production	Fruit, stone nes	tonnes	0	Greece	Area harvested	Fruit, tropical fresh nes	ha	0
Israel	Area harvested	Fruit, tropical fresh nes	ha	2189	Greece	Production	Fruit, tropical fresh nes	tonnes	0
Israel	Production	Fruit, tropical fresh nes	tonnes	222	Greece	Area harvested	Maize	ha	116780
Israel	Area harvested	Maize	ha	3820	Greece	Production	Maize	tonnes	1178050
Israel	Production	Maize	tonnes	73419	Greece	Area harvested	Olives	ha	906020
Israel	Area harvested	Olives	ha	33700	Greece	Area harvested	Rice, paddy	ha	36090
Israel	Production	Olives	tonnes	83000	Greece	Production	Rice, paddy	tonnes	287410
Israel	Area harvested	Vegetables, fresh nes	ha	75045	Greece	Production	Rice, paddy (rice milled equivalent)	tonnes	191702
Israel	Production	Vegetables, fresh nes	tonnes	157794	Greece	Area harvested	Vegetables, fresh nes	ha	960

Table 2. Continued

Area	Element	Item	Unit	Value	Area	Element	Item	Unit	Value
Israel	Area harvested	Vegetables, leguminous nes	ha	26	Greece	Production	Vegetables, fresh nes	tonnes	5700
Israel	Production	Vegetables, leguminous nes	tonnes	284	Greece	Area harvested	Vegetables, leguminous nes	ha	370
Israel	Area harvested	Wheat	ha	50710	Greece	Production	Vegetables, leguminous nes	tonnes	3270
Israel	Production	Wheat	tonnes	116691	Greece	Area harvested	Wheat	ha	355880
Italy	Area harvested	Cereals nes	ha	33910	Greece	Production	Wheat	tonnes	1095150
Italy	Production	Cereals nes	tonnes	109840	Lebanon	Area harvested	Fruit, fresh nes	ha	5715
Italy	Area harvested	Fruit, citrus nes	ha	1580	Lebanon	Production	Fruit, fresh nes	tonnes	49626
Italy	Production	Fruit, citrus nes	tonnes	28270	Lebanon	Area harvested	Maize	ha	1000
Italy	Area harvested	Fruit, pome nes	ha	180	Lebanon	Production	Maize	tonnes	3000
Italy	Production	Fruit, pome nes	tonnes	1590	Lebanon	Area harvested	Olives	ha	62868
Italy	Area harvested	Fruit, stone nes	ha	450	Lebanon	Production	Olives	tonnes	136384
Italy	Production	Fruit, stone nes	tonnes	6070	Lebanon	Area harvested	Vegetables, fresh nes	ha	2900
Italy	Area harvested	Fruit, tropical fresh nes	ha	12160	Lebanon	Production	Vegetables, fresh nes	tonnes	13557
Italy	Production	Fruit, tropical fresh nes	tonnes	214030	Lebanon	Area harvested	Vegetables, leguminous nes	ha	1327
Italy	Area harvested	Maize	ha	602860	Lebanon	Production	Vegetables, leguminous nes	tonnes	13798
Italy	Production	Maize	tonnes	6793130	Lebanon	Area harvested	Wheat	ha	41000
Italy	Area harvested	Olives	ha	1145520	Lebanon	Production	Wheat	tonnes	140000
Italy	Production	Olives	tonnes	2207150	Malta	Area harvested	Cereals nes	ha	0
Italy	Area harvested	Rice, paddy	ha	227320	Malta	Production	Cereals nes	tonnes	0
Italy	Production	Rice, paddy	tonnes	1507490	Malta	Area harvested	Fruit, citrus nes	ha	0
Italy	Production	Rice, paddy (rice milled equivalent)	tonnes	1005496	Malta	Production	Fruit, citrus nes	tonnes	0
Italy	Area harvested	Vegetables, fresh nes	ha	0	Malta	Production	Fruit, pome nes	tonnes	0
Italy	Production	Vegetables, fresh nes	tonnes	0	Malta	Area harvested	Maize	ha	0
Italy	Area harvested	Vegetables, leguminous nes	ha	7370	Malta	Production	Maize	tonnes	0
Italy	Production	Vegetables, leguminous nes	tonnes	46640	Malta	Area harvested	Rice, paddy	ha	0
Italy	Area harvested	Wheat	ha	1711220	Malta	Production	Rice, paddy	tonnes	0
Italy	Production	Wheat	tonnes	6716180	Malta	Production	Rice, paddy (rice milled equivalent)	tonnes	0
Montenegro	Area harvested	Fruit, fresh nes	ha	249	Malta	Production	Vegetables, fresh nes	tonnes	700
Montenegro	Production	Fruit, fresh nes	tonnes	495	Malta	Production	Vegetables, leguminous nes	tonnes	20
Montenegro	Area harvested	Maize	ha	592	Malta	Area harvested	Wheat	ha	0
Montenegro	Production	Maize	tonnes	2687	Malta	Production	Wheat	tonnes	0
Montenegro	Area harvested	Olives	ha	185	Spain	Area harvested	Cereals nes	ha	9310
Montenegro	Production	Olives	tonnes	337	Spain	Production	Cereals nes	tonnes	31750

Table 2. Continued

Area	Element	Item	Unit	Value	Area	Element	Item	Unit	Value
Montenegro	Area harvested	Vegetables, leguminous nes	ha	100	Spain	Area harvested	Fruit, citrus nes	ha	1370
Montenegro	Production	Vegetables, leguminous nes	tonnes	500	Spain	Production	Fruit, citrus nes	tonnes	0
Montenegro	Area harvested	Wheat	ha	763	Spain	Area harvested	Fruit, pome nes	ha	3110
Montenegro	Production	Wheat	tonnes	2339	Spain	Production	Fruit, pome nes	tonnes	33070
Morocco	Area harvested	Cereals nes	ha	3553	Spain	Area harvested	Fruit, stone nes	ha	380
Morocco	Production	Cereals nes	tonnes	2971	Spain	Production	Fruit, stone nes	tonnes	2220
Morocco	Area harvested	Fruit, citrus nes	ha	2007	Spain	Area harvested	Fruit, tropical fresh nes	ha	34200
Morocco	Production	Fruit, citrus nes	tonnes	17719	Spain	Production	Fruit, tropical fresh nes	tonnes	578280
Morocco	Area harvested	Fruit, fresh nes	ha	7783	Spain	Area harvested	Maize	ha	343780
Morocco	Production	Fruit, fresh nes	tonnes	54340	Spain	Production	Maize	tonnes	4214100
Morocco	Area harvested	Fruit, stone nes	ha	34	Spain	Area harvested	Olives	ha	2623720
Morocco	Production	Fruit, stone nes	tonnes	319	Spain	Production	Olives	tonnes	8137810
Morocco	Area harvested	Fruit, tropical fresh nes	ha	70	Spain	Area harvested	Rice, paddy	ha	102060
Morocco	Production	Fruit, tropical fresh nes	tonnes	464	Spain	Production	Rice, paddy	tonnes	739230
Morocco	Area harvested	Maize	ha	71087	Spain	Production	Rice, paddy (rice milled equivalent)	tonnes	493066
Morocco	Production	Maize	tonnes	29888	Spain	Area harvested	Vegetables, fresh nes	ha	7360
Morocco	Area harvested	Olives	ha	1068895	Spain	Production	Vegetables, fresh nes	tonnes	132350
Morocco	Production	Olives	tonnes	1409266	Spain	Area harvested	Vegetables, leguminous nes	ha	6170
Morocco	Area harvested	Rice, paddy	ha	5788	Spain	Production	Vegetables, leguminous nes	tonnes	54070
Morocco	Production	Rice, paddy	tonnes	46275	Spain	Area harvested	Wheat	ha	1914660
Morocco	Production	Rice, paddy (rice milled equivalent)	tonnes	30865	Spain	Production	Wheat	tonnes	8143510
Morocco	Area harvested	Vegetables, fresh nes	ha	9704	Syrian Arab Republic	Area harvested	Fruit, citrus nes	ha	7502
Morocco	Production	Vegetables, fresh nes	tonnes	174477	Syrian Arab Republic	Production	Fruit, citrus nes	tonnes	213584
Morocco	Area harvested	Vegetables, leguminous nes	ha	12942	Syrian Arab Republic	Area harvested	Fruit, fresh nes	ha	6845
Morocco	Production	Vegetables, leguminous nes	tonnes	99274	Syrian Arab Republic	Production	Fruit, fresh nes	tonnes	106506
Morocco	Area harvested	Wheat	ha	2845290	Syrian Arab Republic	Area harvested	Fruit, stone nes	ha	1831
Morocco	Production	Wheat	tonnes	2561898	Syrian Arab Republic	Production	Fruit, stone nes	tonnes	9675
Slovenia	Area harvested	Cereals nes	ha	4700	Syrian Arab Republic	Area harvested	Maize	ha	50393
Slovenia	Production	Cereals nes	tonnes	4440	Syrian Arab Republic	Production	Maize	tonnes	226987
Slovenia	Area harvested	Fruit, citrus nes	ha	0	Syrian Arab Republic	Area harvested	Olives	ha	696363

Table 2. Continued

Area	Element	Item	Unit	Value	Area	Element	Item	Unit	Value
Slovenia	Production	Fruit, citrus nes	tonnes	0	Syrian Arab Republic	Production	Olives	tonnes	781204
Slovenia	Area harvested	Fruit, pome nes	ha	0	Syrian Arab Republic	Area harvested	Rice, paddy	ha	
Slovenia	Production	Fruit, pome nes	tonnes	110	Syrian Arab Republic	Production	Rice, paddy	tonnes	
Slovenia	Area harvested	Fruit, tropical fresh nes	ha	120	Syrian Arab Republic	Production	Rice, paddy (rice milled equivalent)	tonnes	
Slovenia	Production	Fruit, tropical fresh nes	tonnes	1580	Syrian Arab Republic	Area harvested	Vegetables, fresh nes	ha	27459
Slovenia	Area harvested	Maize	ha	39840	Syrian Arab Republic	Production	Vegetables, fresh nes	tonnes	278382
Slovenia	Production	Maize	tonnes	429930	Syrian Arab Republic	Area harvested	Vegetables, leguminous nes	ha	7966
Slovenia	Area harvested	Olives	ha	1420	Syrian Arab Republic	Production	Vegetables, leguminous nes	tonnes	71010
Slovenia	Production	Olives	tonnes	3100	Syrian Arab Republic	Area harvested	Wheat	ha	1350538
Slovenia	Area harvested	Rice, paddy	ha	0	Syrian Arab Republic	Production	Wheat	tonnes	2848472
Slovenia	Production	Rice, paddy	tonnes	0	Turkey	Area harvested	Fruit, citrus nes	ha	251
Slovenia	Production	Rice, paddy (rice milled equivalent)	tonnes	0	Turkey	Production	Fruit, citrus nes	tonnes	2609
Slovenia	Area harvested	Vegetables, fresh nes	ha	1543	Turkey	Area harvested	Fruit, fresh nes	ha	29808
Slovenia	Production	Vegetables, fresh nes	tonnes	29868	Turkey	Production	Fruit, fresh nes	tonnes	521966
Slovenia	Area harvested	Wheat	ha	27280	Turkey	Area harvested	Fruit, stone nes	ha	266
Slovenia	Production	Wheat	tonnes	158130	Turkey	Production	Fruit, stone nes	tonnes	5279
Tunisia	Area harvested	Cereals nes	ha	14025	Turkey	Area harvested	Fruit, tropical fresh nes	ha	
Tunisia	Production	Cereals nes	tonnes	28000	Turkey	Production	Fruit, tropical fresh nes	tonnes	
Tunisia	Area harvested	Fruit, citrus nes	ha	13464	Turkey	Area harvested	Maize	ha	690553
Tunisia	Production	Fruit, citrus nes	tonnes	140579	Turkey	Production	Maize	tonnes	6500000
Tunisia	Area harvested	Fruit, fresh nes	ha	16599	Turkey	Area harvested	Olives	ha	887077
Tunisia	Production	Fruit, fresh nes	tonnes	80530	Turkey	Production	Olives	tonnes	1316626
Tunisia	Area harvested	Fruit, stone nes	ha	600	Turkey	Area harvested	Rice, paddy	ha	125398
Tunisia	Production	Fruit, stone nes	tonnes	5600	Turkey	Production	Rice, paddy	tonnes	980000
Tunisia	Area harvested	Fruit, tropical fresh nes	ha	12085	Turkey	Production	Rice, paddy (rice milled equivalent)	tonnes	653660
Tunisia	Production	Fruit, tropical fresh nes	tonnes	100896	Turkey	Area harvested	Vegetables, fresh nes	ha	29922
Tunisia	Area harvested	Olives	ha	3642569	Turkey	Production	Vegetables, fresh nes	tonnes	555568
Tunisia	Production	Olives	tonnes	2000000	Turkey	Area harvested	Vegetables, leguminous nes	ha	7821
Tunisia	Area harvested	Vegetables, fresh nes	ha	7606	Turkey	Production	Vegetables, leguminous nes	tonnes	54317
Tunisia	Production	Vegetables, fresh nes	tonnes	108793	Turkey	Area harvested	Wheat	ha	6914632

Table 2. Continued

Area	Element	Item	Unit	Value	Area	Element	Item	Unit	Value
Tunisia	Area harvested	Vegetables, leguminous nes	ha	6045	Turkey	Production	Wheat	tonnes	20500000
Tunisia	Production	Vegetables, leguminous nes	tonnes	91593					
Tunisia	Area harvested	Wheat	ha	606000					
Tunisia	Production	Wheat	tonnes	1042000					

Source: FAOSTAT, (n.d.b)

Table 3 shows export quantity and import quantity for some selected cereal products by 2019 among the Mediterranean countries. Sustainable agriculture production is an important to reduce food insecurity in the Mediterranean region. Being close together physically, these countries have an opportunity to import and export food sources within the related region.

Table 3. Total statistics for cereals (2019,tonnes)

Area	Element	Item	Value	Area	Element	Item	Value
Albania	Import Quantity	Barley	6058	Cyprus	Import Quantity	Barley	133575
Albania	Import Quantity	Buckwheat	0	Cyprus	Export Quantity	Barley	0
Albania	Import Quantity	Canary seed	1	Cyprus	Import Quantity	Buckwheat	7
Albania	Import Quantity	Cereals nes	1	Cyprus	Import Quantity	Canary seed	169
Albania	Import Quantity	Maize	93664	Cyprus	Import Quantity	Cereals nes	46
Albania	Export Quantity	Maize	18	Cyprus	Import Quantity	Maize	313348
Albania	Import Quantity	Millet	13	Cyprus	Export Quantity	Maize	26
Albania	Import Quantity	Oats	270	Cyprus	Import Quantity	Millet	32
Albania	Export Quantity	Oats	0	Cyprus	Import Quantity	Oats	32
Albania	Import Quantity	Quinoa	15	Cyprus	Import Quantity	Quinoa	99
Albania	Import Quantity	Rice, paddy	1	Cyprus	Import Quantity	Rice, paddy	80
Albania	Import Quantity	Rye	293	Cyprus	Import Quantity	Rye	2
Albania	Import Quantity	Sorghum	1	Cyprus	Import Quantity	Sorghum	15
Albania	Import Quantity	Triticale	1	Cyprus	Import Quantity	Triticale	266
Albania	Import Quantity	Wheat	236269	Cyprus	Import Quantity	Wheat	61760
Algeria	Import Quantity	Barley	122297	Cyprus	Export Quantity	Wheat	21
Algeria	Import Quantity	Buckwheat	8	Egypt	Import Quantity	Barley	5498
Algeria	Import Quantity	Canary seed	4799	Egypt	Export Quantity	Barley	3
Algeria	Import Quantity	Cereals nes	29	Egypt	Import Quantity	Buckwheat	3
Algeria	Import Quantity	Grain, mixed	24	Egypt	Import Quantity	Canary seed	6047
Algeria	Import Quantity	Maize	4356206	Egypt	Export Quantity	Canary seed	4829
Algeria	Export Quantity	Maize	0	Egypt	Import Quantity	Cereals nes	0
Algeria	Import Quantity	Millet	2840	Egypt	Export Quantity	Cereals nes	239
Algeria	Import Quantity	Oats	448	Egypt	Import Quantity	Maize	8078446
Algeria	Import Quantity	Quinoa	20	Egypt	Export Quantity	Maize	6502
Algeria	Import Quantity	Rice, paddy	1000	Egypt	Import Quantity	Millet	41
Algeria	Import Quantity	Sorghum	18	Egypt	Export Quantity	Millet	1005

Table 3. Continued

Area	Element	Item	Value	Area	Element	Item	Value
Algeria	Import Quantity	Wheat	6775910	Egypt	Import Quantity	Oats	0
Algeria	Export Quantity	Wheat	2	Egypt	Export Quantity	Oats	41
Bosnia and Herzegovina	Import Quantity	Barley	28591	Egypt	Import Quantity	Quinoa	1
Bosnia and Herzegovina	Export Quantity	Barley	24	Egypt	Import Quantity	Rice, paddy	1
Bosnia and Herzegovina	Import Quantity	Buckwheat	275	Egypt	Import Quantity	Sorghum	880
Bosnia and Herzegovina	Import Quantity	Canary seed	1	Egypt	Export Quantity	Sorghum	1643
Bosnia and Herzegovina	Import Quantity	Cereals nes	1	Egypt	Import Quantity	Wheat	10424423
Bosnia and Herzegovina	Import Quantity	Maize	180843	Egypt	Export Quantity	Wheat	2
Bosnia and Herzegovina	Export Quantity	Maize	543	France	Import Quantity	Barley	93466
Bosnia and Herzegovina	Import Quantity	Millet	27	France	Export Quantity	Barley	7171937
Bosnia and Herzegovina	Export Quantity	Millet	0	France	Import Quantity	Buckwheat	4908
Bosnia and Herzegovina	Import Quantity	Oats	3679	France	Export Quantity	Buckwheat	1092
Bosnia and Herzegovina	Import Quantity	Quinoa	8	France	Import Quantity	Canary seed	1545
Bosnia and Herzegovina	Export Quantity	Quinoa	5	France	Export Quantity	Canary seed	2
Bosnia and Herzegovina	Import Quantity	Rice, paddy	24	France	Import Quantity	Cereals nes	6945
Bosnia and Herzegovina	Import Quantity	Rye	1000	France	Export Quantity	Cereals nes	93353
Bosnia and Herzegovina	Export Quantity	Rye	10	France	Import Quantity	Fonio	169
Bosnia and Herzegovina	Import Quantity	Sorghum	6	France	Export Quantity	Fonio	4
Bosnia and Herzegovina	Import Quantity	Triticale	1335	France	Import Quantity	Maize	704995
Bosnia and Herzegovina	Export Quantity	Triticale	78	France	Export Quantity	Maize	3672345
Bosnia and Herzegovina	Import Quantity	Wheat	324580	France	Import Quantity	Millet	2709
Bosnia and Herzegovina	Export Quantity	Wheat	22761	France	Export Quantity	Millet	32779
Bulgaria	Import Quantity	Barley	1713	France	Import Quantity	Oats	12832
Bulgaria	Export Quantity	Barley	267290	France	Export Quantity	Oats	110353
Bulgaria	Import Quantity	Buckwheat	114	France	Import Quantity	Quinoa	9008
Bulgaria	Export Quantity	Buckwheat	87	France	Export Quantity	Quinoa	2042
Bulgaria	Import Quantity	Canary seed	58	France	Import Quantity	Rice, paddy	9308
Bulgaria	Export Quantity	Canary seed	243	France	Export Quantity	Rice, paddy	10525
Bulgaria	Import Quantity	Cereals nes	19	France	Import Quantity	Rye	4634
Bulgaria	Export Quantity	Cereals nes	356	France	Export Quantity	Rye	17886
Bulgaria	Import Quantity	Maize	26524	France	Import Quantity	Sorghum	1871
Bulgaria	Export Quantity	Maize	2588956	France	Export Quantity	Sorghum	125508

Table 3. Continued

Area	Element	Item	Value	Area	Element	Item	Value
Bulgaria	Import Quantity	Millet	1429	France	Import Quantity	Triticale	16501
Bulgaria	Export Quantity	Millet	823	France	Export Quantity	Triticale	61248
Bulgaria	Import Quantity	Oats	152	France	Import Quantity	Wheat	301666
Bulgaria	Export Quantity	Oats	5098	France	Export Quantity	Wheat	19956974
Bulgaria	Import Quantity	Quinoa	398	Greece	Import Quantity	Barley	93759
Bulgaria	Export Quantity	Quinoa	168	Greece	Export Quantity	Barley	1254
Bulgaria	Import Quantity	Rice, paddy	7681	Greece	Import Quantity	Buckwheat	178
Bulgaria	Export Quantity	Rice, paddy	28016	Greece	Export Quantity	Buckwheat	5
Bulgaria	Import Quantity	Rye	741	Greece	Import Quantity	Canary seed	1413
Bulgaria	Export Quantity	Rye	439	Greece	Export Quantity	Canary seed	1
Bulgaria	Import Quantity	Sorghum	98	Greece	Import Quantity	Grain, mixed	29078
Bulgaria	Export Quantity	Sorghum	16171	Greece	Export Quantity	Grain, mixed	669
Bulgaria	Import Quantity	Triticale	96	Greece	Import Quantity	Maize	805728
Bulgaria	Export Quantity	Triticale	756	Greece	Export Quantity	Maize	7214
Bulgaria	Import Quantity	Wheat	56308	Greece	Import Quantity	Millet	404
Bulgaria	Export Quantity	Wheat	4883530	Greece	Import Quantity	Oats	10362
Croatia	Import Quantity	Barley	9976	Greece	Export Quantity	Oats	64
Croatia	Export Quantity	Barley	36714	Greece	Import Quantity	Quinoa	455
Croatia	Import Quantity	Buckwheat	50	Greece	Export Quantity	Quinoa	46
Croatia	Export Quantity	Buckwheat	170	Greece	Import Quantity	Rice, paddy	566
Croatia	Import Quantity	Canary seed	19	Greece	Export Quantity	Rice, paddy	35706
Croatia	Import Quantity	Cereals nes	30	Greece	Import Quantity	Rye	1539
Croatia	Export Quantity	Cereals nes	466	Greece	Export Quantity	Rye	241
Croatia	Import Quantity	Maize	42083	Greece	Import Quantity	Sorghum	634
Croatia	Export Quantity	Maize	882694	Greece	Export Quantity	Sorghum	16
Croatia	Import Quantity	Millet	38	Greece	Import Quantity	Triticale	248
Croatia	Export Quantity	Millet	6	Greece	Export Quantity	Triticale	811
Croatia	Import Quantity	Oats	56	Greece	Import Quantity	Wheat	1038650
Croatia	Export Quantity	Oats	3219	Greece	Export Quantity	Wheat	287034
Croatia	Import Quantity	Quinoa	58	Israel	Import Quantity	Barley	260791
Croatia	Export Quantity	Quinoa	2	Israel	Export Quantity	Barley	36
Croatia	Import Quantity	Rice, paddy	9	Israel	Import Quantity	Buckwheat	235
Croatia	Import Quantity	Rye	1003	Israel	Import Quantity	Canary seed	152
Croatia	Export Quantity	Rye	451	Israel	Import Quantity	Cereals nes	972
Croatia	Import Quantity	Sorghum	26	Israel	Export Quantity	Cereals nes	43
Croatia	Export Quantity	Sorghum	2	Israel	Import Quantity	Maize	1692768
Croatia	Import Quantity	Triticale	220	Israel	Export Quantity	Maize	1223
Croatia	Export Quantity	Triticale	3761	Israel	Import Quantity	Millet	2224
Croatia	Import Quantity	Wheat	122725	Israel	Import Quantity	Oats	318
Croatia	Export Quantity	Wheat	362270	Israel	Import Quantity	Quinoa	1291
Israel	Import Quantity	Rice, paddy	35	Israel	Export Quantity	Quinoa	9
Israel	Import Quantity	Rye	5898	Slovenia	Export Quantity	Oats	47

Table 3. Continued

Area	Element	Item	Value	Area	Element	Item	Value
Israel	Export Quantity	Rye	27	Slovenia	Import Quantity	Quinoa	601
Israel	Import Quantity	Sorghum	26431	Slovenia	Export Quantity	Quinoa	434
Israel	Import Quantity	Wheat	1221666	Slovenia	Import Quantity	Rice, paddy	72
Israel	Export Quantity	Wheat	1841	Slovenia	Export Quantity	Rice, paddy	18
Italy	Import Quantity	Barley	497102	Slovenia	Import Quantity	Rye	709
Italy	Export Quantity	Barley	4104	Slovenia	Export Quantity	Rye	2749
Italy	Import Quantity	Buckwheat	14040	Slovenia	Import Quantity	Sorghum	3
Italy	Export Quantity	Buckwheat	214	Slovenia	Export Quantity	Sorghum	0
Italy	Import Quantity	Canary seed	7646	Slovenia	Import Quantity	Triticale	887
Italy	Export Quantity	Canary seed	209	Slovenia	Export Quantity	Triticale	583
Italy	Import Quantity	Cereals nes	3502	Slovenia	Import Quantity	Wheat	221923
Italy	Export Quantity	Cereals nes	641	Slovenia	Export Quantity	Wheat	105557
Italy	Import Quantity	Fonio	24	Spain	Import Quantity	Barley	1221055
Italy	Import Quantity	Maize	6394217	Spain	Export Quantity	Barley	151462
Italy	Export Quantity	Maize	34220	Spain	Import Quantity	Buckwheat	12151
Italy	Import Quantity	Millet	10333	Spain	Export Quantity	Buckwheat	197
Italy	Export Quantity	Millet	474	Spain	Import Quantity	Canary seed	11926
Italy	Import Quantity	Oats	19247	Spain	Export Quantity	Canary seed	32
Italy	Export Quantity	Oats	7897	Spain	Import Quantity	Cereals nes	19495
Italy	Import Quantity	Quinoa	3079	Spain	Export Quantity	Cereals nes	17178
Italy	Export Quantity	Quinoa	1282	Spain	Export Quantity	Fonio	5
Italy	Import Quantity	Rice, paddy	49476	Spain	Import Quantity	Maize	10012619
Italy	Export Quantity	Rice, paddy	13293	Spain	Export Quantity	Maize	152456
Italy	Import Quantity	Rye	10674	Spain	Import Quantity	Millet	16472
Italy	Export Quantity	Rye	2781	Spain	Export Quantity	Millet	301
Italy	Import Quantity	Sorghum	80596	Spain	Import Quantity	Oats	121710
Italy	Export Quantity	Sorghum	600	Spain	Export Quantity	Oats	33027
Italy	Import Quantity	Triticale	26555	Spain	Import Quantity	Quinoa	3799
Italy	Export Quantity	Triticale	651	Spain	Export Quantity	Quinoa	5088
Italy	Import Quantity	Wheat	7474381	Spain	Import Quantity	Rice, paddy	5418
Italy	Export Quantity	Wheat	65679	Spain	Export Quantity	Rice, paddy	19982
Lebanon	Import Quantity	Barley	107214	Spain	Import Quantity	Rye	117894
Lebanon	Export Quantity	Barley	1935	Spain	Export Quantity	Rye	19845
Lebanon	Import Quantity	Buckwheat	0	Spain	Import Quantity	Sorghum	488557
Lebanon	Export Quantity	Buckwheat	0	Spain	Export Quantity	Sorghum	1528
Lebanon	Import Quantity	Canary seed	483	Spain	Import Quantity	Triticale	15176
Lebanon	Import Quantity	Cereals nes	51	Spain	Export Quantity	Triticale	8476
Lebanon	Export Quantity	Cereals nes	51	Spain	Import Quantity	Wheat	5292711
Lebanon	Import Quantity	Grain, mixed	8	Spain	Export Quantity	Wheat	513222
Lebanon	Export Quantity	Grain, mixed	23	Syrian Arab Republic	Import Quantity	Barley	15937
Lebanon	Import Quantity	Maize	633921	Syrian Arab Republic	Export Quantity	Barley	52500

Table 3. Continued

Area	Element	Item	Value	Area	Element	Item	Value
Lebanon	Export Quantity	Maize	97	Syrian Arab Republic	Import Quantity	Buckwheat	1
Lebanon	Import Quantity	Millet	449	Syrian Arab Republic	Export Quantity	Buckwheat	88
Lebanon	Export Quantity	Millet	0	Syrian Arab Republic	Import Quantity	Canary seed	1930
Lebanon	Import Quantity	Oats	220	Syrian Arab Republic	Import Quantity	Cereals nes	115
Lebanon	Export Quantity	Oats	0	Syrian Arab Republic	Import Quantity	Maize	300746
Lebanon	Import Quantity	Quinoa	244	Syrian Arab Republic	Export Quantity	Maize	1
Lebanon	Export Quantity	Quinoa	9	Syrian Arab Republic	Import Quantity	Millet	1460
Lebanon	Export Quantity	Rice, paddy	54	Syrian Arab Republic	Import Quantity	Oats	6
Lebanon	Import Quantity	Rye	42	Syrian Arab Republic	Import Quantity	Wheat	327576
Lebanon	Import Quantity	Sorghum	49	Tunisia	Import Quantity	Barley	542722
Lebanon	Export Quantity	Sorghum	16	Tunisia	Export Quantity	Barley	375
Lebanon	Import Quantity	Triticale	104	Tunisia	Import Quantity	Canary seed	2390
Lebanon	Import Quantity	Wheat	756819	Tunisia	Import Quantity	Cereals nes	1
Lebanon	Export Quantity	Wheat	11094	Tunisia	Export Quantity	Cereals nes	0
Malta	Import Quantity	Barley	8315	Tunisia	Import Quantity	Maize	1025880
Malta	Import Quantity	Buckwheat	1	Tunisia	Import Quantity	Millet	119
Malta	Import Quantity	Canary seed	770	Tunisia	Import Quantity	Oats	3
Malta	Import Quantity	Cereals nes	3	Tunisia	Import Quantity	Quinoa	3
Malta	Import Quantity	Maize	45570	Tunisia	Import Quantity	Sorghum	6620
Malta	Import Quantity	Millet	34	Tunisia	Export Quantity	Sorghum	2
Malta	Import Quantity	Oats	37	Tunisia	Import Quantity	Triticale	1
Malta	Export Quantity	Oats	322	Tunisia	Import Quantity	Wheat	1848993
Malta	Import Quantity	Quinoa	44	Turkey	Import Quantity	Barley	562777
Malta	Export Quantity	Quinoa	0	Turkey	Export Quantity	Barley	41188
Malta	Import Quantity	Rice, paddy	1	Turkey	Import Quantity	Buckwheat	223
Malta	Import Quantity	Rye	0	Turkey	Export Quantity	Buckwheat	37
Malta	Import Quantity	Sorghum	1	Turkey	Import Quantity	Canary seed	6127
Malta	Import Quantity	Triticale	51	Turkey	Export Quantity	Canary seed	4400
Malta	Import Quantity	Wheat	20431	Turkey	Import Quantity	Cereals nes	0
Montenegro	Import Quantity	Barley	1423	Turkey	Import Quantity	Fonio	1
Montenegro	Import Quantity	Buckwheat	6	Turkey	Import Quantity	Maize	4347475
Montenegro	Import Quantity	Cereals nes	0	Turkey	Export Quantity	Maize	691567
Montenegro	Import Quantity	Grain, mixed	1	Turkey	Import Quantity	Millet	12716
Montenegro	Import Quantity	Maize	30896	Turkey	Export Quantity	Millet	5269
Montenegro	Export Quantity	Maize	32	Turkey	Import Quantity	Oats	0
Montenegro	Import Quantity	Millet	8	Turkey	Export Quantity	Oats	18
Montenegro	Import Quantity	Oats	314	Turkey	Import Quantity	Quinoa	51
Montenegro	Import Quantity	Quinoa	6	Turkey	Export Quantity	Quinoa	69

Table 3. Continued

Area	Element	Item	Value	Area	Element	Item	Value
Montenegro	Import Quantity	Rice, paddy	4	Turkey	Import Quantity	Rice, paddy	101205
Montenegro	Import Quantity	Rye	114	Turkey	Export Quantity	Rice, paddy	381
Montenegro	Import Quantity	Sorghum	5	Turkey	Import Quantity	Rye	0
Montenegro	Import Quantity	Triticale	94	Turkey	Import Quantity	Sorghum	0
Montenegro	Import Quantity	Wheat	9508	Turkey	Export Quantity	Sorghum	103
Montenegro	Export Quantity	Wheat	8	Turkey	Import Quantity	Triticale	3
Slovenia	Import Quantity	Barley	24724	Turkey	Export Quantity	Triticale	107
Slovenia	Export Quantity	Barley	14744	Turkey	Import Quantity	Wheat	10004830
Slovenia	Import Quantity	Buckwheat	1163	Turkey	Export Quantity	Wheat	135157
Slovenia	Export Quantity	Buckwheat	1				
Slovenia	Import Quantity	Canary seed	4				
Slovenia	Export Quantity	Canary seed	1				
Slovenia	Import Quantity	Cereals nes	44				
Slovenia	Export Quantity	Cereals nes	17				
Slovenia	Import Quantity	Maize	703022				
Slovenia	Export Quantity	Maize	763441				
Slovenia	Import Quantity	Millet	87				
Slovenia	Export Quantity	Millet	0				
Slovenia	Import Quantity	Oats	2418				

Source: FAOSTAT, (n.d.c)

Table 4 shows the ranking for sustainable agriculture for Mediterranean countries (top 12). It was seen that Italy, France, Spain, Israel, Portugal and Turkey all performed higher value for sustainable agriculture by 2017 based on FSI. The Food Sustainability Index (FSI) which is developed The Economist Intelligence Unit with the Barilla Center for Food & Nutrition (BCFN), can show the level of sustainable agriculture for Mediterranean countries. According to FSI-2017, it was seen that France was a leader country among FSI-2017 due to its higher value in Nutrition challenge, food loss and waste and sustainable agriculture. Then, Portugal, Italy and Spain were playing an important role in food sustainability in the Mediterranean Basin. Turkey, Israel and Greece also seemed to be important countries for sustainable agriculture (The Economist and Intelligence Unite, 2017).

Table 4. Sustainable agriculture ranking (Food Sustainability Index 2017) for Mediterranean Countries

Ranking	Countries	Sustainable Agriculture*
1	Italy	73
2	France	71.5
3	Spain	69.2
4	Israel	68.9
5	Portugal	68.5
6	Turkey	68.3
7	Greece	67.9
8	Jordan	64.9
9	Lebanon	61.9
10	Egypt	61
11	Morocco	60.6
12	Tunisia	51
*Scores are scaled from 0 to 100, where 100 = the highest sustainability		

Source: adapted from The Economist and Intelligence Unite, 2017: 18

Based on food security indicators from FAO, Table 5 shows the level of food security in the Mediterranean Basin. Each indicator is showed separately for each country in this region as below:

Table 5. Selected food security indicators in the Mediterranean Basin

Area	Item	Year	Unit	Value
Albania	Gross domestic product per capita, PPP, dissemination (constant 2011 international $)	2019	I$	13671.5
Albania	Prevalence of undernourishment (percent) (3-year average)	2018-2020	%	3.9
Albania	Prevalence of moderate or severe food insecurity in the total population (percent) (3-year average)	2018-2020	%	33.8
Albania	Political stability and absence of violence/terrorism (index)	2019	index	0.12
Algeria	Gross domestic product per capita, PPP, dissemination (constant 2011 international $)	2019	I$	11510.6
Algeria	Prevalence of undernourishment (percent) (3-year average)	2018-2020	%	<2.5
Algeria	Prevalence of moderate or severe food insecurity in the total population (percent) (3-year average)	2018-2020	%	17.6
Algeria	Political stability and absence of violence/terrorism (index)	2019	index	-1
Bosnia and Herzegovina	Gross domestic product per capita, PPP, dissemination (constant 2011 international $)	2019	I$	14896.8
Bosnia and Herzegovina	Prevalence of undernourishment (percent) (3-year average)	2018-2020	%	<2.5
Bosnia and Herzegovina	Prevalence of moderate or severe food insecurity in the total population (percent) (3-year average)	2018-2020	%	10
Bosnia and Herzegovina	Political stability and absence of violence/terrorism (index)	2019	index	-0.4
Bulgaria	Gross domestic product per capita, PPP, dissemination (constant 2011 international $)	2019	I$	23191.6
Bulgaria	Prevalence of undernourishment (percent) (3-year average)	2018-2020	%	3
Bulgaria	Prevalence of moderate or severe food insecurity in the total population (percent) (3-year average)	2018-2020	%	13.2
Bulgaria	Political stability and absence of violence/terrorism (index)	2019	index	0.54
Croatia	Gross domestic product per capita, PPP, dissemination (constant 2011 international $)	2019	I$	28753.5
Croatia	Prevalence of undernourishment (percent) (3-year average)	2018-2020	%	<2.5
Croatia	Prevalence of moderate or severe food insecurity in the total population (percent) (3-year average)	2018-2020	%	11

Table 5. Continued

Area	Item	Year	Unit	Value
Croatia	Political stability and absence of violence/terrorism (index)	2019	index	0.76
Cyprus	Gross domestic product per capita, PPP, dissemination (constant 2011 international $)	2019	I$	40226.8
Cyprus	Prevalence of undernourishment (percent) (3-year average)	2018-2020	%	<2.5
Cyprus	Prevalence of moderate or severe food insecurity in the total population (percent) (3-year average)	2018-2020	%	
Cyprus	Political stability and absence of violence/terrorism (index)	2019	index	0.52
France	Gross domestic product per capita, PPP, dissemination (constant 2011 international $)	2019	I$	45834.2
France	Prevalence of undernourishment (percent) (3-year average)	2018-2020	%	<2.5
France	Prevalence of moderate or severe food insecurity in the total population (percent) (3-year average)	2018-2020	%	5.8
France	Political stability and absence of violence/terrorism (index)	2019	index	0.31
Greece	Gross domestic product per capita, PPP, dissemination (constant 2011 international $)	2019	I$	29723.2
Greece	Prevalence of undernourishment (percent) (3-year average)	2018-2020	%	<2.5
Greece	Prevalence of moderate or severe food insecurity in the total population (percent) (3-year average)	2018-2020	%	8.6
Greece	Political stability and absence of violence/terrorism (index)	2019	index	0.29
Israel	Gross domestic product per capita, PPP, dissemination (constant 2011 international $)	2019	I$	40007.3
Israel	Prevalence of undernourishment (percent) (3-year average)	2018-2020	%	<2.5
Israel	Prevalence of moderate or severe food insecurity in the total population (percent) (3-year average)	2018-2020	%	13.7
Israel	Political stability and absence of violence/terrorism (index)	2019	index	-0.78
Italy	Gross domestic product per capita, PPP, dissemination (constant 2011 international $)	2019	I$	42662.5
Italy	Prevalence of undernourishment (percent) (3-year average)	2018-2020	%	<2.5
Italy	Prevalence of moderate or severe food insecurity in the total population (percent) (3-year average)	2018-2020	%	6.7
Italy	Political stability and absence of violence/terrorism (index)	2019	index	0.46
Lebanon	Gross domestic product per capita, PPP, dissemination (constant 2011 international $)	2019	I$	14551.6
Lebanon	Prevalence of undernourishment (percent) (3-year average)	2018-2020	%	9.3
Lebanon	Prevalence of moderate or severe food insecurity in the total population (percent) (3-year average)	2018-2020	%	
Lebanon	Political stability and absence of violence/terrorism (index)	2019	index	-1.64
Malta	Gross domestic product per capita, PPP, dissemination (constant 2011 international $)	2019	I$	43950.6
Malta	Prevalence of undernourishment (percent) (3-year average)	2018-2020	%	<2.5
Malta	Prevalence of moderate or severe food insecurity in the total population (percent) (3-year average)	2018-2020	%	4.3
Malta	Political stability and absence of violence/terrorism (index)	2019	index	1.09
Montenegro	Gross domestic product per capita, PPP, dissemination (constant 2011 international $)	2019	I$	21533.9
Montenegro	Prevalence of undernourishment (percent) (3-year average)	2018-2020	%	<2.5
Montenegro	Prevalence of moderate or severe food insecurity in the total population (percent) (3-year average)	2018-2020	%	13.5
Montenegro	Political stability and absence of violence/terrorism (index)	2019	index	0.01
Morocco	Gross domestic product per capita, PPP, dissemination (constant 2011 international $)	2019	I$	7537.5
Morocco	Prevalence of undernourishment (percent) (3-year average)	2018-2020	%	4.2
Morocco	Prevalence of moderate or severe food insecurity in the total population (percent) (3-year average)	2018-2020	%	28
Morocco	Political stability and absence of violence/terrorism (index)	2019	index	-0.37
Slovenia	Gross domestic product per capita, PPP, dissemination (constant 2011 international $)	2019	I$	38905.5
Slovenia	Prevalence of undernourishment (percent) (3-year average)	2018-2020	%	<2.5
Slovenia	Prevalence of moderate or severe food insecurity in the total population (percent) (3-year average)	2018-2020	%	8.2
Slovenia	Political stability and absence of violence/terrorism (index)	2019	index	0.82
Spain	Gross domestic product per capita, PPP, dissemination (constant 2011 international $)	2019	I$	40805.9

Table 5. Continued

Area	Item	Year	Unit	Value
Spain	Prevalence of undernourishment (percent) (3-year average)	2018-2020	%	<2.5
Spain	Prevalence of moderate or severe food insecurity in the total population (percent) (3-year average)	2018-2020	%	8.8
Spain	Political stability and absence of violence/terrorism (index)	2019	index	0.32
Sudan	Gross domestic product per capita, PPP, dissemination (constant 2011 international $)	2019	I$	4185.6
Sudan	Prevalence of undernourishment (percent) (3-year average)	2018-2020	%	12.3
Sudan	Prevalence of moderate or severe food insecurity in the total population (percent) (3-year average)	2018-2020	%	49.4
Sudan	Political stability and absence of violence/terrorism (index)	2019	index	-1.67
Syrian Arab Republic	Prevalence of undernourishment (percent) (3-year average)	2018-2020	%	
Syrian Arab Republic	Prevalence of moderate or severe food insecurity in the total population (percent) (3-year average)	2018-2020	%	
Syrian Arab Republic	Political stability and absence of violence/terrorism (index)	2019	index	-2.73
Tunisia	Gross domestic product per capita, PPP, dissemination (constant 2011 international $)	2019	I$	10755.6
Tunisia	Prevalence of undernourishment (percent) (3-year average)	2018-2020	%	3
Tunisia	Prevalence of moderate or severe food insecurity in the total population (percent) (3-year average)	2018-2020	%	25.1
Tunisia	Political stability and absence of violence/terrorism (index)	2019	index	-0.83
Turkey	Gross domestic product per capita, PPP, dissemination (constant 2011 international $)	2019	I$	28199.1
Turkey	Prevalence of undernourishment (percent) (3-year average)	2018-2020	%	<2.5
Turkey	Prevalence of moderate or severe food insecurity in the total population (percent) (3-year average)	2018-2020	%	
Turkey	Political stability and absence of violence/terrorism (index)	2019	index	-1.34

Source: FAOSTAT, (n.d.d)

Despite its intense agricultural activities, the Mediterranean region is considered a water-scarce region. Therefore, it can be said that it is a region under threat in terms of exposure to climate change. Since 1970, it has shown an increase in air temperature of about 2 °C in southwestern Europe. The Water Use Index (WEI) (defined as the average annual total freshwater demand divided by the long-term average freshwater resources) shows that Southern countries are among the most water-stressed Mediterranean countries, with many having a higher-than-average WEI. 40%. It is estimated that the Mediterranean population will increase from 180 million today to over 250 million in 20 years (European Environment Agency (EEA), 2015).

A report has been published by CIHEAM showing the consequences of the effects of the Covid-19 pandemic on food safety and agricultural products. In the report, some findings regarding price stability in agricultural products for "*Albania, Algeria, Egypt, Lebanon, Morocco, Tunisia, Turkey*" are shown in Figure 2. It can be said that at the beginning of the epidemic, consumers in most countries experienced some fears caused by the change in food prices. "A pandemic has been observed with large quantities and frequent purchases of storable products such as "*flour, pasta, rice, semolina, pulses, milk powder, sugar and oils*". Therefore, some effects in the form of "temporary famines and consumer price increases" have been observed in countries (CIHEAM, 2020).

Figure 2. Food price changes during the Covid-19 Pandemic (comparing the last year)
Source: adapted from CIHEAM, 2020: 8

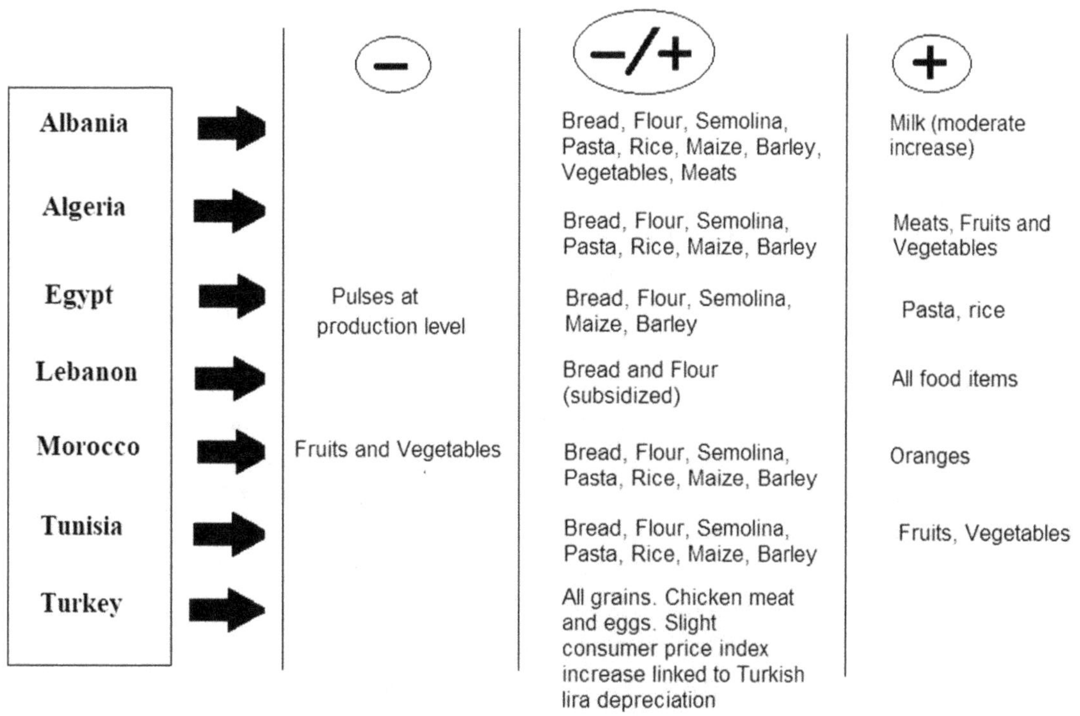

SOLUTIONS AND RECOMMENDATIONS

Agricultural Food Market statistics can help policy makers to see the level of food availability, food utilization, food access and food stability for food security based on agriculture production. Especially, policy makers can develop new strategies or projects to support sustainable agriculture production. In addition, countries can get new measurements to reduce food insecurity based on agriculture food sources. In this context, each food security dimensions should be evaluated separately. Countries should full in the gap between food security dimensions to achieve food security in the long term. Not only one food security dimensions can be enough to reduce food insecurity in a country.

FUTURE RESEARCH DIRECTIONS

This study gives a brief descriptive statistics and profile for agriculture food market in the Mediterranean Basin. The study can list future research directions for the related title as below:

- The future studies can focus on exploring the level of food insecurity based on low agriculture production in the Mediterranean Basin,
- The future studies can focus on examining the importance of working and coordination between Mediterranean countries in the long term.
- There will be helpful if empirical studies show sustainable agriculture production ranking.

- New measurements can be developed to evaluate the level of sustainable agriculture production.
- Agriculture production is related to water security that freshwater sources should be followed as achieving food security in the long term.

CONCLUSION

Food security and water security are both critical issues among today's world and economies. When considering 17 main goals in 2030 SDGs, food security and water security are all seem to be more important for the humanity (Bostancı and Yıldırım, 2022). Due to the effect of climate change, global warming and over wasting, countries had to face with food insecurity and freshwater problems. The literature provides many studies that examining the link between agricultural production, climate change and food security by different approaches (Pawlak and Kolodziejczak, 2020; Lipper et.al., 2014; Spiertz, 2009; Sessitsch and Birgit, 2015; Carvalho, 2006; Ehrlich et.al., 1993; Garrity et.al., 2010; Garrity et.a.l, 2010; Kogo et.al., 2021; de Graaff et.al., 2011; Rosegrant and Ringler, 2000). .

Mediterranean countries have troubled with the results of climate changes in the context of food security. Invasion of alien species in Mediterranean countries and changing weather conditions in this region reduces food sources for people in the long term. the Mediterranean Basin has an ability of agriculture production and aquaculture production to reduce food insecurity in the long time. Accordingly, keeping sustainable agriculture production needs innovative and coordinated projects in the region. The Mediterranean Basin makes each country to work together to achieve food security in the long term. Due to sharing same sea and similar weather conditions, Mediterranean countries can't be ignore the need of working together to reduce food insecurity in the related region.

REFERENCES

Barcelona Convention - Mediterranean 2017 Quality Status Report. (n.d.). *The Mediterranean Marine and Coastal Environment.* https://www.medqsr.org/mediterranean-marine-and-coastal-environment#:~:text=They%20are%20Albania%2C%20Algeria%2C%20Bosnia,scene%20of%20intense%20human%20activity

Bostancı, S. H., & Yıldırım, S. (2021). Sustainable Communities vs. Climate Refugees: Two Opposite Results of Climate Change. In C. Popescu (Ed.), *Handbook of Research on Novel Practices and Current Successes in Achieving the Sustainable Development Goals* (pp. 298–319). IGI Global. doi:10.4018/978-1-7998-8426-2.ch015

Carvalho, F. P. (2006). Agriculture, pesticides, food security and food safety. *Environmental Science & Policy*, *9*(7–8), 685–692. doi:10.1016/j.envsci.2006.08.002

CIHEAM. (2020). *The Covid 19 Pandemic: Threats On Food Security In The Mediterranean Region.* https://www.ciheam.org/publication/the-covid-19-pandemic-threats-on-food-security-in-the-mediterranean-region/

de Graaff, J., Kessler, A., & Nibbering, J. W. (2011). Agriculture and food security in selected countries in Sub-Saharan Africa: Diversity in trends and opportunities. *Food Security*, *3*(2), 195–213. doi:10.100712571-011-0125-4

Echendu, A. J. (2022). Flooding, Food Security and the Sustainable Development Goals in Nigeria: An Assemblage and Systems Thinking Approach. *Social Science*, *2022*(11), 59. doi:10.3390ocsci11020059

Ehrlich, P. R., Ehrlich, A. H., & Daily, G. C. (1993). Food Security, Population and Environment. *Population and Development Review*, *19*(1), 1–32. doi:10.2307/2938383

European Environment Agency (EEA). (2015). *Mediterranean Sea region briefing - The European environment — state and outlook 2015*. Briefing. https://www.eea.europa.eu/soer/2015/countries/mediterranean

FAO. (2018). Transforming Food and Agriculture to Achieve the SDGs: 20 interconnected actions to guide decision-makers. Technical Reference Document. FAO.

FAO. (2021). *World Food and Agriculture - Statistical Yearbook 2021*. doi:10.4060/cb4477en

FAO, IFAD, UNICEF, WFP, & WHO. (2021). *The State of Food Security and Nutrition in the World 2021. Transforming food systems for food security, improved nutrition and affordable healthy diets for all*. FAO. doi:10.4060/cb4474en

FAOSTAT. (n.d.a). *Annual population*, https://www.fao.org/faostat/en/#data/OA

FAOSTAT. (n.d.b). *Crops and livestock products*. https://www.fao.org/faostat/en/#data/QCL

FAOSTAT. (n.d.c). *Supply Utilization Accounts*. https://www.fao.org/faostat/en/#data/SCL

FAOSTAT. (n.d.d). *Suite of Food Security Indicators*. https://www.fao.org/faostat/en/#data/FS

Garrity, D. P., Akinnifesi, F. K., Ajayi, O. C., Weldesemayat, S. G., Mowo, J. G., Kalinganire, A., Larwanou, M., & Bayala, J. (2010). Evergreen Agriculture: A robust approach to sustainable food security in Africa. *Food Security*, *2*(3), 197–214. doi:10.100712571-010-0070-7

Kogo, B. K., Kumar, L., & Koech, R. (2021). Climate change and variability in Kenya: A review of impacts on agriculture and food security. *Environment, Development and Sustainability*, *23*(1), 23–43. doi:10.100710668-020-00589-1

Lipper, L., Thornton, P., Campbell, B., Baedeker, T., Braimoh, A., Bwalya, M., Caron, P., Cattaneo, A., Garrity, D., Henry, K., Hottle, R., Jackson, L., Jarvis, A., Kossam, F., Mann, W., McCarthy, N., Meybeck, A., Neufeldt, H., Remington, T., ... Torquebiau, E. F. (2014). Climate-smart agriculture for food security. *Nature Climate Change*, *4*(12), 1068–1072. doi:10.1038/nclimate2437

Mrabet, R., Savé, R., Toreti, A., Caiola, N., Chentouf, M., Llasat, M. C., Mohamed, A. A. A., Santeramo, F. G., Sanz-Cobena, A., & Tsikliras, A. (2020). Food. In Climate and Environmental Change in the Mediterranean Basin – Current Situation and Risks for the Future. First Mediterranean Assessment Report. Union for the Mediterranean, Plan Bleu, UNEP/MAP.

NASA Global Climate Change. (n.d.). *Scientific Consensus: Earth's Climate Is Warming*. https://climate.nasa.gov/scientific-consensus/

Pawlak, K., & Kołodziejczak, M. (2020). The Role of Agriculture in Ensuring Food Security in Developing Countries: Considerations in the Context of the Problem of Sustainable Food Production. *Sustainability*, *12*(13), 5488. doi:10.3390u12135488

Pérez-Escamilla, R. (2017, July). Food Security and the 2015–2030 Sustainable Development Goals: From Human to Planetary Health: Perspectives and Opinions. *Current Developments in Nutrition*, *1*(7), e000513. doi:10.3945/cdn.117.000513 PMID:29955711

Rosegrant, M. W., & Ringler, C. (2000). Impact on food security and rural development of transferring water out of agriculture. *Water Policy*, *1*(6), 567–586. doi:10.1016/S1366-7017(99)00018-5

Sessitsch, A., & Birgit Mitter, B. (2015, January). 21st century agriculture: Integration of plant microbiomes for improved crop production and food security. *Microbial Biotechnology*, *8*(1), 32–33. doi:10.1111/1751-7915.12180 PMID:25545820

Spiertz, J. H. J. (2009). Nitrogen, Sustainable Agriculture and Food Security: A Review. In E. Lichtfouse, M. Navarrete, P. Debaeke, S. Véronique, & C. Alberola (Eds.), *Sustainable Agriculture*. Springer. doi:10.1007/978-90-481-2666-8_39

The Economist and Intelligence Unite. (2017). *Fixing Food: The Mediterranean Region Building Sustainable Food Systems Through Capacity-Building And Co-Operation*. https://www.barillacfn.com/m/publications/bcfn-fixingfoodthemediterraneanregion2017.pdf

United Nations Department of Economic and Social Affairs. (n.d.). *The 17 Goals*. https://sdgs.un.org/goals

U.S. Department of Agriculture. (n.d.). *Climate Change, Global Food Security, and the U.S. Food System*. https://www.usda.gov/oce/energy-and-environment/food-security#:~:text=Climate%20change%20is%20likely%20to,food%20safety%2C%20among%20other%20causes

Vilar-Compte, M., Burrola-Méndez, S., Lozano-Marrufo, A., Isabel Ferré-Eguiluz, I., Flores, D., Gaitán-Rossi, P., Teruel, G., & Rafael Pérez-Escamilla, R. (2021). Urban poverty and nutrition challenges associated with accessibility to a healthy diet: A global systematic literature review. *International Journal for Equity in Health*, *20*(1), 40. doi:10.118612939-020-01330-0 PMID:33472636

Weber, H. (2017). Politics of 'Leaving No One Behind': Contesting the 2030 Sustainable Development Goals Agenda. *Globalizations, 14*(3), 399-414. doi:10.1080/14747731.2016.1275404

World Atlas. (n.d.). *Mediterranean Countries*. https://www.worldatlas.com/articles/mediterranean-countries.html

World Meteorological Organization (WMO). (2021). *WMO: Climate change threatens sustainable development.* Press Release Number: 22092021. https://public.wmo.int/en/media/press-release/wmo-climate-change-threatens-sustainable-development

Yıldırım, S., & Kaplan, M. (2022). The Threat of Invasive Alien Marine Species to Blue Economy: The Mediterranean Case. In Implications for Entrepreneurship and Enterprise Development in the Blue Economy. IGI Global. doi:10.4018/978-1-6684-3393-5.ch004

Yıldırım, S., & Yıldırım, D. C. (2021). Achieving Seafood Security in the Mediterranean Region: A Case of Turkey. In R. Castanho & J. Martín Gallardo (Eds.), *Management and Conservation of Mediterranean Environments* (pp. 175–195). IGI Global. doi:10.4018/978-1-7998-7391-4.ch011

Yıldırım, S., Yıldırım, D. Ç., & Gedikli, A. (2018). Sustainable Consumption Trends in the World in the Context of Green Economy and Sustainability. In Sustainable Development: Concepts, Methodologies, Tools, and Applications (pp. 1605-1624). IGI Global. doi:10.4018/978-1-5225-3817-2.ch071

ADDITIONAL READING

Yildirim, S. (2020). The Consumer Role for Sustainable Development: How Consumers Contribute Sustainable Development Goals. In V. Chkoniya, A. Madsen, & P. Bukhrashvili (Eds.), *Anthropological Approaches to Understanding Consumption Patterns and Consumer Behavior* (pp. 325–341). IGI Global. doi:10.4018/978-1-7998-3115-0.ch018

Yıldırım, S., & Yıldırım, D. Ç. (2020). Achieving Sustainable Development Through a Green Economy Approach. In S. Patti & G. Trizzino (Eds.), *Advanced Integrated Approaches to Environmental Economics and Policy: Emerging Research and Opportunities* (pp. 1–22). IGI Global. doi:10.4018/978-1-5225-9562-5.ch001

Yıldırım, S., Yıldırım, D. Ç., Aydın, K., & Erdogan, F. (2021). (2021). Regime-dependent effect of tourism on carbon emissions in the Mediterranean countries. *Environmental Science and Pollution Research International*, *28*(39), 54766–54780. Advance online publication. doi:10.100711356-021-14391-7 PMID:34014485

KEY TERMS AND DEFINITION

2030 Sustainable Development Goals (2030 SDGs): 2030 SDGs includes 17 main goals planned to be achieved till 2030 by the members of United Nations (193).

Agricultural Food Market: A food market that includes farming-based products such as crops and livestock.

FAO: Food and Agriculture Organization of the United Nations.

Food Security: Providing sufficient quantity of food sources for everybody with sufficient quality and affordable cost.

Mediterranean Basin: A region includes countries sharing the same weather and Mediterranean Sea's coastal and sea areas.

Chapter 9
Breaking the Vicious Cycle Between Migration and Environmental Degradation:
The Role of Government

Ayfer Gedikli
https://orcid.org/0000-0002-7128-1976
Duzce University, Turkey

Nur Billur Taş
Bilkent University, Turkey

Abdullah Kutalmış Yalçın
Duzce University, Turkey

ABSTRACT

For many decades the world witnessed mass displacements. Migration can be either voluntary or forced depending on the reasons. Mass migration can be also due to climate changes that harden the living conditions. Since there is a bidirectional nexus between migration and environmental degradation, the dramatic increase of refugees, asylum seekers, and immigrants greatly affect the environment and conservation efforts. Globally, the movement of people has caused different types of ecosystem changes including deforestation, water, and air pollution, as well as increased waste. Furthermore, as a two-edged sword for conservation, urbanization and growing population cause increasing per capita demand for energy, goods, and services. In this chapter, migration and sudden-onset and slow-onset hazards, the correlation between migration and environmental degradation, environmental and economic effects on the destination, as well as the role of government during migration processes will be analyzed.

DOI: 10.4018/978-1-6684-4548-8.ch009

INTRODUCTION

In recent years, there has been massive displacement of people on a global scale. For many decades, global migration due to man-made and natural causes has been increasing and breaking records. While the number of immigrants was 60 million in 2000, it reached 82 million in 2013, 258 million in 2017, and 281 million in 2020 (IOM, 2020). Migration is a phenomenon that has links with economic, political, and social factors. Migration causes a deep transformation of the immigrants and has been transformative for societies and economies all over the world. Depending on the reasons for migration, it can be either voluntary or forced. In the forced migration, people can leave their countries for reasons such as widespread violence, political conflicts, violation of human rights, unemployment, worsening living and economic conditions, and lack of income and livelihood. Many studies indicated that people become forced refugees due to similar reasons such as famine, war, or poverty, and massive people movements cause unexpected problems in hosting countries. Besides, when immigrants reach to host country, they face problems such as sheltering, adaptation to the environment, earning for livelihood, and environmental issues (Adelman and Sorenson, 1994; Baker, 1994; Zetter, 1994; Kibreab 1996; Whitaker 1999; Ngai and Koehn, 2002). Forced migration and volunteer migration have different dynamics. The countries which have a high population with respect to their land cause population pressure. People of those countries mostly migrate to expand livelihood opportunities. Converting natural vegetation to agriculture increase the population pressure. Wrong environmental policies increase the pressure on the population and the volunteer migration turns into forced migration (Wageningen University, 2017). Also, as the world is more globalized, the circulation of capital, technology, goods, and services encourages volunteer migration. Due to mass migration, either because of force major reasons or selective reasons, different types of settlements have been formed such as migrant cities, slums, refugee camps, and regions that are occupied by displaced people and irregular migrants (Wageningen University, 2017b).

Migration within the same country and migration to different destinations have different outcomes. Migration in the same country causes regional population density and human capital loss in the sending region whereas international migration leads to more complicated drawbacks both for sending and receiving countries. For example, while sending country loses some part of its human capital, with an increasing number of people due to migration, the receiving country has a more concentrated population in certain regions. Overpopulation leads to overconsumption which causes more pollution and environmental degradation. In the long term, redistribution and reallocation of resources and more consumption may negatively affect the carrying capacity[1] of the environment (O'Lear, 1997:607). Therefore, migration affects sustainability in environmental, demographic, social, and economic dimensions. Migration contributes to sustainability if it is wisely managed and increases well-being and reduces inequality and environmental degradation. Therefore, migration policies are critical for sustainable development incorporating migration. In this context, better integration of ecological, demographic, and sociological environments can lead to sustainable development.

The term sustainable development gained importance in the late 1980s in *The Brundtland Report*[2]. Since then, sustainable development became a goal for not only the development of nations but also for industrial and environmental sustainability and development (The Brundtland Report, 1987:13). The Brundtland report pointed out the urgency of searching for new ways of production without ignoring the sustainable development without ignoring the well-being of the environment, economies, and people as a whole. Besides, sustainable development strategies should have been implemented on a global scale. In addition, sustainable development is complicated progress that has the linkage between the past, today, and the future. What happened in the past, what are the policies today, and what are the outcomes of the future should be taken into account while setting sustainable development strategies (Strange, T. and Bayley, 2008:24, 102). Also, there should be cooperative governance in developed and developing countries to secure their long-term well-being which requires a global contract between advanced and emerging economies to achieve global sustainability (de Lange, Wise and Nahman, 2010).

Sustainable development involves economic, social, and environmental aspects (Ukko et al., 2019). Since sustainable development provides a linkage between the well-being of the current generation and future generations, the concept of capital can be used to create this linkage. Financial capital, human capital, social capital, and natural capital are different forms of capital. Among all, natural capital is related to environmental sustainability. Protecting the environment stands for preserving the flora and fauna as well as establishing sustainability for the next generations (Fuchs, Raulino, and Guerra, 2020:511; Khan et al., 2019).

Since there has been a mass migration process in recent years, the relationship between migration and sustainability should be analyzed in detail. Sustainable development is highly correlated with a sustainable environment. The fact is that there is strong bidirectional causality between migration and environmental sustainability (Black et al., 2011; Morrissey, 2013; IOM, 2008). Existing economic, social, political, and technical systems cannot satisfy the needs, thus they need fine-tuning emergency assistance and development strategies in accordance with disaster relief. The complex problems of migration need to have a holistic and integrated approach involving green, circular, environmentally friendly, and food secure settlements (Wageningen, 2017b). Therefore, in order to have a sustainable environment and economic growth in the age of global migration, policymakers need to implement a wide range of policies. It is a fact that there is no single solution, and a wide range of measures and regulations are required. Besides, equal priority should be given to policies to promote the long-term resilience of cities and societies.

This chapter explores some of the main ways in which environmental change and migration have been linked. Considering the complexity and multi-dimensionality of the subject, this paper seeks to illustrate first how migration impacts environment quality and economic development. Accordingly, sustainable development and sustainable environment vulnerabilities are aimed to be analyzed. The factors such as socio-economic development, economic growth, demographic factors, poverty, resource scarcity, governance frameworks, and population growth due to migration will be some of the main points. In the last part of the study, some policy suggestions will be put forth for the government to manage migration processes with sustainable economic performance and with a sustainable environment.

Theoretical and Conceptual Framework

Changing a permanent or temporary place of residence is called migration. In the definition of migration, there is no criterion related to the place, voluntary or involuntary, or whether it is relocation within the country or outside the country. In other words, theoretically, all types of spatial mobility fall under the definition of migration and any kind of relocation in a short or long distance, voluntary or involuntary, within the country or abroad is defined as migration. Whether local or international, short-term or long-term, or trouble-free; every displacement contains different levels of life threats, obstacles, and social and psychological struggles (Lee, 1966:49-50).

Immigration does not lead to an increase in the global population but increases the concentration of the population in a certain area resulting in more resource consumption and pollution. In the long period, the redistribution of resources may decrease the carrying capacity of the environment. It is also impossible to redistribute the people to the empty or less populated areas since these regions cannot offer sufficient caring capacity (O'Lear, 607).

Despite different types of displacements being called migration, each of them has different features. In addition, migration processes do not immediately become permanent. Depending on the type of the migration whether it is forced or volunteer, there are different stages. Four-stages of volunteer migration are as follows (Castles and Miller, 1998:28):

i. Temporary labor migration of young workers. They send the remittance to their families and are mostly determined to turn back to their homeland when there is enough income is provided.
ii. After a while, they adapt to their social and economic environment and they prefer to stay and develop their social networks in the new environment.
iii. When they feel safe, they intend to bring their families. The family reunion comes after growing consciousness of long-term settlement and growing orientation towards the receiving country.
iv. The permanent settlement comes if the government of receiving country gives permission through a legal framework. This may be through eventual citizenship due to political exclusion, socio-economic marginalization, or the formation of ethnic minorities

The stages of forced migration are as follows:

i. The decision to migrate after a sudden event such as disaster, political conflict, or war.
ii. Asylum seeking and being a refugee are alternatives to migration.
iii. Legal permission provides safe transfers of migrants to the host country.
iv. Illegal ways can be applied to irregular migrations
v. Staying at camps and waiting for acceptance by the host country
vi. The permanent settlement comes after getting permission and legal acceptance.

The magnitude of the migration varies depending on certain factors explained below (Lee, 1966:52-54):

- If there are vital differences between the home and targeted regions, there may be large-scale migration. If the target region offers higher opportunities in both working and social life, it accelerates the decision to migrate.
- The magnitude of migration is also directly related to the differentiation of people. There are similarities between groups of people living in the same regions in terms of ethnicity, religion, education, or income level. As the level of sameness and similarity increases within the group, the level of decision-making increases and vice versa. However, if there are different ethnical, racial, or religious groups living within the same region this is a clear indicator of multicultural immigrant societies. The USA is a good example of that. The country defines itself as a *"melting pot"* of different groups of people. In addition, immigrant groups concentrate on certain occupational groups and try to maintain their existence in that sector with the help of their ethnic identities. In many countries, the laundry industry is in the hands of immigrant Chinese. Greeks in the USA are very active in the food and beverage sector. Differentiation between groups of people inevitably results in the superiority of one group over others on racial, ethnic, etc. grounds. The process of superiority can sometimes go as far as racial discrimination, as in the reaction movements against black people in the USA. Ethnic discrimination ends with the assimilation of minorities.

- The magnitude of migration is also related to the level of overcoming obstacles. In particular, the absence of physical difficulties or legal barriers that may prevent migration may lead to large migrations toward the destination country.
- The volume of migration is highly correlated with the fluctuation of the home country's economic conditions. Specific factors such as the economic crisis, especially during the depression periods, the restriction of job opportunities, the rapid increase in unemployment, and the worsening of living conditions encourage individuals to migrate.
- Unless legal restrictions and physical barriers, migration accelerates and increases toward the destination countries. The increase in differences between regions, the emergence of economic and social level differences between individuals, and the failure to take measures to slow down migration accelerate migration.
- The development of a country also affects the magnitude of migration. In an economically developing country, with industrial and social investments, the differences in the level of development between regions gradually decrease. Hence, differences in quality of life and education among individuals gradually decrease. In cases where social justice and social peace are ensured and individuals feel economically and socially safe, they no longer intend to migrate.

Some characteristics of migration have been changing for decades. They are (Castles and Miller, 1998:8-9):

a. *Globalization of migration:* There is an increasing tendency that more and more countries are included in the migratory movements. Besides, the diversity of the areas of the origin and receiving countries increases in economic, social, and cultural backgrounds.
b. *Acceleration of migration:* The volume of migration has been increasing all over the world. This acceleration leads to initiate urgent governmental policies which may bring an extra burden on both financial and social policies.
c. *Differentiation of migration:* There is no specific type of migration since there are different motives for it. Security, job possibilities, better living conditions, and education are some of the reasons for the migration. Accordingly, refugee, labor migration, and permanent settlements are different types of migrations that need different policies to handle the problem. Indeed, one type of migration calls for the other one. For example, labor migration brings permanent migration after a while.
d. *Feminization of migration:* In the 1960s, most of the migration movements were for job seeking and refugee movements. Those movements were male-dominated, and women were included in the migration through family reunions. In recent years, females and children are also included in the movements. Females play a major role in international migration in many countries such as Filipinos to the Middle East, and Thais to Japan.
e. *Politicization of migration:* Not only domestic policies but also bilateral and international relations are designed on the security policies of states that are deeply affected by international migration.

Indeed, migration is an adventure to the unknown. The longer the journey, the higher the risk and uncertainty. There are economic, social, and environmental consequences of migration both for sending and receiving countries.

Table 1. Economic, social, and environmental consequences and impacts on governmental policies of migration

	Receiving (host) country	Sending (home) country
Economic consequences	• Increased younger labor force • Impacts on local labor market • More unskilled/low skill people • Increasing unemployment • Impact on macroeconomic performance	• Remittance flow • Loss of labor young labor force • Impact on macroeconomic performance • Less production
Social consequences	• Brain drain • Cultural diversity • Increasing inequality • Cultural adaptation • Education • Language • Increased crime	• Brain drain • Decreasing population and loss of young labor force
Environmental consequences	• Increased pollution • Increased population • Increased consumption • Increased suburbs • Increased demand for energy and natural resources	• Decreased pollution • Decreased population • Less consumption • Decreased demand for energy and natural resources
Effects on Governmental policies	• Setting effective immigration policies • Additional laws for adaptation • Transfers and supports • Support to find a job • Protect the environment by legal precautions • Levying more tax to cover expenditures for newcomers • Preparing camps for the unexpected and huge number of immigrants • Legal issues: citizenship, ID card, driver's license	• Finding solutions to control the migration -safety and security -job possibilities -social rights -more income

Source: Gavonel et al., 2021:102; authors' own explanations

Obstacles to Managing the Migration

Migration is very difficult both for sending and receiving countries with its multidimensional problems. Some of the obstacles are explained as follows:

Vulnerability and fragility: Vulnerability represents the low level of socio-economic development. In this vein, vulnerability can be categorized as the vulnerability of the population subject to migration and vulnerability of the country/region. For example, low literacy rates, lack of skills, poverty, low Human Development Index, high unemployment, poor macroeconomic conditions of the country, economic crisis, poor social security system, and high fragility are the determinants of vulnerability. OECD (2016:21) describes fragility as *"the combination of exposure to risk and insufficient coping capacity of the state, system and/or communities to manage, absorb or mitigate those risks"*. Fragility has five dimensions: Economic, environmental, political, security, and social as shown in Table-2.

Lack of financial support: Financial support is limited in developing countries. Hence, with their vulnerable economic conditions, they cannot provide enough humanitarian assistance for immediate relief and recovery.

Table 2. Dimensions of fragility

Subtitles	Description
Economic	Vulnerability due to the weakness of economic institutions and human capital, macroeconomic instability, income inequality, low economic growth rate, high unemployment
Environmental	Vulnerability is due to environmental and climatic threats that impact people's lives. Environmental pollution, disasters, and epidemics
Political	Vulnerability due to political instability such as political inclusiveness, transparency, corruption, lack of social support by the government
Societal	Vulnerability due to vertical and horizontal inequalities among social groups; cleavages among social groups due to ethnical, religious, racial differences

Source: OECD (2016:23)

Immigrants are not homogenous: Refugees and migrants are not homogenous either in ethical, financial, or cultural backgrounds. While some of them can speak a foreign language, some of them cannot express themselves. While some of them have skills or education, some others are unskilled or illiterate. Thus, it is hard to create a unique policy suitable for all of them.

Excessive displacement in a short time: If there is a war starts suddenly, or a disaster happens in a few hours, there may be sudden massive displacement. In this case, the situation is very challenging for both the home and host countries.

Lack of cooperation among the host and home countries: This problem occurs mostly if the sending country is in a civil war or political conflict. Also, if the host and home countries are the two parties in conflict, cooperation becomes even harder.

Poor employment of migrants: Lack of skills, lack of knowledge about the job, and language problems deprive migrants of finding a job in host countries. Furthermore, the labor market may be problematic both for local people and migrants in host countries. If migrants cannot find a job for their livelihood, migration becomes even more suffering for them. Figure-1 represents the problems of migrants in employment in host countries. Lack of skill seems to be the most important reason for the unemployment of migrants in selected EU countries. While language is the greatest problem in Austria, unable to learn the requirements of a job is one of the common problems of immigrants in EU countries.

Current Situation of Migration

International migration is actually a transnational revolution that reshapes societies and governmental policies. Both receiving and sending countries are affected by migration but in different ways. Countries such as the USA, Canada, Australia, and New Zealand are the *"classical countries of immigration"*. Western and Northern regions of Europe are exposed to massive migration movements between World War II and the 1970s. Although some of the Southern European states such as Greece, Italy, and Spain used to be the zones of emigration, in recent years they became immigration areas. Hungary, Poland, and the Czech Republic also became immigration lands. In the Middle East, Turkey is an exceptional country being both subject to emigration and immigration. In the 1960s and 1970s, thousands of Turkish workers migrated to Western European countries to work. Recently, Turkey is a haven for Syrian and Iraqi refugees. Natural resource-rich Gulf countries such as Saudi Arabia, Kuwait, Qatar, and the United Arabic Emirates became major magnets to immigrants for Arabic and Asian immigrants. In Africa, colonialism and white settlement caused the establishment of a labor migrant system for mines and plantations. Former French colonies would like to migrate to France from Algeria, Tunisia, and other African countries. Particularly in Sub-Saharan countries, there are millions of people who try to migrate to reach prosperity and better living conditions (Castles and Miller, 1998:6-8). In recent years, particularly in the Middle East and Africa, mass migration movements occur because of political struggles, and civil war which causes an increasing number of refugees all over the world. At the end of 2020, 48 million people internationally displaced. In mid-2021, 84 million people were forcibly displaced and 4,4 million people are asylum-seekers. 42% (approximately 35 million people) of forcibly displaced people are children under 18 years of age, and around 1 million babies were born in refugee camps in the 2018-2020 period. Approximately 85% of refugees are hosted in developing countries. As illustrated in Table-2, 68% of all refugees originated from five countries composed of Syria, Venezuela, Afghanistan, South Sudan, and Myanmar. 39% of all refugees are accepted by Turkey, Colombia, Uganda, Pakistan, and Germany. Among them, Turkey hosts the largest number of refugees with 3,7 million people (UNHCR, 2021).

Figure 1. The reasons for unemployment of migrants in sample host countries
Source: de Lima et al. (2016:21)

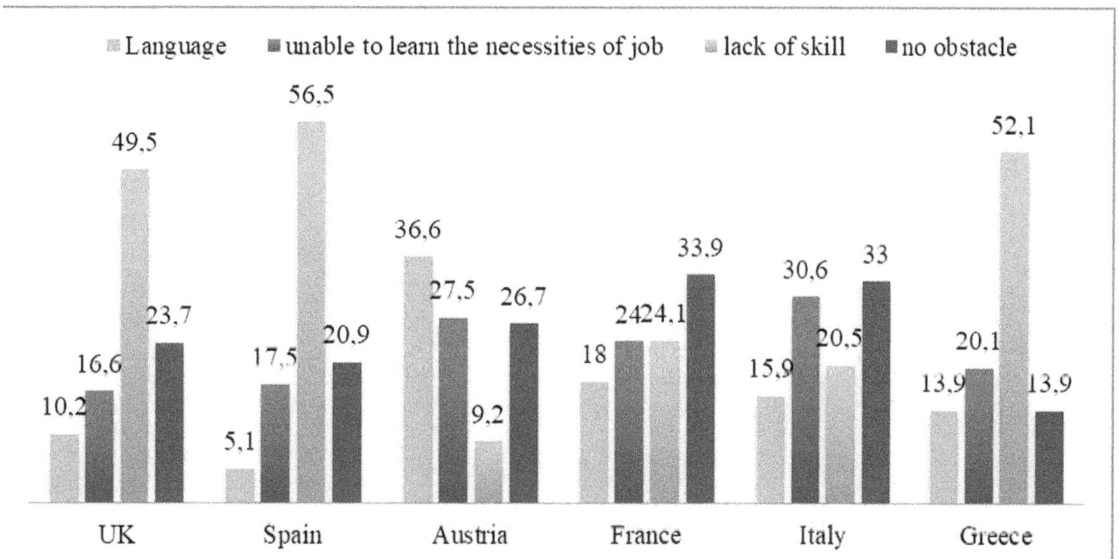

Table 2. Number of people who migrated from/to in the top five sending and hosting countries

Sending countries (million people)		Hosting countries (million people)	
Syrian Arab Rep.	6,8	Turkey	3,7
Venezuela	4,1	Colombia	1,7
Afghanistan	2,6	Uganda	1,5
South Sudan	2,2	Pakistan	1,4
Myanmar	1,1	Germany	1,2

Source: UNHCR, 2021

In 2020, Syria was the top country with 6.6 million people internally displaced and more than 6.7 million refugees in neighboring countries. Turkey hosted 3.6 million Syrian registered and 350.000 refugees of other nationalities such as Afghans, Iraqis, and Pakistanis. Although Turkey used to be a transit country because of its geographical position, in recent years, the country became a destination country. In 2020, while there was an increase in Syrians under temporary protection, there was a decline recorded in irregular migrants on land and sea. Because of the controls imposed by the pandemic, Turkey froze all registrations, refugee status determinations, and resettlements (ICMDP, 2020:3-4). As of March 31, 2022, the total number of Syrian people of concern for asylum is 5.724.230. As shown in Figure-2, 65,7% of them are in Turkey (3.763.565 people); 14,7% are in Lebonan (839.086 people); and 14,7% are in Jordan (674.268 people) (UNHCR, 2022). In Turkey, 50.639 Syrian refugees are hosted in temporary refugee centers and 3.712.047 of them are out of temporary refugee centers (Ministry of Interior Presidency of Migration Management, 2022)

Figure 2. Syrian asylum population and their distribution among the hosting countries
Source: UNHCR (2022)

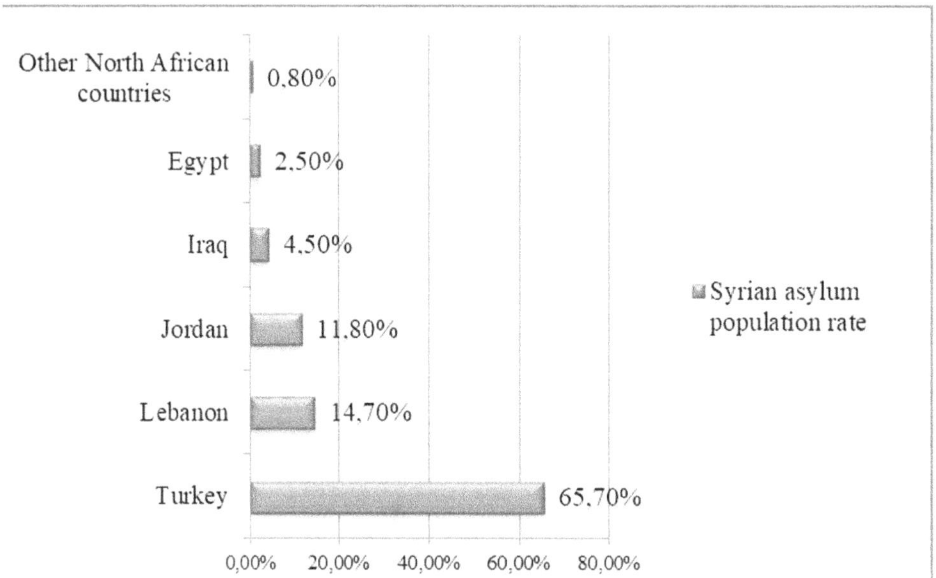

The highest number of Syrian refugees is hosted in Turkey. Figure-3 represents the number of Syrian refugees in Turkey spanning the period 2011-April 28, 2022. As shown in Figure-2, there is a sharp incline of fleeing to Turkey starting in 2013. Starting from this date, the number of refugees from Syria increased exponentially toward Turkey. After 2019, Syrian refugees stayed almost the same due to the Covid-19 pandemic precautions.

Figure 3. Number of Syrians under temporary protection in Turkey (2011-2022)
Source: Ministry of Inferior Presidency of Migration Management (2022)

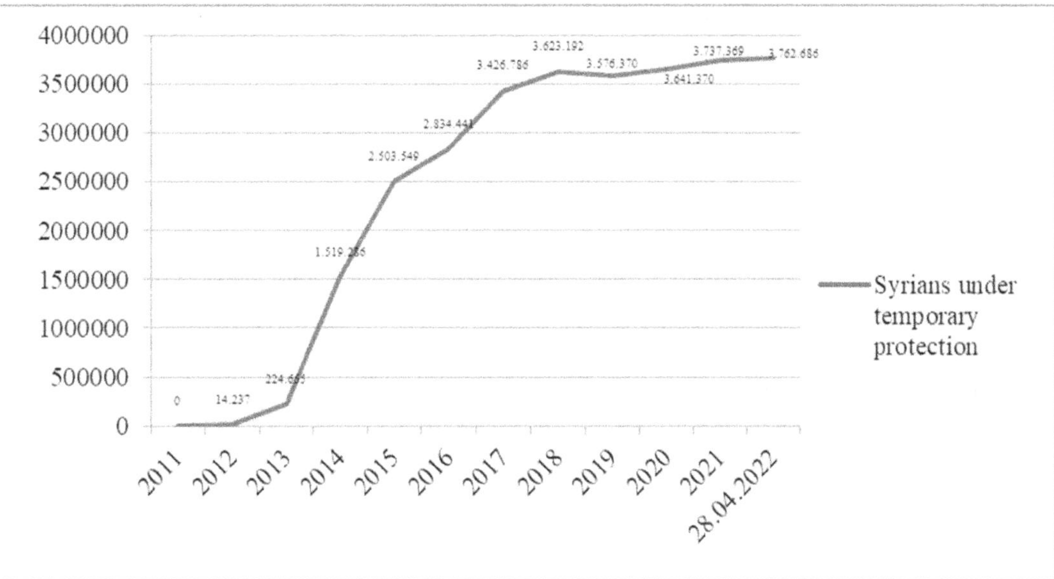

The Economic Consequences of Migration for Home and Host Countries

The negative impacts of migration on the receiving country's economy are diversion of mainstream development, supply shocks, excessive consumption of scarce resources, need assistance, and increasing public expenditures. However, while there is a disequilibrium between the massive movement of people and natural disasters, there may be specific advantages inferred from this unpleasant process. Having a cheap labor force of immigrants and greater demand for the construction sector and sheltering can support the relevant sectors in the host country. Although constructing shelters may be harmful to the environment, it may have a positive contribution to the housing and construction sectors which supports national economic growth. In the literature, this process is handled in two different ways: First, a high level of external intervention is represented as a *"relief model"* considering the negative consequences of disasters, and a *"development after the disaster"* model considers it as finding opportunities from the threats. If there is enough land for refugees, legal and political considerations about the status of refugees can be practically solved. National governments' involvement contributes different dimensions to the process and provides effective assistance (Zetter, 1994:85-86)

The impact of migration on home and host countries' economic performance can be categorized as follows in Table-3 (OECD, 2017:24):

As shown in Table-3, the effects of migration are different for home and host countries. While it causes a shortage in the labor market in the home country, it leads to an excess labor supply in the host country. As more low-skilled people are employed in the agricultural sector, both countries' agricultural sectors are impacted by migration. Since some of the people employed in the agricultural sector leave the country, there are vacancies and new employment areas in sending country. Also, more people are employed in the agricultural sector in the host country. Besides, there will be more possibility to find cheaper labor supply in the host country. The migration of high-educated and high-skilled people to their home country causes a brain drain, and most of the high-skilled immigrants do not intend to turn back even though conditions become better in their home country. Immigrant families get less government support from the host country, but they tend to increase the informal market. Remittances are supportive both for immigrant families and the country's budget. Many low and low-middle-income countries benefit from remittances to cover budget deficits. Some of the immigrants open their businesses in urban areas of their homeland after some capital accumulation (OECD, 2017:24).

As illustrated in Figure-4, there is an inverse relationship between the income level of the countries and the rate of personal remittances in their GDPs. The higher the economic performance, the fewer the people migrate for working for livelihood. Thus, the ratio of personal remittances is the highest in the least developed countries in the 2010-2020 period. The World Bank revealed that although personal remittances declined by 1.7% in 2020 due to the Covid-19 pandemic and global recession, there is a sharp increase in remittances by 7.3% to reach $589 billion in low- and middle-income countries in 2021. The World Bank (2021) also estimated that for a second consecutive year, remittance flow to low-and middle-income countries (excluding China) would exceed the foreign direct investment and overseas development assistance.

Table 3. Impacts of migration on home and host country's economies

	Labor market	**Agriculture**	**Education**	**Financial services and investment**	**Health and Social support and protection**
Emigration (for the home country)	-Labor shortage in certain sectors -less pressure on the labor market -reduction in household labor supply	-New employment by other workers replacing the emigrants	-Emigration of well-educated people negatively affect human capital (brain drain) -Low-skilled emigration may lead the young generation to drop out of education		
Remittance	-Reduces household labor supply -Helps to stimulate the self-employment -support to cover budget deficits of the home country	-more investment in agriculture	-remittance receiver households spend more on education	-more business owners in urban areas -more investment in real estate	-not on social expenditures but more on health facilities
Back migration	-more self-employment -enrich the skills in the home country	-more investment in agriculture -more diversification even within the agricultural sector	-Few high-skilled people return, but they support to improve of the quality of local human capital	-households run the business more than non-migrants	-return-migrants get less benefit than non-migrants from government transfers
Immigration (host country)	-additional labor supply by immigrants particularly in the low skilled labor market	-more employment in the agricultural sector by immigrants to hire-in-labor and sell their products	-less interest in schooling than non-immigrant children	-immigrant families intend to have more a non-agricultural business	-immigrants get fewer government transfers -they are mostly employed in informal sectors and thus they pay less tax

Figure 4. Personal remittances of % of GDP in a different group of countries (Least developed, low and middle-income countries)
Source: The World Bank Data (2022)

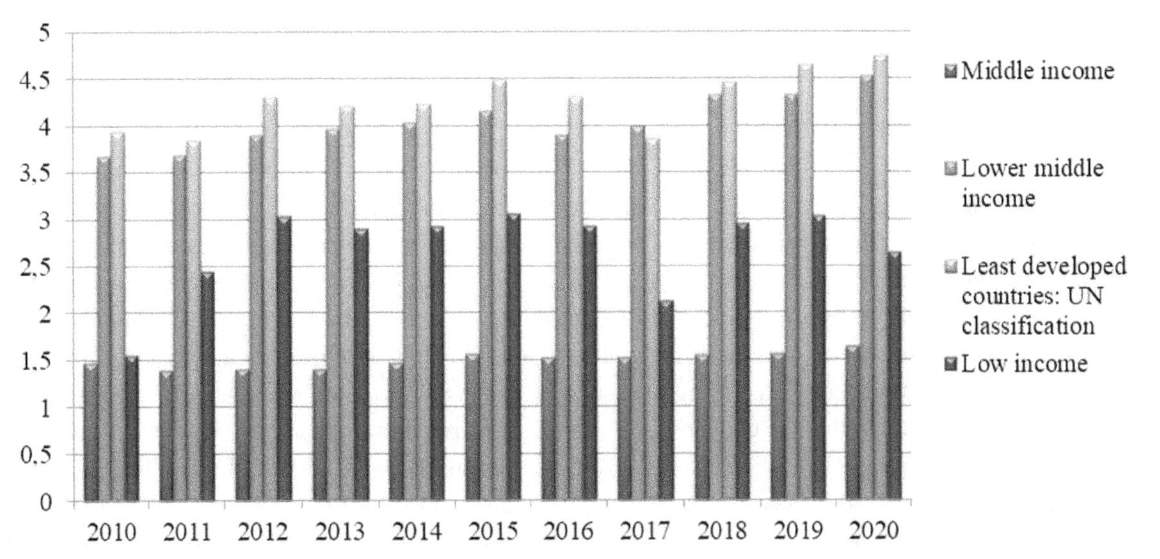

Figure 5. The contribution of remittances to GDP of selected developing countries (2010-2020)
Source: The World Bank Data (2022)

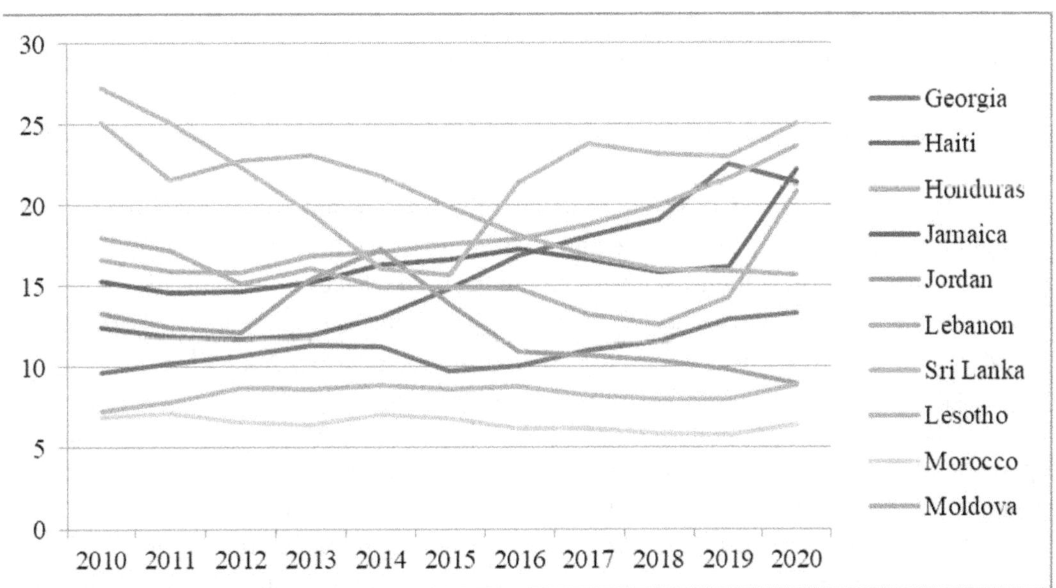

As shown in Figure-5, the share of remittances in GDP is the highest in Lesotho during this period. While it was 27% of GDP in 2010, it decreased to 15% in 2015 and rose to 25% again in 2020. The second highest personal remittances belong to Moldova with 21,57% in 2010. This rate declined to 15% in 2020 due to the pandemic precautions. Honduras showed the highest acceleration in personal remittance during this period. While it was 16% in 2010, it reached 23,6% of GDP in 2020.

Migration and Environment

Actually, the relationship between migration and environmental degradation is bidirectional. Both of them are the reasons and the consequences of each other. Migration for economic or political struggles directly or indirectly impacts the environment. Even the migration for touristic activities can endanger the environmental quality as demographic shifts can cause overpopulation even for a short period of time. Moreover, the construction of camps or suburbs for refugees or migrants can be harmful to the environment if water is scarce. Consuming excessive water in a short time may cause competition for water use (Véron, 2014). In the last decades, the intensity of forced migration and population displacement has caused the growth of environmental problems. Drought and floods, deforestation, sea-level rise, and security reasons are some of the reasons for migration. In addition to climate and geographical hazards, the Covid-19 pandemic caused an unprecedented humanitarian crisis all over the world in 2020. Especially developing countries were badly affected by the pandemic due to weak macroeconomic conditions, problematic health systems, and a lack of funds to allocate to fight the outbreak. The World Bank revealed that an additional 115 million people in 2020 and 150 additional people in 2021 have to fight extreme poverty because of the outbreak. Since the pandemic caused excessive expenditures for health support, it increased the indebtedness of the developing countries with vulnerable economic conditions. Health crises combined with economic problems that cause miserable living conditions urge people to migrate. Moreover, the Red Cross pointed out that during January 2020- September 2020 period, there were 132 extreme weather events, and 90 of them happened together with the pandemic (Eckstein, Küzel and Schafer, 2021:23).

In particular, many pieces of research confirmed the relationship between social conflicts, war, forced migration, and environmental degradation. Cities' deforestation, damage to irrigation systems, damage to wildlife, and loss of forests are very common consequences of wars. Radioactive wastes are also used as weapons whose devastating effects last decades. Oil fires spilled in the sea during the Gulf war, radioactive waste leakage was tested in Afghanistan by the former Soviet Union, and the Syrian government applied chemical weapons against its citizens (Black, 2014:9).

It is widely accepted that immigration and refugee settlements may have stressed the natural resources, leading to both environmental and social consequences. As the number of dwellers increases within the city, resource demand increases dramatically. This leads to accelerated conversion of forests to agricultural areas, extraction of more natural resources, and consuming more natural resources, minerals, and water. There will be more fishing, deforestation, and hunting. In addition, there will be more waste production that exacerbates threats to human health and environmental quality (Martin, 2005:332). In their research, Black and Sessay (1997) investigated the negative effects of forced migration to the Senegal River Valley on environmental quality. The authors confirmed the relationship between migration and environmental degradation nexus. They concluded that when immigrants could not afford to buy wood fuel, they started to cut the forest to get the fuel that results in deforestation.

For sustainable environmental and economic quality, the following principles can be taken into consideration (Wageningen, 2017b):

- Green areas and water quality of life of refugees should be ensured
- Food security should be ensured
- Circular solution for refugee camps and suburbs (energy needs, waste management, and watering)
- Including NGOs, public authorities, businesses, and communities in the policy selection processes
- Helping to create jobs for livelihood
- Training and facilitation of sustainable land use in rural areas
- Training and facilitation of environmental protective solutions in urban areas

Refugee Shelters

Whether in the formal or informal economy, the housing sector has a multiplier effect on both the employment and building materials sector within the industry. The cheaper labor force of immigrants and local people with lower opportunity costs stimulates the income transfers and supports the capital formation of the national economy. These positive effects are observed more in developing countries with low-income households. In many developing countries, housing investments may reach up to 20% of GNP and this sector is the single largest type of fixed capital investment accounting for up to 50%. It should be noted that a sharp increase in low-skilled labor supply in the labor market in the host country may bring unexpected employment problems. The informal sector can grow very fast and there may be a huge number of immigrants employed in low-skilled businesses instead of local people. While this process brings a cheaper labor force for the employers, it also brings labor marker deteriorations which result in more unemployment among the local people of the host country since immigrants are ready to work with lower wages than the minimum wage and informally. However, as sheltering is one of the basic needs of refugees, a large scale of migration and unplanned rapid urbanization due to fast settlements can be faced in many developing countries. Rapid urbanization leads to deforestation, environmental pollution, and degradation. To keep the refugees in a safe and healthy place and protect the environment, governments take the responsibility for spatial planning. To make such planning, governments need to check investment capital, management, coordination, logistics, and shelter. Besides, during shelter construction, environment-related policies should be taken into account. Although in many developing countries construction sector are one of the pioneer sectors of economic development, it also brings environmental consequences (Zetter, 1994:86-88). Especially refugee camps and slums are unfavorable transient settlements. Yet the number of people living in those places is often equal to regular cities, living conditions are most miserable in camps. Thus, an integrated approach that involves natural-based environmental factors may provide protective and adequate quality of living standards for refugees.

Environmental Change and Migration

Climate change is one of the reasons for migration since it implies risks for jobs and livelihoods, particularly for disadvantageous groups such as women, persons with disabilities, indigenous people, tribal people, and the people who already have been challenged with economic and social vulnerabilities in the region (ILO, 2021:5). The following effects of climate changes lead to the displacement of dwellers (Véron, 2014):

- Increasing intensity and frequency of natural disasters
- Worsening living conditions due to weather conditions, lack of food, decreasing water supplies, diseases because of changes in the ecosystem, etc.
- Rising sea levels that affect agricultural production and living conditions
- Extinction of natural resources or more competition for them because of scarcity

Climate changes, global warming, and environmental degradation have not only environmental but also socioeconomic and humanitarian consequences. Hurricanes, floods, earthquakes, and other disasters may also cause great movements. In particular, lower social classes are more vulnerable to disasters and they have the tendency to migrate. As an example, during Hurricane Katrina, displacement patterns were observed among the economically disadvantaged groups. South Africa is another region highly affected by climate change, and every single country is exposed to multiple climate-related problems. Among other regional countries, India, Pakistan, Sri Lanka, Bangladesh, and the Maldives were more affected by the environmental consequences of climate change such as rising sea levels, floods, heat, drought, and changing rainfall patterns. Between 2010-2019, 700 million people have been affected by at least one climate-related disaster. Just in 2019, 7.8 million people had to displace due to recurrent natural events. In the same year, India had to experience eight tropical storms that caused 5 million new displacements. Based on these data, it is roughly estimated that 143 million people will displace internally all over the world just because of climate change (ILO, 2021:5). Figure-6 shows the reasons for migration: conflicts, violence, or disasters.

Figure-6 indicates that while disasters were the pioneer reason for migration in 2008-2012 period, conflicts and violence became more important reasons for displacement in 2018. In 2018 number of people who migrated due to violence or conflict was 10,8 million and the number of people who migrated due to disasters was 17,2 million people.

Figure 6. Displacements due to conflict/violence or disasters (million people, 2008-2018)
Source: OECD (2019:6)

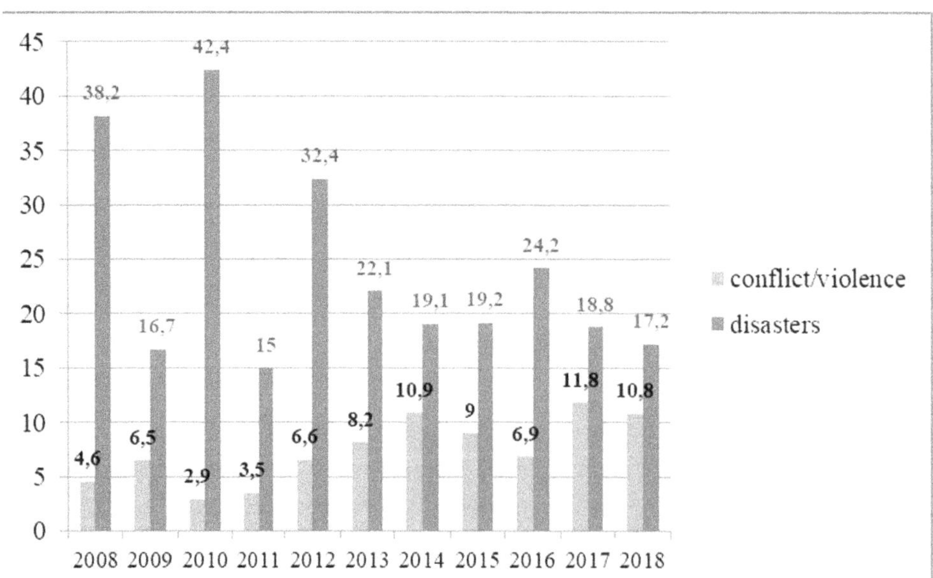

As illustrated in Table-4, Global Climate Risk Index (GCRI) has ranked among the most affected 10 countries over the period 2000-2019. According to Table-4, during this period, Puerto Rico first, Myanmar second, and Haiti third among the top ten countries of the climate risk index. In Table-4, it is also shown that the human development index rankings of the riskiest countries are bad. Mozambique is 181st, Haiti is 170th, and Pakistan is 154th in HDI ranking.

Table 4.

Climate Risk Index (CRI) 2000-2019	Country	Fatalities	Losses (million $)	Losses per unit GDP (%)	Number of Events	Human Development Index Ranking (2019)
1	Puerto Rico	149.85	4 149.98	3.66	24	Not available
2	Myanmar	7 056.45	1 512.11	0.8	57	147
3	Haiti	274.05	392.54	2.3	80	170
4	Philippines	859.35	3 179.12	0.54	317	107
5	Mozambique	125.4	303.03	1.33	57	181
6	The Bahamas	5.35	426.88	3.81	13	58
7	Bangladesh	572.5	1 860.04	0.41	185	133
8	Pakistan	502.45	3 771.91	0.52	173	154
9	Thailand	137.75	7 719.15	0.82	146	79
10	Nepal	217.15	233.06	0.39	191	142

Source: Eckstein, Küzel and Schafer (2021); UNDP (2022)

The Climate Risk Index indicates that the impacts of climate change have been felt severely all over the world and that frequent extreme climate events are some of the major drivers of disasters. The index also indicates that the most affected countries are low and lower-middle-income countries. Climate impacts that threaten lives and livelihood combine with the poor macroeconomic conditions of these developing countries. The result is extreme poverty, lack of human security, food, and natural resource scarcity, and slower economic growth (Eckstein, Küzel and Schafer, 2021:23). Hence, affected people need to flee from the impacted region. However, during the disasters, more vulnerable, poorer groups of the population that represent the economically bottom of the society are unable to flee and have to survive under miserable conditions. Indeed people need capital to move to escape from environmental hazards. If there is not enough financial source (capital for migration, wealth, access to credit) social network (family, friend, community, or other social ties in the destination area), or human capital (education, skill, language, etc.) they cannot dare to leave. Environmental migrants are sometimes subject to conflicts in communities that may depend on the timeframe of the migration, the presence of other conflict drives, and governmental policies. Thus, it can be concluded that if a large number of people migrate in a short time and there are existing challenges in the receiving area, the conflicts may be more severe. Furthermore, compared to the general migration and refugee flows to neighboring countries, migration due to disasters has less potential for conflict since people are mostly displaced within their own countries. As the population increases and urbanization rise, more people are exposed to sudden and unexpected environmental hazards. Therefore, the prevalence of the immobile population in vulnerable areas needs more concern from governmental policies and support (Migration Data Portal, 2020).

Depending on whether the disaster is sudden or slow, the mobility response of people may change. If the disaster is sudden-onset, displacement risk is dependent on the hazard itself (type, frequency, or intensity), exposure (people who are in hazard-prone regions), and vulnerability (the degree of sensitivity to hazard impacts). Besides, climate change also increases the frequency and intensity of natural disasters such as storms, floods, famines, or heavy rains. The cities in low-lying coastal areas are at greater risk since a higher number of people live in coastal areas. Also, there are a growing number of people migrating to urban areas in low and middle-income countries. If the disaster is slow-onset, it also causes displacement. The degree of human vulnerability is the determinant of the movement. For example, sea-level rise due to climate changes may impact human mobility in different ways. The severity of the reasons such as coastal erosion, the salinization of groundwater in the hazard areas, floods, and salinization of agricultural lands increase the possibility of migration. In certain conditions, people are immobile despite the hazardous effects of disasters. This immobility can be categorized as voluntary immobility and involuntary immobility which is called a trapped population. As the severity of environmental and climate change increases, the population can be trapped in areas where they are vulnerable to environmental effects. On the contrary, there may be a massive number of people fleeing from the environmentally degraded areas. In this case, high human mobility itself becomes a "threat multiplier" (Flavell et al., 2020:9-10). In this regard, governments have a great responsibility to minimize environmental degradation and environmental stressors. Particularly in rural areas, effective precautions should be taken to keep people from irregular migration.

Role of Governments in Migration Policies

Since migration has been increasing, regulations for immigration are inevitable policies in all hosting countries. Migration policies are parts of migration governance. The governance is mostly implemented by the ministries or related departments of hosting countries. Immigration policies of governments are defined by Bjerre et al. (2015) as *"government's statements of what it intends to do or not do (including laws, regulations, decisions or orders) in regards to the selection, admission, settlement and deportation of foreign citizens residing in the country"*. The immigration policy of an individual country is a combination of different laws. Migration policies of governments and legal systems cover different areas including humanitarian/asylum strategies, family union, irregular migration, emigration, travel and temporary mobility of people, nationality, integration of immigrants to the host culture, economic and social development, and labor market struggles of the host country. (Migration Data Portal, 2021). Besides, there are regulations on one hand and control mechanisms on the other hand. Regulations provide legal provisions that define rights, and controls are the mechanism that monitors if the regulations are well-established and implemented. For example, while a regulation points out that immigrants should have a work permit to be employed, control mechanisms follow the sanctions for illegal immigrant employment (Bjerre et al., 2015:559, 563). Governments initiate necessary laws and policies for migration in accordance with Sustainable Development Goals (SDGs). In SDG targets 10.7, UN member countries commit to facilitating the *"orderly safe and responsible migration and mobility of people, including through implementation of planned and well-managed migration policies"*. It was also stated in target 29 as *"international migration is a multidimensional reality of major relevance for the development of countries of origin, transit, and destination, which requires coherent and comprehensive responses. We will cooperate internationally to ensure safe, orderly, and regular migration involving full respect for human rights and the humane treatment of migrants regardless of migration status, of refugees, and of displaced persons. Such cooperation should also strengthen the resilience of communities hosting refugees, particularly in developing countries. We underline the right of migrants to return to their country of citizenship, and recall that States must ensure that their returning nationals are duly received."* Thus, governments have a great role and responsibility in managing migration processes.

Furthermore, politicians should not approach the problem as a matter of local political maneuver against their rivals. The problem should be handled with long-term wise international policies. Governance of migration can be done through setting stable policies, launching regulations and measures, and implementing and controlling them. Policy recommendations to the government for migration governance are as follows (OECD, 2016:137-142; OECD, 2017; authors' own suggestions)

1. Recognizing all dimensions of migration

These dimensions include the different types of risks and the lack of capacity to absorb those risks affecting both receiving and sending countries. States and societies confront a combination of risks and if they cannot manage those risks, or mitigate the consequences, then the situation becomes more fragile. Economic, governmental, security, environmental, and societal dimensions should be revised and provided a new framework to consider how migration is framed and assessed. Recognizing the dimensions of migration can help policymakers design better policies. First of all, policymakers need to have a more holistic analysis of the context. The better analysis will bring better policy design, ensuring that policies will support the capacity of state and society systems to manage, absorb and mitigate potential conflicts due to migration. A multidimensional framework will help to analyze the problem in different dimensions. Furthermore, this will help to implement new strategic plans, design new policies, and monitor the consequences.

2. More attention to migration

More attention should be paid to all consequences of migration. Migration includes risks, uncertainties, and economic and political impacts. The fact is that migration has been increasing over the decades. It is getting more complex, and getting risky. Thus, not only collaborations among the government to initiate global policies but also the international community needs to dedicate more resources and attention to humanitarian and settlement problems.

3. Extirpate the negative perceptions of migrants.

Governments need to proactively plan to extirpate negative perceptions of migrants to overcome potential conflicts with the local communities. It is also necessary to avoid the concentration of migrants in a few specific and small areas (ICMDP, 2021:18).

4. Changes in existing paradigms

Problematic areas of migration are elaborate and differ from region to region. Immigrants should not be considered victims or occupants. Both domestic and international policies should be harmonized and complementary to each other. In addition, the international community should support education, job, and cultural integration programs. Besides, immigrant receiver countries should be supported financially by international institutions.

5. Invest in the reduction of migration

Well-designed policies to keep the people in their homeland save lives, resources, and money. It is both logical and effective. One of the determinants of success is analyzing the reasons for global migration. It may be beneficial to offer alternative policies both for domestic migration and international migration.

6. Integration of domestic policies with international policies

Actions should start at home, first. Domestic policies can contribute to the solution, marginalization, and capacity to support migration around the world.

7. *Development strategies of sending and hosting countries should include immigrants.* In this vein, policymakers and public authorities both in origin and hosting countries should coordinate their strategies.
8. Introducing the strategic programs and policies to minimize the cost of migration and maximize the benefit for both migrant and hosting country
9. The harmonization between the public policies should be ensured for the development and migration strategies
10. Improving coordination between public authorities, NGOs, and other non-state actors on migration policies
11. Host and home countries should have cooperation such as bilateral and regional migration agreements that guarantee the protection of migrants' rights and social benefits.
12. More effective domestic and global environmental policies that increase the public awareness should be initiated
13. As the number of immigrants increases in city centers, there is a growing demand for energy. There should be transparency about both the number of immigrants and purchasing resources such as oil, and minerals.

If actual data are shared, the limits can be evaluated more accurately. In addition, these data can be used to predict long-term access to markets and future needs for energy, shelter, and food. In this vein, reducing deforestation and climate change, providing incentives to increase agricultural production, and protecting diversity come to the fore to protect the environment. Besides, investing in improving the capacity and living conditions of camps, providing financial support to immigrants, incentives and leverage investments on new strategies, stimulating financial flow from public and private sources to protect the environmental quality and strengthen the infrastructure of more crowded cities can be useful to protect the environment.

14. More support for disadvantageous groups

Children, women, handicapped, and elderly people are more disadvantageous than young males in the migration process. Children and women expose to harmful attacks such as child abuse, prostitution, human trafficking, and organ mafia. Therefore, to protect them, migration offices and the international community should work and implement international policies on these disadvantageous groups.

15. Developing supportive financial strategies

The availability of adequate and appropriate funds and resources is crucial to addressing the root causes and drives of migration. The efficiency of financial resources will provide a dynamic international financial environment and consensus among the parties.

16. Legal framework

The legal framework should be revised in improved to involve the rights and responsibilities of immigrants.

CONCLUSION

This paper traces the relationship between migration, environmental degradation, and governmental policies. It also reflects on how policy is provided to societies affected by migration, and how policies should be reformulated by the governments to keep the environment clean, and people more stable. The fact is that migration wreaks destruction on human lives and societies. Irregular and massive migration obstructs the development of both receiving and sending countries compounds the risk of more fragility and feeds devastating effects on environmental quality. Irregular migration may obstruct economic development; compound the risks of fragility in macroeconomic performance and feeds vicious cycles of poverty and environmental degradation. Through mass migration, fragility may become more exponential and there may be more possible to face more stresses on environmental quality, climate change, fragile cities, and overpopulated urban areas. This process will not just leave the status quo in place, but if there are no further precautions, it will be even worse. Thus, policies to alleviate the toll of irregular migration and environmental degradation are crucial to fighting these problems. In this process, governments have a great responsibility. And, they should have both national policies and international cooperation to solve the problem.

REFERENCES

Adelman, H., & Sorenson, J. (1994). African refugees: Development aid and repatriation. Westview Press.

Baker, J. (1995). Refugee and labor movement in Sub-Saharan Africa. In *Studies on Emergencies and Disaster Relief, No.2* (pp. 6–28). Nordiska Afrikainstitutet.

Bjerre, L., Helbling, M., Römer, F., & Zobel, M. (2015, Fall). Conceptualizing and Measuring Immigration Policies: A Comparative Perspective. *The International Migration Review*, *49*(3), 555–600. doi:10.1111/imre.12100

Black, R. (2014). *Refugees, environment, and development*. Routledge.

Black, R., Adger, N., Dercon, S., Geddes, A., & Thomas, D. (2011). *Migration and Global Environmental Change-Future Challenges and Opportunities.* Government Office of Science, Final Project Report. https://www.un.org/development/desa/pd/sites/www.un.org.development.desa.pd/files/unpd-cm10201202-11-1116-migration-and-global-environmental-change.pdf

Black, R., & Sessay, M. F. (1997). Forced migration, environmental change and woodfuel issues in the Senegal River Valley. *Environmental Conservation*, *24*(3), 251–260. doi:10.1017/S0376892997000337

Castless, S., & Miller, M. J. (1998). *The age of migration-International population movements in the modern world* (2nd ed.). Macmillan Press Limited.

De Lange, W. J., Wise, R. W., & Nahman, A. (2010, November/December). Securing a sustainable future through a new global contract between rich and poor. *Sustainable Development (Bradford)*, *18*(6), 374–384. doi:10.1002d.413

De Lima, P., Bernabe, S., & Bubbico, R. L. (2016). *Migration and the EU, Challenges, Opportunities, the Role of EIB*. European Investment Bank.

Eckstein, D., Küzel, V., & Schafer, L. (2021). Global Climate Risk Index. *Germanwatch, 1*. https://www.germanwatch.org/en/19777

Flavell, A., Melde, S., & Milan, A. (2020). *Migration, environment and climate change: Impacts*. Second report in the "Migration, environment and climate change" series. Texte, 43/2020, Project No. (FKZ) 3717181060, Report No. FB000245/2,ENG.

Fuchs, P. G., Raulino, C. E., & Guerra, J. B. O. S de A. (2020). Green business in the context of the sustainable development. In W. L. Filho, A. M. Azul, L. Brandli, A. L. Salvia, & T. Wall (Eds.), *Decent Work and Economic Growth, Encyclopedia of the UN Sustainable Development Goals* (pp. 507–518). Springer. doi:10.1007/978-3-319-71058-7_12-1

Ganovel, M. F., Adger, W. N., de Campos, R. S., Boyd, E., Carr, E. R., Fabos, A., Fransen, S., Jolivet, D., Zickgraf, C., Codjoe, S. N. A., Abu, M., & Siddiqui, T. (2021). The migration-sustainability paradox: Transformations in mobile worlds. *Environmental Sustainability*, *49*, 98–109. doi:10.1016/j.cosust.2021.03.013

ICMDP. (2021). *Regional Migration Outlook 2021-Turkey & Western Balkans*. ICMDP Turkey & Western Balkans Regional Coordination Office. https://www.icmpd.org/file/download/51375/file/Turkey0and-0Western0Balkans0Regional0Migration0Outlook.pdf

ILO. (2021). *Impact of COVID-19 on nexus between climate change and labour migration in selected South Asian countries: An exploratory study*. https://www.ilo.org/wcmsp5/groups/public/---ed_protect/---protrav/---migrant/documents/publication/wcms_822838.pdf

IOM. (2008). *Migration, development, and environment.* IOM International Organization of Migration, No. 35. https://publications.iom.int/system/files/pdf/mrs_35_1.pdf

IOM. (2020). *World migration report 2020.* https://publications.iom.int/system/files/pdf/wmr_2020.pdf

Khan, S. A. R., Sharif, A., Golpira, H., & Kumar, A. (2019, November/December). A green ideology in Asian emerging economies: From environmental policy and sustainable development. *Sustainable Development, 27*(6), 1063–1075. doi:10.1002d.1958

Kibreab, G. (1996, December). Eritrean and Ethiopian urban refugees in Khartoum: What the eye refuses to see. *African Studies Review, 39*(3), 131–178. doi:10.2307/524946

Lee, S. E. (1966). A Theory of Migration. *Demography, 3*(1), 47–57. doi:10.2307/2060063

Martin, A. (2005, May). Environmental Conflict between Refugee and Host Communities. *Journal of Peace Research, 42*(3), 329–346. doi:10.1177/0022343305052015

Migration Data Portal. (2020). *What you need to know about the impact of environmental change on migration.* https://www.migrationdataportal.org/blog/what-you-need-know-about-impact-environmental-change-migration

Migration Data Portal. (2021). *Migration policies and governance.* https://www.migrationdataportal.org/ar/themes/migrationspolitik-und-regierungsfuehrung

Ministry of Interior Presidency of Migration Management. (2022). *Temporary protection.* https://www.goc.gov.tr/gecici-koruma5638

Morrissey, J. W. (2013, December). Understanding the relationship between environmental change and migration: The development of an effects framework based on the case of northern Ethiopia. *Global Environmental Change, 23*(6), 1501–1510. doi:10.1016/j.gloenvcha.2013.07.021

Ngai, P. B., & Kochn, P. H. (2002). *Organizational communication in refugee camp situations.* UNHCR The UN Refugee Agency, New Issues in Refugee Research, Working Paper No. 71.

O'Lear, S. (1997, June). Migration and the Environment: A Review of Recent Literature. *Social Science Quarterly, 78*(2), 606–618.

OECD. (2016). *States of fragility 2016: Undestanding violence.* . https://www.oecd-ilibrary.org/sites/9789264267213-5-en/index.html?itemId=/content/component/9789264267213-5-en#:~:text=Fragility%20is%20defined%20as%20the,absorb%20or%20mitigate%20those%20risks doi:10.1787/9789264267213-5-en

OECD. (2017). *Interrelations between Public Policies, Migration and Development.* OECD Development Center, OECD Publishing. https://www.oecd-ilibrary.org/docserver/9789264265615-en.pdf?expires=1651525981&id=id&accname=guest&checksum=580CCE6E6B161AA7868D7D49F3B0DDE6

Strange, T., & Bayley, A. (2008). *Sustainable development-Linking economy, society, environment.* OECD Publication. doi:10.1787/9789264055742-en

The World Bank. (2021, November 17). Remittance Flows Register Robust 7.3 Percent. *Growth, 2021*. https://www.worldbank.org/en/news/press-release/2021/11/17/remittance-flows-register-robust-7-3-percent-growth-in-2021

The World Bank Data. (2022). *Personal remittances, received (% GDP).* https://data.worldbank.org/indicator/BX.TRF.PWKR.DT.GD.ZS?locations=HT-AM-GE-PH-MA

Ukko, J., Saunila, M., Rantala, T., & Havukainen, J. (2019, May/June). Sustainable development: Implications and definition for open sustainability. *Sustainable Development (Bradford)*, *27*(3), 321–336. doi:10.1002d.1904

UN. (2015). *Transforming our world: the 2030 Agenda for Sustainable Development- Resolution adopted by the General Assembly on 25 September 2015. 17th Session Ageda items 15 and 116. A/RES/70/1.* https://www.un.org/ga/search/view_doc.asp?symbol=A/RES/70/1&Lang=E

UNHCR. (2021). *Refugee data finder*. https://www.unhcr.org/refugee-statistics/

UNHCR. (2022). *Operational Data Portal-Refugee situations*. https://data2.unhcr.org/en/situations/syria

Véron, J. (2014). *How are migration and the environment connected?* https://www.ined.fr/en/everything_about_population/demographic-facts-sheets/focus-on/migrations_environment/

Wageningen University. (2017a). *Rural migration and environmental degradation: A vicious cycle?* NOW VENI. https://www.wur.nl/en/project/Rural_migration_and_environmental_degradation.htmq

Wageningen University. (2017b). *Innovation to improve the fate of refugees and their hosts*. https://www.wur.nl/en/Research-Results/Research-Institutes/Environmental-Research/Programmes/Green-Cities/Migration-and-settlements.htm

Whitaker, B. E. (1999). *Changing opportunities: Refugees and host communities in western Tanzania.* New Issues In Refugee Research, Working Paper No. 11.

Zetter, R. (1994). Shelter provision and settlement policies for refugees: A state of the art review. In *Studies on Emergencies and Disaster Relief, No.2* (pp. 29–98). Nordiska Afrikainstitutet.

ENDNOTES

1 Carrying capacity can be defined as the capacity of the environment to provide the resources, food, clothing and shelter people need for a reasonable quality of life.

2 See The World Commission on Environment and Development (1987).

Chapter 10
Environmental Analysis of Azerbaijan and Its Development:
Environmental Factors Effecting Azerbaijan

Bahruz Balakishiyev
WSB University, Poland

ABSTRACT

The purpose of local environmental programs of municipalities is to involve the entire local population in the solution of environmental issues that are not provided for in the state environmental programs or in addition to them. These programs include measures to maintain the existing ecological balance in the local environment; greening and landscaping of the municipality; collection, transportation, neutralization, processing, and other issues in this area; protection of water, air, and land from all kinds of pollution together with neighboring municipalities, implementation of environmental measures, and other measures of local importance.

INTRODUCTION

Environmental policy has always been the basis of state policy. A clear example of this is the document on "Azerbaijan 2030: National Priorities for Socio-Economic Development" approved by the Decree of the President of the Republic of Azerbaijan Ilham Aliyev dated February 2, 2021, according to which the country's socio-economic development. The implementation of five National Priorities is envisaged, one of which is the Clean Environment and Green Growth Country Priority. The main goal for the successful implementation of this priority should be to create a high-quality environment, green energy space. According to this priority, along with the perspective economic development of the country, it is necessary to ensure the improvement of the environment, rapid restoration and growth of greenery, efficient use of water resources and sustainable energy sources (Mammadov & Khalilov, 2005).

DOI: 10.4018/978-1-6684-4548-8.ch010

Legislative acts adopted in the field of ecology define the rights, responsibilities and powers of municipalities in the field of ecology in local government.

Thus, "On the status of municipalities", "On environmental protection", "On protection of greenery", "On water economy of municipalities", "On soil fertility", "On specially protected natural areas and objects", "Environmental education of the population and the Law of the Republic of Azerbaijan "On Enlightenment" and the provisions of the Land Code of the Republic of Azerbaijan on municipalities.

Increasing and preserving greenery is also very important in protecting and improving the environment (Mammadov et al., 2009).

ENVİRONMENTAL FACTORS AFFECTİNG AZERBAİJAN

New industrial development and economic growth have presented a lot of environmental challenges to Azerbaijan. Some of those challenges are degradation of nature, scarcity of water, deforestation, climate change, marine pollution, and air pollution. Now, the government realizes the environmental challenges, and it's planning to lower the pollution rate (Mammadov & Khalilov, 2005).

The rapid development of all sectors of the economy over the last century has resulted in an increase in the negative impact of human activities on the environment and the overuse of natural resources. Maintaining the ecological balance, rational use of natural resources, protection of water, soil and air from pollution has become a universal problem. In addition to demographic growth, concerns such as the sharp increase in consumption and the depletion of natural resources that cannot be recovered in the future have also affected environmental thinking and activities. Maintaining the necessary balance between the economy, society and the environment in the context of the global environmental crisis can only be achieved through the formation of a new environmentally safe and economically optimal model of development - sustainable development. Although each country has its own environmental strategy and policy, global goals are aimed at ensuring that people live in a healthy environment, protecting and developing the environmental values of society, and improving the quality of the environment (Asian Development Bank, 2005).

Taking appropriate measures to address environmental problems in Azerbaijan, the state policy aimed at improving the environmental situation is an important part of the country's sustainable development strategy. To ensure environmental security, minimize and improve environmental pollution, rational use of natural resources to meet the needs of present and future generations, use alternative energy sources and achieve energy efficiency, assess needs at the national level on global environmental issues, identify solutions, international Expanding relations with organizations is one of the main directions of the country's environmental policy.

The country's environmental strategy is aimed at protecting natural resources at the national, international and regional levels, strengthening the coordination of activities in the field of environmental protection, applying the principles of science-based development, ensuring sustainability in the use of economic and human resources. Back in 2003, Azerbaijan adopted the National Program for Environmentally Sustainable Socio-Economic Development (Mammadov & Khalilov, 2005).

Rehabilitation of polluted areas, increasing the scale of forested areas, protection of biodiversity, fertilization and landscaping of arid lands, protection of the marine environment of the Caspian Sea, mitigation of climate change, modernization of hydrometeorological monitoring systems and monitoring networks, improvement of legislation, international cooperation, Raising the level of environmental awareness has been one of the activities. At present, in order to ecologically restore the areas contaminated with oil and oil products, "Improvement of the ecological condition and restoration of the landscape of the former iodine-bromine plant in Surakhani region," Inventory and assessment of contaminated lands in the Absheron Peninsula" and so on. projects are being implemented. At the same time, appropriate measures are being taken to manage waste, including plastic waste, and campaigns are being organized to recycle waste (International Eco-Energy Association, 2002).

As a result of afforestation and reforestation measures taken to increase the scale of forested areas, the area of forested areas now accounts for 12 percent of the country's territory. Over the past 17 years, 152.4 thousand hectares of afforestation and reforestation measures have been carried out, 96.445 million trees have been planted. A total of 5,386,000 trees were planted on 4,438 hectares of land in arid areas as part of landscaping measures on non-forest lands. To ensure the sustainability of landscaping projects, the areas have been provided with drip irrigation systems. Irrigation uses reusable water sources - desalinated seawater, as well as wastewater. In addition, fruit trees, including olive and other fruit trees, are preferred. The total area of specially protected natural areas increased from 478,000 hectares in 2003 to 893,000 hectares, reaching 10.3% of the country's territory (10 national parks, 10 state nature reserves and 24 state nature reserves). In order to increase the biological resources in the water basins, about 7 billion fish, including sturgeon and salmon fry, have been raised and released into the water basins in the last 17 years (Ministry of Environment and Natural Resources, 2003).

Azerbaijan is a country that is highly sensitive to the effects of climate change, and the country is already aware of the negative effects of climate change - floods, mudslides, droughts, thermal stresses, etc. feels the increase in the number of natural phenomena such as these highlights adequate adaptation measures (Ministry of Environment and Natural Resources, 2003).

Recently, new technologies have been applied in the field of hydrometeorology to strengthen control over dangerous hydrometeorological events caused by climate change, improve the forecasting and early warning system. At the same time, the monitoring system has been improved, new technologies have been introduced in the field of assessment and forecasting of environmental processes, a complex of automatic stations has been set up for continuous monitoring of air pollution, and information on air condition has been provided to the public online. In recent years, Azerbaijan has taken measures to improve its legislation and bring existing regulations into line with the provisions of international treaties to which the country is a party, as well as EU legislation. A number of new laws have been adopted, amendments and additions have been made to improve environmental legislation. The legislation mainly covers the areas of atmospheric air, water resources, domestic and industrial wastes, water bioresources, and biodiversity protection. Along with the adoption of new laws, amendments and additions to existing laws, as well as to ensure their implementation, relevant regulations and rules have been developed and approved. To address the existing problems, the Ministry of Ecology and Natural Resources attaches special importance to expanding relations with international organizations and donor countries. Thus, UN Development, Environment, Industrial Development Programs, NATO, Global Environment Facility, European Economic Cooperation and Development Organization, Economic Cooperation Organization, World Bank, Asian Bank, World Wildlife Fund, etc. cooperation with organizations continues. At the same time, bilateral cooperation is being established with developed countries on the basis of relevant agreements. Great attention is paid to acceding to international agreements in the field of environment. Thus, to date, the Republic of Azerbaijan has acceded to 21 conventions and signed relevant protocols (Asian Development Bank, 2005).

One of the priorities of Azerbaijan's development strategy is to ensure the highest level of nature protection, improvement of the ecological situation and living of citizens in a healthy natural environment.

On the initiative of the President of the Republic of Azerbaijan, Mr. Ilham Aliyev, 2010 has been declared the "Year of Ecology" in Azerbaijan. Under the leadership of President Ilham Aliyev, important work is being done in our country to solve environmental problems, protect the environment and restore ecological balance. The Heydar Aliyev Foundation's "Let's plant a tree" project, which serves to protect the ecology and the environment, has also made great contributions to this area (Ministry of Environment and Natural Resources, 2003).

REFERENCES

Asian Development Bank. (2005). *Country Environmental Analysis: Azerbaijan*. Asian Development Bank.

International Eco-Energy Association. (2002). *Proceedings of the 6th Baku International Congress "Energy, Ecology, Economy."* Author.

Mammadov, G., & Khalilov, M. (2005). *Ecology and Protection of the Environment. Elm* Press.

Mammadov, G. S., Khalilov, M. Y., & Mammadova, S. Z. (2009). Ecology Atlas. Academic Press.

Ministry of Environment and Natural Resources. (2003). *National Caspian Action Plan of Azerbaijan*. Author.

Chapter 11
Valuing Forest Ecosystem Services in Portugal:
A Literature Review

José Victorino Cristiano do Rosário
https://orcid.org/0000-0003-3007-0125
Centre for Transdisciplinary Development Studies (CETRAD), University of Trás-os-Montes e Alto Douro, Portugal

Lívia Madureira
Centre for Transdisciplinary Development Studies (CETRAD), University of Trás-os-Montes e Alto Douro, Portugal

ABSTRACT

The national literature on forest ecosystem valuation is scarce and little is known about how important the valuation of forest ecosystem services and their internalization in low density regions of Portugal are. Hence, there is a need for technicians, academics, and researchers to mitigate this knowledge gap through further research in this area. The chapter is a literature review with the objective of systematizing and synthesizing the knowledge produced in the period between 1992 and 2021 with regard to estimates of the economic value of forest ecosystem services in Portugal as well as finding evidence that relates the mechanisms of internalization of externalities in the sustainable development of low-density regions. A meta-regression was estimated, and the results indicate 220 international dollar/hectare/year in 2019 (190 euros/hectare/year) for forest ecosystem services in Portugal. Payment mechanisms for non-market forest ecosystem are still at an embryonic stage, which does not allow an accurate measurement of their real contribution to the sustainability of low-density regions.

DOI: 10.4018/978-1-6684-4548-8.ch011

INTRODUCTION

As the main land use in Portugal, the forest has the potential to tackle the sustainable transition challenge if we take into account the sustainable management of forestry goods and services and the probability of population settlement and valorisation of territories with forestry production. The present book chapter explores the recent research trends and dynamics on the economic valuation of Forest Ecosystem Services (FES) in Portugal to deliver clear guidelines to decision-makers and minimise the knowledge gap identified in previous research related to the scarcity of the national literature on forest ecosystem valuation. Portugal is a country where numerous regions are considered low-density regions, and the valuation of the forest ecosystem can be considered a strategy toward sustainability in these regions. Economic valuation expresses nature's contribution to monetary value (Farber et al., 2002), appraising both use and non-use values and allowing decision-makers and policymakers to identify, evaluate, and estimate trade-offs with other development goals (Balmford et al., 2002; Christie et al., 2012). The evaluation of FE S can contribute to the conservation and management of forest resources by improve the sustainable use of resources and convincing policymakers about the importance of forest management strategies (Acharya et al., 2019; Aslaksen et al., 2015; Baral et al., 2016; Häyhä et al., 2015). The primary role of economic analysis is to support the decision making of managers and policy makers (Khalfaoui et al., 2020). Therefore, the assessment process offers incentives to managers for sustainable forest management. In this sense, knowledge of the economic value of the multifunctionality of FES is necessary for those who can define, implement and evaluate public policies and public initiatives capable of capturing a favour in sustainable forest management and solving market failures (Bingham et al., 1995; Garcia et al., 2018).

The value of FES reflects the different ways they satisfy human needs (Marta-Pedroso et al., 2014). As highlighted by Costanza et al. (1997), research on the economic valuation of forest ecosystem goods and services is relevant, transdisciplinary and it was carried out in the scope of disciplines such as silviculture, environmental management, natural resources, conservation of biodiversity and the environment, accounting, among others. The national literature on forest ecosystem valuation is scarce, and little is known about how important the valuation of FES is and their internalisation in low-density regions of Portugal. Hence, technicians, academics, and researchers need to narrow this knowledge gap through further research. The present work is a literature review with meta-analysis aiming to systematise and synthesise the knowledge produced between 1992 and 2021 concerning estimates of the economic value of FES in Portugal and finding evidence that relates the mechanisms of internalisation of externalities in the sustainable development of low-density regions.

BACKGROUND

The first forest assessment studies appeared after the 1992 World Earth Summit and the 1992 Biological Diversity Convention in Rio de Janeiro. Later, Costanza et al. (1997) reported the first estimated value of worldwide ecosystem services at US 33 trillion in 1995, almost 1.2 times the total gross domestic product (GDP) globally. This value is still an indicative number that express the researcher´s intention to incorporate natural capital into the economic process. This objective was successfully achieved if we consider the evolution, research trends, and the number of citations reported on this topic (Costanza et al., 2014).

Studies show that human-induced pressure such as population growth, climate change, and increased demand for agricultural and forest products are the main reasons for attenuated forest values and related ecosystem services. For instance, Tilman et al. (2001) stated that land for agriculture expansion comes from the forest, grassland and other natural ecosystems; and if current global trends continue, a net loss of natural ecosystems to agriculture will amount to 10^9 ha by 2050, larger than the total area of the USA.

Although the importance of forests is recognised, including global benefits due to the protection of watersheds, soil and climate regulation, forests and their potential contribution to human well-being are still underestimated by society, and the value of forest ecosystem service is not incorporated into decision making, particularly those without market prices (Bernués et al., 2014; Ninan & Kontoleon, 2016; Sil et al., 2017). FES received considerable attention after the innovative work of the Millennium Economic Assessment (MEA) in 2005, and the main contribution of this initiative was the establishment of ecosystem services as a political tool for the management of sustainable natural resources and boosted evaluation studies of ecosystem services (Acharya et al., 2019; Fisher et al., 2009).

According to Raum (2018), in a study carried out in the United Kingdom, the motivations for the growing interest in the economic (monetary) valuation of FES are mainly and not exclusively the following: (1) it is required by international obligations; (2) is in line with the dominant political philosophy of the market; (3) keeps its promise to include the environment more fully in current economic decision-making processes; (4) can help to draw more attention to biodiversity conservation; (5) keeps the promise of new sources of income from public and private sources; and (6) can be used as a convenient argument for promoting tree planting.

The Portuguese forest is predominantly managed under private ownership, corresponding to 84.2% of the total forest area, 6.5% of which belongs to industrial companies. The remaining 15.8% correspond to public areas, where only 2% are in the private domain of the State, the lowest percentage in Europe (ICNF, 2015). Portugal is also characterised by the predominance of small agroforestry properties, especially in the country's northern and central regions, and the main negative externality in the sector is forest fires. For instance, from 2013 to 2017, the burnt area expanded from 38,961.5 ha to 541,491 ha, respectively or about an 86.5% increase (ICNF, 2019a).

The study of the economic valuation of the Portuguese forest is important because the forest is the dominant land use in mainland Portugal, occupying 36.2% of the territory, according to the last forest inventory carried out in 2010 (IFN6, 2015).

The first attempt to estimate the total economic value (TEV) of forests in the Mediterranean region was edited by Croitoru and Merlo (2005), in collaboration with several authors from the Mediterranean region, where Mendes (2005) for the case of Portugal stands out.

According to Mendes (2005) (cited by Lopes, 2013), about 83.5% of the forest's value comes from the benefits of ecosystem services that are not valued in the market. More recent research estimates this value at around 50% of the total economic value. Still, the national literature on forest ecosystem valuation is scarce, and little is known about the importance of FES valuation and its internalisation into low-density regions of Portugal.

The next section will present the methodology employed, followed by the key findings' interpretations and conclusions.

METHODOLOGY

A survey was made of the scientific and technical works that quantified the TEV of Portugal's FES. Two databases were built: one for market forest ecosystem services and goods and the other for nonmarket services and goods. The SCOPUS and Web of Science databases and other secondary data sources such as the Institute for Nature and Forest Conservation (ICNF), Instituto Nacional de Estatística (INE), statistics from the Food and Agriculture Organization of the United Nations (FAOSTAT), Information System on Quotations of Forest Products in Production (SICOP) and at EU level (EUROSTAT) were used as sources for constructing the databases.

The SCOPUS database searched for the keywords: (economic) AND (valuation) OR (assessment) AND TITLE-ABS-KEY (forest) AND TITLE-ABS-KEY (ecosystem) OR TITLE -ABS-KEY (recreation) OR TITLE-ABS-KEY (forestry) OR TITLE-ABS-KEY (cork) AND TITLE-ABS-KEY (Portugal) in the titles, abstracts, and keywords. After analysing the hits and applying the inclusion criteria, considering only articles with estimates of the economic value of FES in Portugal, 31 articles remained. In a second phase, other technical documents, doctoral theses and master's dissertations that present estimated values were also analysed. The two databases with the described biophysical, monetary, and temporal units facilitated the production dynamics and dissemination of knowledge analyses for the valuation of FES in Portugal. The selected literature delimited the theoretical and conceptual framework of the economic valuation of FES in Portugal. Moreover, it revealed notable gaps in the valuation studies, including information about valuation methodologies used over the years and the most valued types of ecosystem services. Furthermore, the literature demonstrated the need for improvements in the economic value calculations for these services, which would provide reliable information based on a rigorous methodology to the public decision-makers.

Overall, 118 sources were consulted, including scientific articles, project reports, doctoral theses, master's dissertations, and technical notebooks. Despite an increased number of studies on the economic valuation of FES from 1996 to 2021, many were of methodological scope or literature reviews, and only 34 reported estimated values. The values were classified according to the type of ecosystem valued and the ecosystem service of the forest it represents. The Common International Classification of Ecosystem Services (CICES v.5.1) was utilised and included 48 ecosystem services divided into 20 groups, eight divisions and three sections. There is enormous diversity in the way values are reported in each study.

Additionally, comparisons of reported values estimated in different years were standardised to correct for inflation, with 2019 being the base year. which was necessary because the economy's general price levels rise over time so that any amount of money has less value in terms of the goods and services it can buy over time. Standardisation was performed using GDP deflators that measure the annual rate of change in prices in an economy. GDP deflators for Portugal were obtained from the World Bank's World Development Indicators, following the price standardisation methodology suggested by the ESVD (2020). The estimated values expressed in escudos and dollars were converted into euros using the Bank of Portugal exchange rate for the reference year. The values were then converted into 2019 International Dollars, considering the methodology used by the ESVD (2020). The standardised price level was calculated using the following formula:

$$V_{2019} = V_t(D_{2019}/D_t),$$

where $\mathbf{V}_{2019}$ is the value at the price level in 2019, $\mathbf{V}_t$ is the value at the price level of the study year, $\mathbf{D}_{2019}$ indicates the GDP deflator for the base year 2019, and $\mathbf{D}_t$ is the GDP deflator index of the study year.

The value in international dollars was calculated using the following formula:

$$Vint\$=VLc\times FXppp$$

where **Vint$** is the value in international dollars, **VLc** is the observed value in the local currency, and **FXppp** is the exchange rate adjusted to the purchasing power parity between the local currency and the USD. The GDP deflator data and the exchange rate adjusted for purchasing power provided by the World Bank and OECD (ESVD, 2020) made these calculations possible.

The constructed database made it possible to estimate a meta-regression combining original research studies, generalising the results and estimating the relationship between dependent variables and a set of explanatory variables.

The values of different estimates of FES carried out in different study areas of Portugal, including national studies, were combined to estimate a meta-equation. This meta-equation showed a statistical relationship between forest values and variables, including ecosystem service type (e.g., carbon sequestration, timber production, hydrological regime, forest fire protection services, landscape biodiversity and forest recreation) and whether the study was conducted in a protected area.

RESULTS

Available Studies on the Economic Valuation of FES in the Portuguese Case

One of the first studies on the economic valuation of FES and goods in Portugal was about recreation in the forest in the Portuguese context. This doctoral thesis by Mendes (1996) quantified the value of protected areas, namely the Peneda-Gerês National Park, using a combination of the Travel Cost and Contingent Valuation method and the consumer surplus and willingness to pay as economic measures. This work sought to provide a reference on management policies for protected areas, suggesting a price for a good without a market price, using the willingness to pay as a proxy, after having estimated the consumer surplus associated with the use of the protected area for recreation (Mendes, 1996).

In 2005, after nine years, the work by Mendes (2005) was published and reported the total economic value of FES for mainland Portugal at 1,193,236,000 euros per year at 2001 prices. This work served as the basis for determining the economic value of forest areas for the national forest as a whole, which was carried out in 2006 by the Directorate General of Forestry Resources and constituted a relevant component of the definition of the forest strategy, giving rise to the structuring matrix that it reflected the different value components, given the distribution across different forest types.

Subsequently, several works on the economic valuation of ecosystem services performed at the local level emerged, with the moment of inflexion or a considerable increase in publications in 2010, including the study by Mendes and Proença (2011) that estimated the average consumer surplus per day in the Peneda-Gerês National Park at 194 euros at 2005 prices (217 euros at 2019 prices) using the travel cost method. This study served as a tool to improve government policy and promote nature tourism.

The work of Simões (2012) stands out in Portugal when analysing the demand for forest recreation. This study combined declared and revealed preferences methods to examine the effect of forest fires on visits to forests and woods. Their results confirmed that visitors are sensitive to price and quality changes. Additionally, the authors questioned the lack of knowledge about recreation in the forest, the demand for these services, the preferred forest characteristics and the visitors' motivations and characteristics. Consequently, these questions prompted further investigations and served as references that made it possible to compare the results of these investigations. The study also estimated that the intended number of trips would be reduced by 47%, and the loss of well-being would be around 118.16 euros per year in a forest fire scenario. The results of studies valuing forest recreation are presented in Table 1.

Table 1. Studies valuing forest recreation, estimates and methods. The case of Portugal.

ID	Authors	Publication Year	Estimated value	Currency	Unit	Temporal Unity	Values in euros of 2019	Values in Int$ Dollars of 2019	Employed Methods
1	Mendes, I.	1996	14,961	escudos	visitor	day	186.86	106.70	Travel cost
1	Mendes, I.	1996	16,635	escudos	visitor	day	207.77	118.64	Travel cost
1	Mendes, I.	1996	11,763	escudos	visitor	day	146.92	83.89	Travel cost
1	Mendes, I.	1996	21,375,975	escudos		year	266,988.42	152,450.39	Travel cost
2	Mendes, I.	2005	123	€	visitor	day	178.47	101.91	Travel cost
3	Mendes, A.	2005	2.75	€	visitor	day	3.99	6.27	Travel cost
4	Kastenholz E. & Rodrigues A.	2007	10.00	€	visitor	day	14.51	8.29	Expenditure levels
4	Kastenholz E. & Rodrigues A.	2007	25.00	€	visitor	day	36.27	20.71	Expenditure levels
5	Antunes, Sandra et al.	2010	3.31	€	visitor	day	4.80	2.74	Cost valuation
6	Pacheco J.	2011	4.73	€	visitor	day	5.29	3.02	Contingent valuation
6	Pacheco J.	2011	576,447.42	€		year	644,793.09	368,176.86	Travel cost
6	Pacheco J.	2011	760,700.02	€		year	850,891.34	485,858.96	Contingent valuation
7	Mendes & Proença	2011	194	€	hectare	year	217.00	123.91	Contingent valuation
8	Simões, P	2012	102.75	€	visitor	year	114.93	65.63	Contingent valuation
8	Simões, P	2012	251.73	€	visitor	year	281.58	160.78	Travel cost
8	Simões, P	2012	134	€	visitor	year	149.41	85.31	Travel cost
9	Simões, P	2013	48	€	visitor		52.71	30.10	Count Data Models
9	Simões, P	2013	37	€	visitor	year	40.63	23.20	Count data
10	Simões, P	2013	201.5	€	visitor	year	221.29	126.36	Count data
10	Simões, P	2013	106.92	€	visitor	year	117.42	67.05	Count data
10	Simões, P	2013	82.24	€	visitor	year	90.32	51.57	Travel cost
11	Rodrigues et al.	2015	0.798	€	hectare	year	0.89	0.51	Market price
11	Rodrigues et al.	2015	64.517	€	hectare	year	72.17	41.21	Market price
11	Rodrigues et al.	2015	65.315	€	hectare	year	73.06	41.72	Market price
12	Lopes, L.	2015	10	€	hectare	year	11.19	6.39	Review
13	Madeira A.	2016	288.5	€	visitor	year	303.19	173.12	Travel cost

Table 1. Continued

ID	Authors	Publication Year	Estimated value	Currency	Unit	Temporal Unity	Values in euros of 2019	Values in Int$ Dollars of 2019	Employed Methods
13	Madeira A.	2016	511.91	€	hunter	year	537.97	307.18	Travel cost
14	Oliveira F. et al.	2017	4	€	visitor	year	4.14	2.36	Contingent valuation
15	Amaral B.	2019	13.66	€	visitor	day	13.89	7.93	Travel cost
16	Mota, A.	2019	5	€	visitor	day	5.08	2.90	Contingent valuation
17	Gomes, F.	2013	3.99	€	visitor	day			Contingent valuation

Analysing the history of citations, we can see that the work of Mendes (2005) influenced later studies as it was the first to cover the entire Portuguese forest. In 2007, two works by Croitoru (2007) were published. In the first, the author explores the question of the economic value of non-wood products covering the entire Mediterranean, highlighting the value of non-woody forest products such as honey, cork, wild mushrooms, and grazing, estimated based on Mendes (2005), in 206 euros/hectare at 2005 prices (298.90 euros at 2019 prices). In the second work, the author, continuing to analyse the Mediterranean, opens the spectrum of analysis to the entire forest, also presenting references to Portugal provided by Mendes (2005). A notable contribution was that this study revealed that wood products account for a small part of the forests' benefits, suggesting the benefits of watershed protection is much more important. Additionally, recreation was already starting to be important in the North Mediterranean with a tendency to grow throughout the region; thus greater incorporation of the multifunctionality of the forest was recommended in these countries' forest policies. In 2011, Madureira et al. (2011) focused on strategic forest management by employing the Contingent Valuation Method (CVM) in supporting decision support systems to characterise an alternative forest management strategy in detail. The authors present an innovation in the method that consisted of combining CVM with advanced techniques to facilitate the construction of assessment scenarios with a presentation of the visual impacts of alternative forest management strategies (conservation, cutting allowing natural regeneration and clear-cutting) that contributed to the literature with a methodological procedure that came to be used in other subsequent studies. Sil (2014), using InVEST models, estimated carbon sequestration in the upper Sabor river basin at US$ 248.64. Very recently, a scientific report of the ECOFOR.PT project (PDR2020-2023-045916) was published, which presents for the period from 2014 to 2019, not counting the exceptional year of 2017 because of severe forest fire occurred, the average of the Total Economic Value of the forestry space in Portugal, in a default estimate, at 2016 prices, as 2.240 billion euros (Mendes et al., 2021). These results will be important to consider for future forest policy design in Portugal and at the EU level.

Subtracting the social costs of forest fires (prevention, combat and loss of goods and services), the average Total Economic Value, without the year 2017, in the period from 2014 to 2019, was 1.926 billion euros. The undermentioned study estimated that the Total Economic Value could have negative values with large fires, such as in 2017 when it dropped to -77.795 million euros (Mendes et al., 2021).

As shown in Figure 1, the authors' affiliation investigating forest ecosystem valuation is led by the University of Lisbon; however, other institutions are committed to this theme.

The analysis of the entities that funded research in this area is interesting because it provides an enlightened view of the entities that support this research. Indeed, the entire evolution of knowledge production depends on funding. Thus, the various programmes of the European Union and the Portuguese Foundation for Science and Technology have made a great contribution (Figure 2).

Table 2. Studies and estimates on the economic value of other ecosystem services in Portugal

ID	Authors	Publication Year	Estimated Value	Currency	Unity	Temporal Unity	Values in Euros of 2019	Values in Int$ dollars of 2019	Ecosystem Service	Method
4	Marta-Pedroso et al.	2018	502,234	€	hectare	Year	547,708.00	312,741.26	Carbon sequestration	Value Transfer
4	Marta-Pedroso et al.	2018	6,734,829	€	hectare	Year	7,344,623.70	4,193,780.14	Erosion regulation	Value Transfer
17	Marta-Pedroso et al.	2018	579,241	€	hectare	Year	631,687.48	360,693.55	Biodiversity	Value Transfer
18	Marta-Pedroso et al.	2020	154	€	ha^{-1}	Year	159.43	91.03	Carbon sequestration	Value Transfer
18	Marta-Pedroso et al.	2020	112	€	ha^{-1}	Year	115.95	66.20	Carbon sequestration	Value Transfer
18	Marta-Pedroso et al.	2020	194	€	ha^{-1}	Year	281.49	160.73	Carbon sequestration	Value Transfer
18	Marta-Pedroso et al.	2020	236	€	ha^{-1}	Year	244.32	139.51	Carbon sequestration	Value Transfer
19	Mendes A.	2005	1,193,232,000	€		Year	1,731,372,793.06	988,613,864.80	All forest ecosystems	Market price
20	Sil A. et al.	2014	248.64	USD	hectare	Year	336.01	191.86	Carbon sequestration	InVEST
21	Von Essen et al.	2019	3,500,000	€		Year	3,558,881.50	2,032,121.33	Cork production	InVEST models
21	Von Essen et al.	2019	7,790,000	€		Year	7,921,053.39	4,522,921.48	Carbon sequestration	InVEST models
22	Vale M.	2014	64,000,000	€	tons CO_2	Year	71,588,069.49	40,876,787.68	Carbon sequestration	Substitution market
23	Madureira L. et al.	2011	11	€		Year	15.96	9.11	Forest landscape protection	Contingent Valuation
23	Madureira L., et al.	2011	19	€		Year	27.57	15.74	Forest landscape protection	Contingent Valuation
23	Madureira L., et al.	2011	20	€		Year	29.02	16.57	Forest landscape protection	Contingent Valuation
23	Madureira L., et al.	2011	15	€		Year	21.76	12.43	Forest landscape protection	Contingent Valuation
23	Madureira L., et al.	2011	54	€		Year	78.35	44.73	Forest landscape protection	Contingent Valuation
24	Croitoru, L.	2007	206	€	hectare	Year	298.90	170.67	Non-woody products	Market price
25	Croitoru, L.	2007	70	€	hectare	Year	101.57	57.99	Timber production	Market price
25	Croitoru, L.	2007	130	€	hectare	Year	188.63	107.71	Timber production	Market price
25	Croitoru, L.	2007	23	€	hectare	Year	33.37	19.06	Non-woody products	Market price
25	Croitoru, L.	2007	143	€	hectare	Year	207.49	118.48	Non-woody products	Market price
25	Croitoru, L.	2007	8	€	hectare	Year	11.61	6.63	Hunting	Market price

META-ANALYSIS

The meta-analysis concept implies combining original research studies, generalising the results and estimating the relationship between dependent variables and a set of explanatory variables. Herein, 118 scientific articles, project reports, doctoral theses, master's dissertations and technical notebooks were consulted. However, only 34 studies have fulfilled the meta-analyses requirements. Overall, there were 133 observations of estimates of economic values of goods and services of the forest ecosystem in Portugal among these documents. Each value was classified according to the type of valued ecosystem and the FES that it represents.

Figure 1. Publications by the authors' affiliation institution

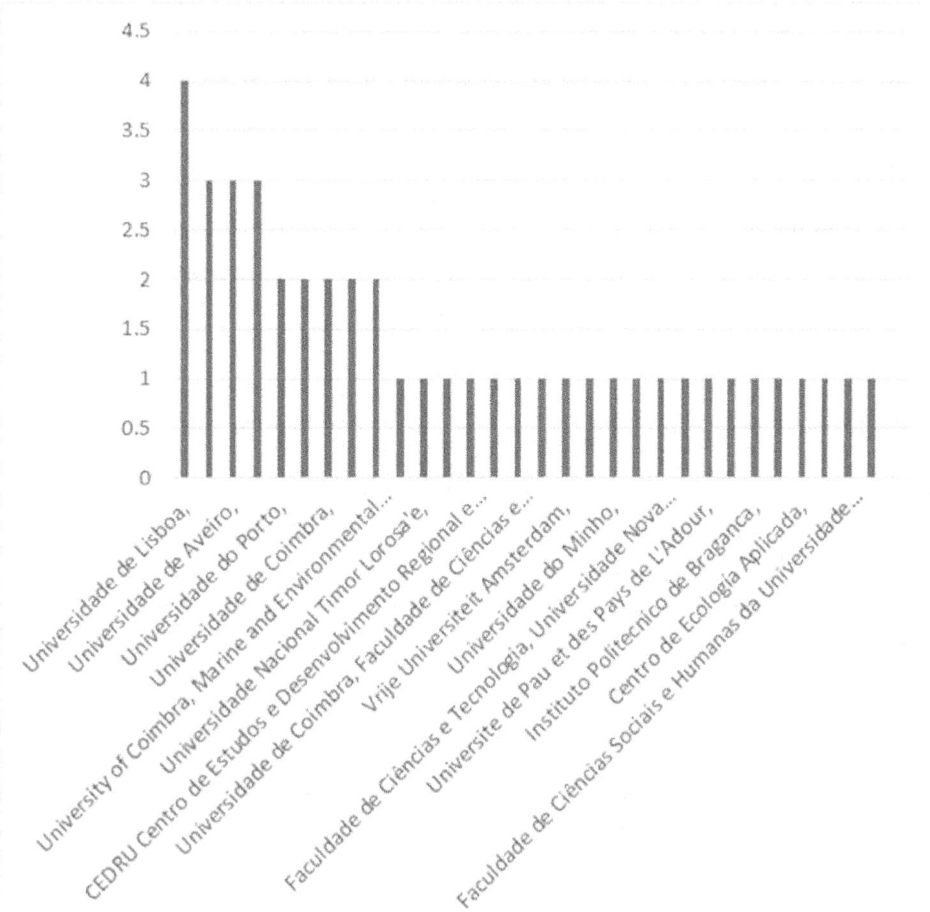

Figure 2. Publications by funding entity

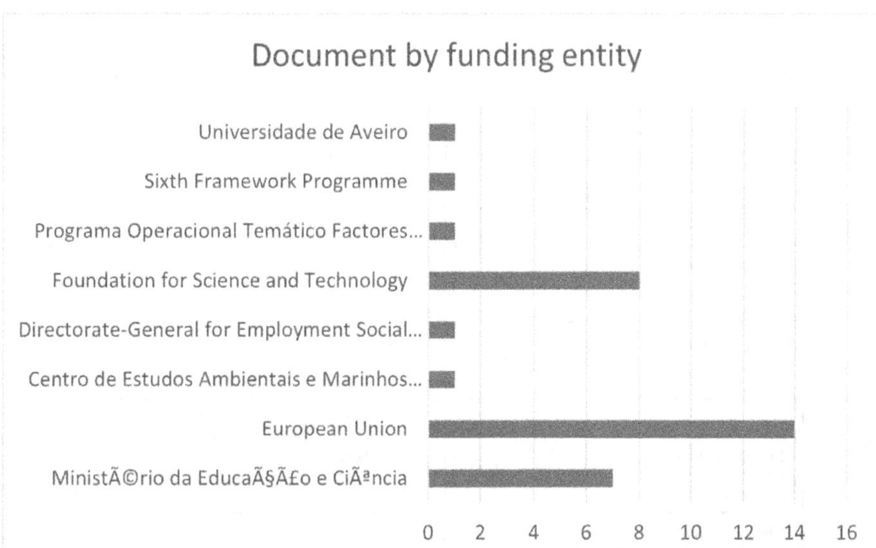

Concerning different currencies of several years reported in the studies, all euro values were converted to international dollars in December 2019, considering a previously employed methodology (ESVD, 2020). With this initial set of data, we specify an econometric model with a dependent variable, the economic value of the FES (expressed in international dollars 2019 per hectare per year), and independent variables to indicators (dummies) that distinguish the method employed for evaluation.

In this case, the values of the different FES estimates made in different areas of continental Portugal studies, including national studies, were combined to estimate a meta-equation that shows a statistical relationship between forest values and a set of explanatory variables, including the type of ecosystem service (e.g., carbon sequestration, wood production, hydrological regime, forest fire protection services, landscape and forest recreation biodiversity) and if the study was carried out in an area considered protected. The resulting model is as follows:

$$Yi = \alpha + x1\beta1 + x2\beta2 + \ldots\ldots xn\beta n + \mathcal{E}$$

where Yi - α represents the constant term, x- β refers to the coefficients associated with their explanatory variables to be estimated, and $\mathcal{E}$ is a vector of residues distributed independently and identically (Romo-Lozano et al., 2019; TiParpa et al., 2019)

Specification of the regression model: Yi = α + β1 Method + β2 Ecosystem service + β3 Study area + β4 scope.

Table 3. Variables considered in the model

	VARIABLE	DETAILS	TYPE OF VARIABLE
Valuation methods used in the study	Market price	(=1, 0 otherwise)	dummy
	Value transfer	(=1, 0 otherwise)	dummy
	Travel cost	(=1, 0 otherwise)	dummy
	Contingent valuation	(=1, 0 otherwise)	dummy
	Other methods	(=1, 0 otherwise)	dummy
Characteristics of the study area	Latitude		Continuous variable
	Area of the study site in hectares		Continuous variable
	Protected zone	(=1, 0 otherwise)	dummy
Forest ecosystem services	Biodiversity	(=1, 0 otherwise)	dummy
	Forest recreation	(=1, 0 otherwise)	dummy
	Wood production	(=1, 0 otherwise)	dummy
	Non-woody Products	(=1, 0 otherwise)	dummy
	Carbon sequestration	(=1, 0 otherwise)	dummy
	Forest_fire_regimes	(=1, 0 otherwise)	dummy
	Hydrological regime	(=1, 0 otherwise)	dummy
Characteristics of the studies	Scope of the study	(1=Local, 2-National)	
	Article	(=1, 0 otherwise)	Dummy
	Other documents	(=1, 0 otherwise)	
	Sample size		Continuous variable
	Publication year		Continuous variable

The model explains 42% of the variation for the dependent variable as indicated by R^2 with R^2 adjusted to 37%.

According to the last Forest Inventory (2015), the area of forest in Portugal is 3,305,300 hectares, it can be concluded that the value of FES in mainland Portugal is approximately 220.85 international dollars per hectare per year.

The variables that explain the value of FES that are statistically significant and positive are carbon sequestration with a coefficient of 9.23, which means that an increase of one unit in the amount of carbon sequestered increases 9.23 international dollars average the value of FES in hectares per year. Other significant and positive variables are recreation in the forest (t-value 2.52), Biodiversity, and hydrological regime. The forest fire protection service, although not statistically significant, is a component to be considered in the Portuguese context. The Market Price method is associated with high values of FES, followed by travel cost, contingent valuation and other methods, all of which are statistically significant at 95%.

Regarding the prerequisites of meta-regression, tests were performed where the Absence of Multicollinearity (VIF < 10) was performed. Also Homoscedasticity was tested and null hypothesis was rejected. (Breusch-Pagan test).

Value transfer example: $Yi = \alpha + \beta 1$ Method$+\beta 2$ Ecosystem service$+\beta 3$ study area$+\beta 4$ protected zone.

Table 4. Results of meta-analysis

Source	SS	df	MS	Number of obs	= 133
				F(11, 121)	= 8.23
Model	6.8908e17	11	6.2644e+16	Prob > F	= 0.0000
Residual	9.2053e17	121	7.6077e+15	R-squared	= 0.4281
				Adj R-squared	= 0.3761
Total	1.6096e18	132	1.2194e+16	Root MSE	= 8.7e+07

Table 5 Meta-analysis results

Estimated value in IntUSD	Coef.	Std. Err.	t	P>t	[95%Conf. Interval]	
Market price	2.39e+08	4.35e+07	5.50	0.000	1.53e+08	3.25e+08
Value transfer	1.02e+08	5.29e+07	1.94	0.055	-2272118	2.07e+08
Travel cost	1.34e+08	4.80e+07	2.79	0.006	3.88e+07	2.29e+08
Contingent valuation	1.01e+08	4.28e+07	2.36	0.020	1.64e+07	1.86e+08
Other_methods	1.40e+08	4.44e+07	3.16	0.002	5.22e+07	2.28e+08
Protected Zone	1.88e+07	1.98e+07	0.95	-0.344	2.04e+07	5.81e+07
Biodiversity	5.83e+07	2.47e+07	2.36	0.020	9405807	1.07e+08
Forest recreation	5.44e+07	2.16e+07	2.52	0.013	1.17e+07	9.71e+07
Carbon sequestration	9.23e+07	2.95e+07	3.12	0.002	3.38e+07	1.51e+08
Forest fire regime	4.87e+07	4.64e+07	1.05	0.295	-4.30e+07	1.41e+08
Hydrologic regime	1.18e+08	4.36e+07	2.71	0.008	3.21e+07	2.05e+08
_cons	-2.07e+08	4.44e+07	-4.65	0.000	-2.95e+08	-1.19e+08

Table 6. Value transfer example

Estimated value Intdollars	Coef.	Estimated variable value	Result
Market price	2.39e+08	1	2.39e+08
Travel cost	1.34e+08	1	1.34e+08
Other methods	1.40e+08	1	1.40e+08
Biodiversity	5.83e+08	1	5.83e+08
Forest recreation	5.44e+07	1	5.44e+07
Carbon sequestration	9.23e+07	1	9.23e+07
Hydrological regime	1.18e+08	1	1.18e+08
Const	-2.07e+08	1	-2.07e+08
Total			730000000

How can the internalisation mechanisms of the value of nonmarket goods and services contribute to the development of low-density regions in Portugal?

The mechanism for internalising the value of nonmarket goods and services in forest areas can be part of the equation for the sustainable development of low-density regions. As internalisation implies a situation in which a forestry producer is compensated monetarily or otherwise by those that benefit from a positive externality, which mitigates the common tendency of loss of forest area in favour of bushland and transforms forestry into a more attractive activity, capable of increasing the value of low-density inland regions. The most used nonmarket goods and services of forest areas in Portugal, although not yet widespread to satisfactory levels, is the typology of internalisation of goods and services called Payments for Environmental Services (PES), which are contracts between owners of areas producing these services and public or private entities that pay them to manage these areas to produce the intended ecosystem services (Mendes et al. 2021).

Some evidence of payment mechanisms for ecosystem services in forest spaces in Portugal, both public and private, can be analysed to assess their real contribution to the country and low-density regions' sustainability in particular: The first public mechanism is the Permanent Forest Fund (PFF) that was created by the Basic Law of Forest Policy of 17 August 1996 (Law No. 33/96) and had other objectives: Financing projects for reforestation of areas affected by fires; Compensate economically the owners of sensitive ecosystems for the damages that result from restrictions imposed by the need for their conservation; Fund specific research actions, favouring the form of programme contracts; Institute a subsidised forest credit system. However, although this fund could be the most important public mechanism for paying for forest ecosystem services in Portugal, it is far from being able to fulfil these functions due to limitations in financial allocations. For example, a good part of this fund's resources have gone to public entities of the central and local administrations and not to forest producers and their organisations, even when the public forest represents only about 2% of the total forest area in Portugal (Mendes et al., 2021). The second is the Environmental Fund created by Decree-Law No. 42-A/2016 on 12 August. However, its scope of action is very wide, with the forestry sector having a residual role. The third is the programmes that encourage forestry investment co-financed by the European Union. Furthermore, public funding granted within the scope of measures to encourage forestry investment within the framework of Rural Development Programmes co-financed by the European Union can also be justified as a form of payment for ecosystem services in forest areas.

Except for measures that support agricultural land afforestation, all these programmes have suffered from a problem, securing investments. However, it no longer supports management when the area is the object of this investment, which allows us to conclude that the potential mechanisms for internalising the externalities of forestry activity in Portugal still has a long way to go towards sustainable forest management and low-density territories sustainability. The fourth public programme is the Landscape Transformation Programme which was established by the Council of Ministers Resolution No. 49/2020, of 24 June, combines funding from the European Rural Development Fund and the Environmental Fund to support not only forestry investment but also the maintenance and management in the medium term, as well as the remuneration of ecosystem services. The Terra Prima weed control project to promote carbon sequestration is one example of public-private mechanisms. In partnership with UNAC – União da Floresta Mediterrânica and its associates and with financial support from the Portuguese Carbon Fund, Terra Prima developed a control project forests to promote carbon sequestration between January 2011 and December 2014. Payment mechanisms such as the Project “In the heart of the Montado”53 is a good example in the private sector. “In the heart of the cork” (Green Heart of Cork) (Bugalho & Silva, 2014) is a project by the Associação Natureza de Portugal and WWF in partnership with APFC – Associação de Produtores Florestais do Concelho de Coruche e Limítrofes, and ANSUB – Associação de Produtores Florestais do Vale do Sado.

Multifunctional management project for burned forest areas in Arganil co-financed by the Jerónimo Martins Group.

In the Portuguese case, and even in other European countries, the mechanisms for internalising the value of non-market goods and services in forest areas are still very far from being achieved. However, capturing a substantial share of this value for forestry producers constitutes a starting point for the sustainable management of forest spaces in low-density regions. According to recent studies, the value of nonmarket goods and services is approximately 50% of the total economic value of forest ecosystem services in Portugal, compared to the 1.062 billion euros of value. By default, the production of nonmarket goods and services produced by forest areas with the approximately 25 million euros of the tax on forest products earmarked for the Permanent Forest Fund could be an important instrument of internalisation of that value for the benefit of forestry producers. The distance is enormous, even if it is argued that not all the value of 1.062 million euros should be internalised for the benefit of forestry producers and even if the following is added to the Permanent Forest Fund funds: public expenditure for the conservation of biodiversity in forest spaces; silvo-environmental measures and support for forestry investment from the Rural Development Programme; the few experiences of private contracting of payment for ecosystem services (Mendes et al., 2021).

FUTURE RESEARCH DIRECTIONS

The present book chapter, when reviewing the available studies on the economic valuation of forest services and goods in Portugal and trying to find evidence on the importance of payment mechanisms for ecosystem services, leaves several open questions that are basically just directions for future research: Among these questions for future research are, for example, what other methods, in addition to the literature review, would help us to measure the contribution of payment mechanisms for FES to the sustainability of low-density regions? Looking at the variation in the total economic value of forest ecosystem services per hectare, we see that this value in 2001 was 344 euros per hectare, and in 2019 it changed to 173.52 euros per hectare, now including the social costs of forest fires. These numbers are essential in the formulation of policies for forestry activity sustainability and impel us to carry out continuous studies to arrive at a clearer understanding that allows us to address updated recommendations to public decision-makers and managers in the forestry sector.

CONCLUSION

The study's main objective was to survey the studies related to the valuation of forests in Portugal. The available data enabled us to conclude that there is an increase in the literature. The existing literature, although scarce, has responded to the needs of society and public decision-makers; that is, the existing literature has always been able to provide data for the creation of forestry strategies and policies. Doctoral theses, master's dissertations, scientific reports and scientific articles were reviewed in both SCOPUS and Web of Science databases, and according to the most recent publication, the total economic value of Portugal's FES and goods went from 344 euros per hectare in 2001 to 173.52 euros per hectare in 2019 at 2016 prices. The Gross Economic Value of forestry production in Portugal is approximately €53.45/hectare, without considering the externalities of forest fires or intermediate consumption.

These values are comparable to the meta-analysis results, which indicate 220.85 international dollars per hectare per year (190 euros/hectare/year).

Notably, the 45.5% of the gross value of forest production in Portugal corresponds to non-market or public goods and services. Payment mechanisms for the nonmarket forest ecosystem are still at an embryonic stage, which does not permit accurate measurements of their real contribution to the sustainability of low-density regions. It is recommended that future studies analyse the status quo regarding the implementation of a strategic forestry plan in Portugal, addressing the progress in sustainable forest management and the valuation of FES. The combination of methods, utilising both market and nonmarket prices for a comprehensive FES valuation in Portugal captures better the considerations for future research in this area.

The limitations and weaknesses of the work are related to the fact that the transfer of value is problematic insofar as many studies reviewed cover the entire territory and others are related to a specific part of a site area of study. Another weakness has to do with the possibility of the omission of recent research in this area is one of the limitations of the present research. Even so, the work has practical relevance, it can be used for consultation by students and scholars in the area. According to reviewed studies there is a need to increase support in terms of continuous investment to solidify the mechanisms of internalization, so that producers really benefit from the positive externalities they provide to society. Thus, this chapter provides clear guidelines for decision-makers insofar as it systematizes the knowledge produced on the economic valuation of services in the forest ecosystem of Portugal, giving concrete clues as to the value to be charged or paid for a walk in the forest, the carbon sequestration by the forest as well as the importance of the forest for the development of low-density regions.

ACKNOWLEDGMENT

This research was supported by the Rural Development Program 2020, that financially supported this application registered under PDR No. [2020-2023-045913], Operation 20.2.3, the Portuguese Foundation for Science and Technology (FCT), Ministry of Science, Technology, and Higher Education (MCTES), European Social Fund (FSE) through NORTE 2020 (North Regional Operational Program 2014/2020) and European Union (EU) through the Grant: 2020.07852.BD [grant number 2020.07852] to José Rosário.

REFERENCES

Acharya, R., Maraseni, T., & Cockfield, G. (2019). Global trend of forest ecosystem services valuation – An analysis of publications. In *Ecosystem Services* (Vol. 39). Elsevier B.V. doi:10.1016/j.ecoser.2019.100979

Acharya, R. P., Maraseni, T., & Cockfield, G. (2019). Global trend of forest ecosystem services valuation – An analysis of publications. In Ecosystem Services (Vol. 39). Elsevier B.V. doi:10.1016/j.ecoser.2019.100979

Acharya, R. P., Maraseni, T. N., & Cockfield, G. (2019). Local users and other stakeholders' perceptions of the identification and prioritization of ecosystem services in fragile mountains: A case study of Chure region of Nepal. *Forests*, *10*(5), 421. Advance online publication. doi:10.3390/f10050421

Alamgir, M., Pert, P. L., & Turton, S. M. (2014). A review of ecosystem services research in Australia reveals a gap in integrating climate change and impacts on ecosystem services. In International Journal of Biodiversity Science, Ecosystem Services and Management (Vol. 10, Issue 2, pp. 112–127). Taylor and Francis Ltd. doi:10.1080/21513732.2014.919961

Aslaksen, I., Nybø, S., Framstad, E., Garnåsjordet, P. A., & Skarpaas, O. (2015). Biodiversity and ecosystem services: The Nature Index for Norway. *Ecosystem Services*, *12*, 108–116. doi:10.1016/j.ecoser.2014.11.002

Baral, S., Basnyat, B., Khanal, R., & Gauli, K. (2016). A Total Economic Valuation of Wetland Ecosystem Services: An Evidence from Jagadishpur Ramsar Site, Nepal. *TheScientificWorldJournal*, *2016*, 1–9. Advance online publication. doi:10.1155/2016/2605609 PMID:27830175

Bartczak, A., Lindhjem, H., Navrud, S., Zandersen, M., & Zylicz, T. (2008). Valuing forest recreation on the national level in a transition economy: The case of Poland. *Forest Policy and Economics*, *10*(7–8), 467–472. doi:10.1016/j.forpol.2008.04.002

Barth, N. C., & Döll, P. (2016). Assessing the ecosystem service flood protection of a riparian forest by applying a cascade approach. *Ecosystem Services*, *21*, 39–52. doi:10.1016/j.ecoser.2016.07.012

Bastian, O., Syrbe, R. U., Slavik, J., Moravec, J., Louda, J., Kochan, B., Kochan, N., Stutzriemer, S., & Berens, A. (2017). Ecosystem services of characteristic biotope types in the Ore Mountains (Germany/Czech Republic). *The International Journal of Biodiversity Science, Ecosystem Services & Management*, *13*(1), 51–71. doi:10.1080/21513732.2016.1248865

Bernués, A., Rodríguez-Ortega, T., Ripoll-Bosch, R., & Alfnes, F. (2014). Socio-cultural and economic valuation of ecosystem services provided by Mediterranean mountain agroecosystems. *PLoS One*, *9*(7), e102479. Advance online publication. doi:10.1371/journal.pone.0102479 PMID:25036276

Bertram, C., & Larondelle, N. (2017). Going to the Woods Is Going Home: Recreational Benefits of a Larger Urban Forest Site — A Travel Cost Analysis for Berlin, Germany. *Ecological Economics*, *132*, 255–263. doi:10.1016/j.ecolecon.2016.10.017

Bingham, G., Clark, E., Cooper, W., Costanza, R., Hale, T., Hayden, G., Kellert, S., Norgaard, R., Norton, B., Payne, J., Russell, C., & Suter, G. (1995). *ECOLOGICAL ECONOMICS Analysis Issues in ecosystem valuation: improving information for decision making*. Academic Press.

Botelho, A., Lourenço-Gomes, L., Pinto, L., Sousa, S., & Valente, M. (2016). Using stated preference methods to assess environmental impacts of forest biomass power plants in Portugal. *Environment, Development and Sustainability*, *18*(5), 1323–1337. doi:10.100710668-016-9795-6

CELPA. (2018). *Boletim estatístico indústria Papeleira Portuguesa*. Portuguese Paper Industry Statistical Bulletin.

Christie, M., Hanley, N., & Hynes, S. (2007). Valuing enhancements to forest recreation using choice experiment and contingent behaviour methods. *Journal of Forest Economics*, *13*(2–3), 75–102. doi:10.1016/j.jfe.2007.02.005

Corona, P., Quatrini, V., Schirru, M., Dettori, S., & Puletti, N. (2018). Towards the economic valuation of ecosystem production from cork oak forests in sardinia (Italy). *IForest (Viterbo)*, *11*(5), 660–667. doi:10.3832/ifor2558-011

Costanza, R., de Groot, R., Farberk, S., Grasso, M., Hannon, B., Limburg, K., & Naeem, S. O, R. v, Paruelo, J., Raskin, R. G., Suttonkk, P., & van den Belt, M. (1997). The value of the world's ecosystem services and natural capital. Nature, 387.

Costanza, R., de Groot, R., Sutton, P., van der Ploeg, S., Anderson, S. J., Kubiszewski, I., Farber, S., & Turner, R. K. (2014). Changes in the global value of ecosystem services. *Global Environmental Change*, *26*(1), 152–158. doi:10.1016/j.gloenvcha.2014.04.002

Croitoru, L. (2007). Valuing the non-timber forest products in the Mediterranean region. *Ecological Economics*, *63*(4), 768–775. doi:10.1016/j.ecolecon.2007.01.014

Czajkowski, M., Buszko-Briggs, M., & Hanley, N. (2009). Valuing changes in forest biodiversity. *Ecological Economics*, *68*(12), 2910–2917. doi:10.1016/j.ecolecon.2009.06.016

D'Amato, D., Rekola, M., Li, N., & Toppinen, A. (2016). Monetary valuation of forest ecosystem services in China: A literature review and identification of future research needs. In *Ecological Economics* (Vol. 121, pp. 75–84). Elsevier. doi:10.1016/j.ecolecon.2015.11.009

Einecker, R., & Kirby, A. (2020). Climate Change: A Bibliometric Study of Adaptation, Mitigation and Resilience. *Sustainability*, *12*(17), 6935. doi:10.3390u12176935

Engelman, M., Lagerkvist, C.-J., & Gren, I.-M. (2018). Hunters' trade-off in valuation of different game animals in Sweden. *Forest Policy and Economics*, *92*, 73–81. doi:10.1016/j.forpol.2018.04.004

Ezebilo, E. E. (2016). Economic value of a non-market ecosystem service: An application of the travel cost method to nature recreation in Sweden. *The International Journal of Biodiversity Science, Ecosystem Services & Management, 12*(4), 314–327. doi:10.1080/21513732.2016.1202322

Fisher, B., Turner, R. K., & Morling, P. (2009). Defining and classifying ecosystem services for decision making. *Ecological Economics, 68*(3), 643–653. doi:10.1016/j.ecolecon.2008.09.014

Förster, J., Schmidt, S., Bartkowski, B., Lienhoop, N., Albert, C., & Wittmer, H. (2019). Incorporating environmental costs of ecosystem service loss in political decision making: A synthesis of monetary values for Germany. *PLoS One, 14*(2), e0211419. Advance online publication. doi:10.1371/journal.pone.0211419 PMID:30759137

Garcia, S., Abildtrup, J., & Stenger, A. (2018). How does economic research contribute to the management of forest ecosystem services? In Annals of Forest Science (Vol. 75, Issue 2). Springer-Verlag France. doi:10.100713595-018-0733-7

Getzner, M., Meyerhoff, J., & Schläpfer, F. (2018). Willingness to pay for nature conservation policies in state-owned forests: An austrian case study. *Forests, 9*(9), 537. Advance online publication. doi:10.3390/f9090537

Grilli, G., Paletto, A., & de Meo, I. (2014). Economic valuation of forest recreation in an alpine valley. *Baltic Forestry, 20*(1), 167–175.

Häyhä, T., Franzese, P. P., Paletto, A., & Fath, B. D. (2015). Assessing, valuing, and mapping ecosystem services in Alpine forests. *Ecosystem Services, 14*, 12–23. doi:10.1016/j.ecoser.2015.03.001

Häyhä, T., Franzese, P. P., Paletto, A., & Fath, B. D. (2015). Assessing, valuing, and mapping ecosystem services in Alpine forests. *Ecosystem Services, 14*, 12–23. doi:10.1016/j.ecoser.2015.03.001

Hein, L. (2011). Economic Benefits Generated by Protected Areas: The Case of the Hoge Veluwe Forest, the Netherlands. *Ecology and Society, 16*(2), art13. doi:10.5751/ES-04119-160213

Hlaváčková, P., & Šafařík, D. (2016). Quantification of the utility value of the recreational function of forests from the aspect of valuation practice. *Journal of Forest Science, 62*(8), 345–356. doi:10.17221/50/2016-JFS

Hougner, C., Colding, J., & Söderqvist, T. (2005). *Economic valuation of a seed dispersal service in the Stockholm National Urban Park.*, doi:10.1016/j.ecolecon.2005

ICNF, Instituto da Conservação da Natureza e das Florestas. (2006). *Estratégia nacional para as florestas* [National strategy for forests]. Author.

ICNF, Instituto da Conservação da Natureza e das Florestas. (2019a). *IFN 6*. Author.

ICNF, Instituto da Conservação da Natureza e das Florestas. (2019b). *Programa Regional de Ordenamento Florestal, PROF da Região de Trás-os-Montes e Alto Douro. Documento Estratégico. CAP. A B e C* [Regional Program for Forest Management, PROF of the Trás-os-Montes and Alto Douro Region. STRATEGIC DOCUMENT. Chap. A B and C]. Author.

INE. (2019). *Instituto Nacional de Estatísticas* [National Statistical Institute]. Estatísticas Agrícolas.

Ingaramo, R., & Salizzoni, E. (2017). Valuing Forest Ecosystem Services for Spatial and Landscape Planning and Design. *Journal Valori e Valutazioni*, *19*, 65–78.

Khalfaoui, M., Daly-Hassen, H., Stiti, B., & Jebari, S. (2020). Toward decision-making support: Valuation and mapping of new management scenarios for tunisian cork oak forests. *Forests*, *11*(2), 197. Advance online publication. doi:10.3390/f11020197

Kibria, A. S. M. G., Behie, A., Costanza, R., Groves, C., & Farrell, T. (2017). The value of ecosystem services obtained from the protected forest of Cambodia: The case of Veun Sai-Siem Pang National Park. *Ecosystem Services*, *26*, 27–36. doi:10.1016/j.ecoser.2017.05.008

Kountouris, Y., & Remoundou, K. (2011). *Valuing the Welfare Cost of Forest Fires: a Life Satisfaction Approach*. Academic Press.

Lankia, T., Neuvonen, M., Pouta, E., Sievänen, T., & Torvelainen, J. (2020). Outdoor recreation in ecosystem service accounting: Pilot accounts from Finland. *Scandinavian Journal of Forest Research*, *35*(3–4), 186–197. doi:10.1080/02827581.2020.1760342

Lienhoop, N., & Brouwer, R. (2015). Agri-environmental policy valuation: Farmers' contract design preferences for afforestation schemes. *Land Use Policy*, *42*, 568–577. doi:10.1016/j.landusepol.2014.09.017

Lopes, L. F. G., dos Santos Bento, J. M. R., Arede Correia Cristovão, A. F., & Baptista, F. O. (2015). Exploring the effect of land use on ecosystem services: The distributive issues. *Land Use Policy*, *45*, 141–149. doi:10.1016/j.landusepol.2014.12.008

Madureira, L., Nunes, L. C., Borges, J. G., & Falcão, A. O. (2011). Assessing forest management strategies using a contingent valuation approach and advanced visualisation techniques: A Portuguese case study. *Journal of Forest Economics*, *17*(4), 399–414. doi:10.1016/j.jfe.2011.04.001

Marini Govigli, V., Górriz-Mifsud, E., & Varela, E. (2019). Zonal travel cost approaches to assess recreational wild mushroom picking value: Trade-offs between online and onsite data collection strategies. *Forest Policy and Economics*, *102*, 51–65. doi:10.1016/j.forpol.2019.02.003

Marta-Pedroso, C., Laporta, L., Gama, I., & Domingos, T. (2018). Economic valuation and mapping of ecosystem services in the context of protected area management (Natural park of Serra de São Mamede, Portugal). *One Ecosystem*, *3*, e26722. Advance online publication. doi:10.3897/oneeco.3.e26722

Marta-Pedroso, C., Laporta, L., Proença, V., Azevedo, J. C., & Domingos, T. (2014). Changes in the ecosystem services provided by forests and their economic valuation: A review. In Forest Landscapes and Global Change: Challenges for Research and Management (pp. 107–137). Springer. doi:10.1007/978-1-4939-0953-7_5

Masiero, M., Pettenella, D. M., & Secco, L. (2016). From failure to value: Economic valuation for a selected set of products and services from Mediterranean forests. *Forest Systems*, *25*(1), 051. Advance online publication. doi:10.5424/fs/2016251-08160

Matsiori, S., Anagnos, N., & Soutsas, K. (2012). Economic valuation of forest recreation: The case of the University Forest of Pertouli in Greece. *Journal of Food Agriculture and Environment*, *10*(2), 866–870.

Mendes, A. M. S. C. (2005). Portugal. In T. E. V. Towards, M. Merlo, & L. Croitoru (Eds.), *Valuing Mediterranean Forests* (pp. 331–352). CAB International. doi:10.1079/9780851999975.0331

Mendes, A. M. S. C., Madureira, L., Sottomayor, M., Alves, R., & Rosário, J. V. C. (2021). *ECOFOR. PT – Valorização Económica dos Bens e Serviços dos Ecossistemas Florestais de Portugal. Relatório Científico. Operação 20.2.3 – Assistência Técnica RRN – Área 3*. Candidatura Nº PDR 2020-2023-045913. https://www.researchgate.net/publication/358646726_ECOFORPT_-_Valorizacao_Economica_dos_Bens_e_Servicos_dos_Ecossistemas_Florestais_de_Portugal_Relatorio_Cientifico

Merlo, M. L., Croitoru, Merlo, M., Croitoru, L. (Eds.). (2005). Valuing Mediterranean forests_ towards total economic value-CABI. CABI Publishing.

Meya, J. N. (2020). Environmental Inequality and Economic Valuation. *Environmental and Resource Economics*, *76*(2–3), 235–270. doi:10.100710640-020-00423-2

Millennium Ecosystem Assessment. (2005). *Ecosystems and human well-being : synthesis*. Island Press.

Ninan, K. N., & Kontoleon, A. (2016). Valuing forest ecosystem services and disservices - Case study of a protected area in India. *Ecosystem Services*, *20*, 1–14. doi:10.1016/j.ecoser.2016.05.001

Notaro, S., & Paletto, A. (2012). The economic valuation of natural hazards in mountain forests: An approach based on the replacement cost method. *Journal of Forest Economics*, *18*(4), 318–328. doi:10.1016/j.jfe.2012.06.002

Oliveira, F., Pintassilgo, P., Pinto, P., Mendes, I., & Silva, J. A. (2017). Segmenting visitors based on willingness to pay for recreational benefits: The case of Leiria National Forest. *Tourism Economics*, *23*(3), 680–691. doi:10.5367/te.2015.0526

Paletto, A., de Meo, I., Grilli, G., & Nikodinoska, N. (2017). Effects of different thinning systems on the economic value of ecosystem services: A case-study in a black pine peri-urban forest in Central Italy. *Annals of Forest Research*, *60*(2), 313–326. doi:10.15287/afr.2017.799

Paletto, A., Geitner, C., Grilli, G., Hastik, R., Pastorella, F., & Garcìa, L. R. (2015). Mapping the value of ecosystem services: A case study from the Austrian Alps. *Annals of Forest Research*, *58*(1), 157–175. doi:10.15287/afr.2015.335

Popa, B., Borz, S. A., Nita, M. D., Ioras, F., Iordache, E., Borlea, F., Pache, R., & Abrudan, I. V. (2016). Forest ecosystem services valuation in different management scenarios: A case study of the maramures mountains. *Baltic Forestry*, *22*(2), 327–340.

Ratisurakarn, T. (2019). The Valuation of Forest Ecological Services: A Meta-Analysis. *Southeast Asian Journal of Economics*, *7*(2), 61–84.

Raum, S. (2018). Reasons for Adoption and Advocacy of the Ecosystem Services Concept in UK Forestry. *Ecological Economics*, *143*, 47–54. doi:10.1016/j.ecolecon.2017.07.001

Riccioli, F., Fratini, R., Fagarazzi, C., Cozzi, M., Viccaro, M., Romano, S., Rocchini, D., Diaz, S. E., & Tattoni, C. (2020). Mapping the recreational value of coppices' management systems in Tuscany. *Sustainability (Switzerland)*, *12*(19), 1–18. doi:10.3390u12198039

Sil, Â., Fernandes, P. M., Rodrigues, A. P., Alonso, J. M., Honrado, J. P., Perera, A., & Azevedo, J. C. (2019). Farmland abandonment decreases the fire regulation capacity and the fire protection ecosystem service in mountain landscapes. *Ecosystem Services*, *36*, 100908. Advance online publication. doi:10.1016/j.ecoser.2019.100908

Sil, Â., Fonseca, F., Gonçalves, J., Honrado, J., Marta-Pedroso, C., Alonso, J., Ramos, M., & Azevedo, J. C. (2017). Analysing carbon sequestration and storage dynamics in a changing mountain landscape in Portugal: Insights for management and planning. *The International Journal of Biodiversity Science, Ecosystem Services & Management*, *13*(2), 82–104. doi:10.1080/21513732.2017.1297331

Soares, A. L., Rego, F. C., McPherson, E. G., Simpson, J. R., Peper, P. J., & Xiao, Q. (2011). Benefits and costs of street trees in Lisbon, Portugal. *Urban Forestry & Urban Greening*, *10*(2), 69–78. doi:10.1016/j.ufug.2010.12.001

Soliman, T., Mourits, M. C. M., van der Werf, W., Hengeveld, G. M., Robinet, C., & Lansink, A. G. J. M. O. (2012). Framework for Modelling Economic Impacts of Invasive Species, Applied to Pine Wood Nematode in Europe. *PLoS One*, *7*(9), e45505. Advance online publication. doi:10.1371/journal.pone.0045505 PMID:23029059

Tang, M., Liao, H., Wan, Z., Herrera-Viedma, E., & Rosen, M. A. (2018). Ten years of Sustainability (2009 to 2018): A bibliometric overview. *Sustainability (Switzerland)*, *10*(5), 1655. Advance online publication. doi:10.3390u10051655

Teodoro, A. C., & Duarte, L. (2013). Forest fire risk maps: A GIS open source application - a case study in Norwest of Portugal. *International Journal of Geographical Information Science*, *27*(4), 699–720. doi:10.1080/13658816.2012.721554

Tilman, D., Fargione, J., Wolf, B., Dantonio, C., Dobson, A., Owarth, R., & Schindler, D. (2001). Forecasting Agriculturally Driven Global Environmental Change. *Science*, *292*(5515), 281–284. doi:10.1126cience.1057544 PMID:11303102

Vecchiato, D., & Tempesta, T. (2013). Valuing the benefits of an afforestation project in a peri-urban area with choice experiments. *Forest Policy and Economics*, *26*, 111–120. doi:10.1016/j.forpol.2012.10.001

von Essen, M., do Rosário, I. T., Santos-Reis, M., & Nicholas, K. A. (2019). Valuing and mapping cork and carbon across land use scenarios in a Portuguese montado landscape. *PLoS One*, *14*(3), e0212174. Advance online publication. doi:10.1371/journal.pone.0212174 PMID:30845222

KEY TERMS AND DEFINITIONS

Declared Preference Methods: The declared preference methods apply to situations where the goods or services to be valued do not have a direct relationship with market goods or services. In these cases, these methods consist of placing the consumers of the goods or services to be valued in hypothetical situations in which they would have to exchange expenditure on other goods or services for these. From the respondents' answers, the values of their willingness to pay for the consumption of the good or service to be valued or their willingness to accept compensation to give up this consumption are deducted (Mendes et al., 2021).

Economic Valuation of Ecosystem Services: Process of expressing nature's contribution in monetary value (Farber et al., 2002).

Economic Value: Gain or loss in human well-being, when there is an increase or decrease in service provision (Madureira et al., 2013).

Ecosystem Services: Direct, indirect, and passive benefits that human populations derive from ecosystems.

Internalization: In the case of a negative externality, internalization consists of one or two of the following actions: the agent generating the externality monetarily or otherwise compensates whoever is affected by the externality. This agent takes steps to reduce or eliminate the externality (for example, installs effluent treatment equipment that sends it to a watercourse. In the case of a positive externality, there is internalization when the agent that generates the externality is compensated monetarily or otherwise by those who benefit from the externality (Mendes et al., 2021).

Low Density Regions: Regions displaying low demographic density (population per square kilometer) as well as weak economic density (GDP per capita, economic developments levels), both below national and European levels. Low density regions often combine these characteristics with physical and relational distance to centers of decision-making.

Meta-Regression: Is a value function, only estimated based on determinations of the type of value in question made in several contexts and not in a single context. Therefore, the dependent variable is the valuation of the ecosystem service in question and the explanatory variables include characteristics of the situations in which the value of each service was estimated (e.g., geographic area, protected area or not, plant species present in the forest space, relevant characteristics of the population, etc.) and the valuation method that was used in each situation (Mendes et al., 2021).

Revealed Preference Methods: Draw statistical inferences on values from actual choices people make within markets. Estimation of the values people place on environmental amenities and disamenities proceeds by specifying a theoretical framework and conducting data analyses from purchase decisions (prices paid and quantities purchased) according to this conceptual framework.

Chapter 12
Analysis of the Quality of Vegetation on the Island of Palma (Spain)

José Manuel Naranjo Gómez
https://orcid.org/0000-0001-7998-9154
School of Agricultural Engineering, University of Extremadura, Spain

Jacinto Garrido Velarde
University of Extremadura, Spain

ABSTRACT

Natural resources are of vital importance in any territory. This is the case of the ultra-peripheral and low-density region such as the island of Palma, where the vegetation is enormously rich and varied, and also together with the water plays a transcendental role in the sustainability of the island. For this reason, the quality of the vegetation and its spatial distribution on the island were analysed. The water content in plants and soils was also determined. These objectives could be met using images from the Landsat satellite and the management of this raster information with a geographic information system. From the results obtained according to the spatial distribution, contrasts were identified between areas of the island where there was vegetation with great vigorousness and abundant water content, concerning others with vegetation of worse quality and lower water content.

INTRODUCTION

Starting in the 1990s, there was a notable boost in the field of regulations and the development of environmental management instruments. The territory begins to be understood as that space in which social and economic relations are produced (Romero González, 2005).

More integrated, more holistic, or systemic visions are now addressed. These ideas also reach the rural environment, where agriculture is no longer the queen activity, almost exclusive to rural space and society, (Pisan 1994 cited in Mora Aliseda, J., and Castellano Álvarez, FJ, 2005), and where new

DOI: 10.4018/978-1-6684-4548-8.ch012

policies with a multidisciplinary and multisectoral approach are beginning to make their way. Thus, the readjustment of the agricultural sector is combined with economic diversification, the management of natural resources, the improvement of environmental functions, culture, and leisure activities (Mora and Castellano, 2005). This vision of joint management implies the concept of sustainable development and therefore the need to also protect the landscape in an integrated way (Plieninger, et al., 2006).

On the other hand, citizens also demand a greater degree of information and participation in territorial policies, which makes it necessary to govern more horizontally, that is, closer to the citizen or ultimately more democratic (Romero González, 2005). Therefore, the need for a new political and territorial culture began to emerge, which during the second half of the 1990s culminated in the approval of the European Territorial Strategy, Postdam 1999. Its text very well summarizes the objectives of this new focus on land use planning, and that indirectly affects the landscape as a mirror of it. Thus, in article 3 of the text, it is established that territorial policies must achieve balanced and sustainable development, having as basic objectives for this: economic and social cohesion, conservation and management of natural resources and cultural heritage, and competitiveness. most balanced of the European territory. Objectives that suppose a triangle in the territorial management of society, economy, and environment.

But without a doubt, the great leap in terms of landscape, and its conception as a legal asset that must have a specific regulation, is reached a year later in the European Landscape Convention (CEP), developed in Florence on October 20, 2000. The fundamental objective of this convention is to sensitize the Member States to the idea that the protection and improvement of landscapes must begin by providing them with sufficient legal coverage. Until now, landscape, as we have seen, has been frequently cited in-laws, although in most cases it has remained a legally indeterminate concept. In a Europe where landscapes are beginning to be more valued by society and where at the same time they are suffering from progressive degradation, this Landscape Convention represents an important starting point (Zoido, 2001). First of all, the convention defines in article 1. that landscape shall be understood as "any part of the territory as perceived by the population, whose character is the result of the action and interaction of natural and/or human factors". In this sense, this definition already gives the landscape its character in a much more global vision of it when considering the human-nature interaction. In addition, the agreement clearly defines, among other commitments, the legal recognition of the landscape as a fundamental element of the human environment, the formulation of its own or specific policies that regulate it, as well as other perhaps more innovative commitments such as the participation of the population in the identification and assessment of their landscapes, the teaching, and training of specialists, or the declaration of transboundary landscapes and their joint management (Zoido, 2000).

The territory is the base on which all human activities are based, which in turn shape and alter it based on its technological development or its aptitudes (vocation of the territory for certain agricultural, mining, forestry activities...), economic model, cultural schemes, and demographic densities. It is the object of political intervention to promote, control or reorient the spatial dynamics tending towards an improvement in the living conditions of citizens, including environmental preservation through the compatibility between the economy and the rational use of natural resources (Sheppard, et al., 1989; Montero, et al., 2005; Jaraiz, et al., 2018; Mora, et al., 2018).

From both the continental and the national or regional-local scale, the information obtained from satellite images and the management of this raster information with a Geographic Information System stands as a good study instrument for the evolution of landscapes. at the different levels already indicated above from the supranational (Feranec, et al., 2010; Castanho, et al., 2018) to the local (Díaz-Pacheco and Gutiérrez, 2014; Corbau, et al., 2019), passing by the national (Feranec, et al., 2007; Kucsicsa, et al., 2019) in very diverse aspects such as environmental, agro-livestock, land planning or socioeconomic.

Therefore, in the present study, Landsat satellite images are used to analyze the quality of the vegetation and its spatial distribution on the Island of La Palma, where plant diversity and its water resources play a transcendental role in the development of tourism and as a consequence, revitalization of the economy of the regions.

METHODOLOGY

Initially from the United States Geological Survey (USGS.gov) spatial data infrastructure, the different available images that could be used for analysis on the island of La Palma were observed. After discarding those images that had a large number of clouds, and those that had been obtained long before today, the image to be analysed was selected.

The selected image was obtained on 09-16-2021, just before the eruption of the Cumbre Vieja volcano on 09-19-2021. In addition, this was captured by the Landsat 8 satellite, which has two main sensors. The Operational Land Imager (OLI) produces 9 spectral bands from bands 1 to 9, with a resolution of 15, 30, and 60 meters. Likewise, the Thermal Infrared Sensor (TIRS) has two thermal bands with a spatial resolution of 100. As a consequence, the image used is made up of several bands of different resolutions and wavelengths (Table 1).

Table 1. Operational Land Imager (OLI) that produces 9 spectral bands from band 1 to 9 and the Thermal Infrared Sensor (TIRS)

Bands	Wavelength (micrometres)	Resolution (meters)
Band 1 - Coastal aerosol	0.43-0.45	30
Band 2 - Blue	0.45-0.51	30
Band 3 - Green	0.53-0.59	30
Band 4 - Red	0.64-0.67	30
Band 5 - Near Infrared (NIR)	0.85-0.88	30
Band 6 - SWIR 1	1.57-1.65	30
Band 7 - SWIR 2	2.11-2.29	30
Band 8 - Panchromatic	0.50-0.68	15
Band 9 - Cirrus	1.36-1.38	30
Band 10 - Thermal Infrared (TIRS) 1	10.6-11.19	100
Band 11 - Thermal Infrared (TIRS) 2	11.50-12.51	100

Source: https://www.usgs.gov/faqs/what-are-band-designations-landsat-satellites

Next, from the Download Centre of the National Geographic Information of Spain (CNIG), the administrative limits of each of the islands that make up the Canary archipelago were obtained.

All the above information, both raster, and vector, was managed through a Geographic Information System in QuantumGIS 3.10. In the first place, all the information bands were clipped according to the limit of the island of La Palma. In this way, a greater speed was achieved in obtaining the results obtained, since fewer resources were required in general, and specifically of the internal memory of the computer equipment.

Subsequently, different bands were combined to obtain the true-colour image, the representative image of the island's geology, the infrared image, and the representative image of agriculture. Then, the Normalized Difference Vegetation Index (NDVI) and Green Normalized Difference Vegetation Index (GNDVI) were calculated to analyse the vegetation, and the Normalized Difference Water Index (NDWI) and Moisture Stress Index (MSI) to analyse the moisture content, and water in plants and soil.

Regarding the combination of bands, to obtain natural colour, colour infrared, short-wave infrared, agriculture, and geology images, bands 4, 3, and 2 were combined; 5,4, and 3; 7,6, and 4; 6.5, and 2; 7,6, and 2; respectively.

The true-colour image replicates the display of an image as closely as possible to the appreciation that the human eye can make when looking at the Earth.

However, because chlorophyll reflects near-infrared light, colour infrared images are more useful for analysing vegetation. This appears red, and the healthier it is, the brighter this colour is. Also, dark areas represent water and urban areas are white. Even compared to a real colour image, these images offer the possibility of more easily differentiating the different types of vegetation.

As for short-wave infrared images, they offer results similar to traditional colour infrared photography and are commonly used to monitor soil and drainage patterns. Vegetation is shown in different shades of red. Thus, the lighter shades represent areas with little vegetation or grasslands. Conversely, darker shades of green show healthier vegetation. Also, the ground appears in different shades, but brown. Also, cyan blue is used to show urban areas, while light cyan or even white is used to represent ice, snow, and clouds.

Regarding the agriculture image, it is used for monitoring crops, since the healthier the vegetation, it appears in darker green. Likewise, the vegetation differs from the bare earth, since the latter is represented in different shades of magenta.

The geology image is used to identify various geological formations that may be of interest for analysis due to their geological characteristics. This is the case of faults and volcanoes.

Regarding the determined indices, those directly related to the vegetation, the NDVI, and GNDVI, were initially calculated.

The NDVI is used to quantify the vigour of the vegetation. For this reason, it is useful for understanding the quality and density of vegetation. As a consequence, it is possible to evaluate the changes in the health of the plants. The mathematical formulation is as follows:

$$NDVI = \frac{(Band5 - Band4)}{(Band5 + Band4)}$$

The possible values that can be obtained in the different parts of an analysed image can be between -1.0 and 1.0. Negative values correspond to water and clouds. Values less than 0.1 represent waste rock, sand, or snow. Likewise, moderate values between 0.2 and 0.5 correspond to scarce vegetation, such as grasslands, senescent crops, or shrubs. Higher values between 0.6 and 0.9 represent dense vegetation typical of temperate and tropical forests, or crops but in their maximum growth stage.

The GNDVI index supposes a variation of the NDVI so that the values obtained are more sensitive to the variation of the chlorophyll content of the vegetation. This index is used to estimate photosynthetic activity, and to determine the nitrogen and water consumption of the plant cover. It was introduced by Anatoly Gitelson, Yoran Kaufman, and Mark Merzlyak in 1996. The mathematical formulation is as follows:

$$GNDVI = \frac{(Band5 - Band3)}{(Band5 + Band3)}$$

The values obtained by this index range between -1.0 and 1.0. Thus, values less than 0 represent water or bare soil.

Subsequently, the NDWI and MSI indices, related to plant and soil moisture, were calculated.

The NDWI shows the moisture content in the plants and the soil, being able to determine the water stress of the vegetation. The main advantage over NDVI is that it eliminates variations induced by the internal structure of plant tissues and their dry matter content, improving the precision in determining the water content of vegetation. This index was introduced by Gao in 1996 using the following equation:

$$NDWI = \frac{(Band3 - Band5)}{(Band3 + Band5)}$$

The values obtained will oscillate in a range between -1.0 and 1.0. Thus, values greater than 0, the closer they are to 1, show higher water content.

The MSI is used to perform biophysical modelling, analyse canopy stress, and predict productivity. This index was introduced by Pauline Welikhe, Joseph Essamuah–Quansah, Souleymane Fall, and Wendell McElhenney in 2017. The mathematical formulation is:

$$MSI = \frac{Band6}{Band5}$$

The lower values obtained represent a higher moisture content of the soil, and as a consequence, lower water stress of the plants. In this regard, the representation of the vegetation usually occupies a range of values between 0 and 3. The range of values between 0.2 and 2.0 corresponds to green vegetation.

RESULTS

From the methodology used previously, the corresponding images were obtained for the real colour image (Figure 1), the infrared image (Figure 2), the short-wave infrared image (Figure 3), and the descriptive images of agriculture. (Figure 4) and geology (Figure 5) of the island of La Palma.

Figure 1. A real colour image of the island of La Palma
Source: Own elaboration from the Landsat 8 satellite image

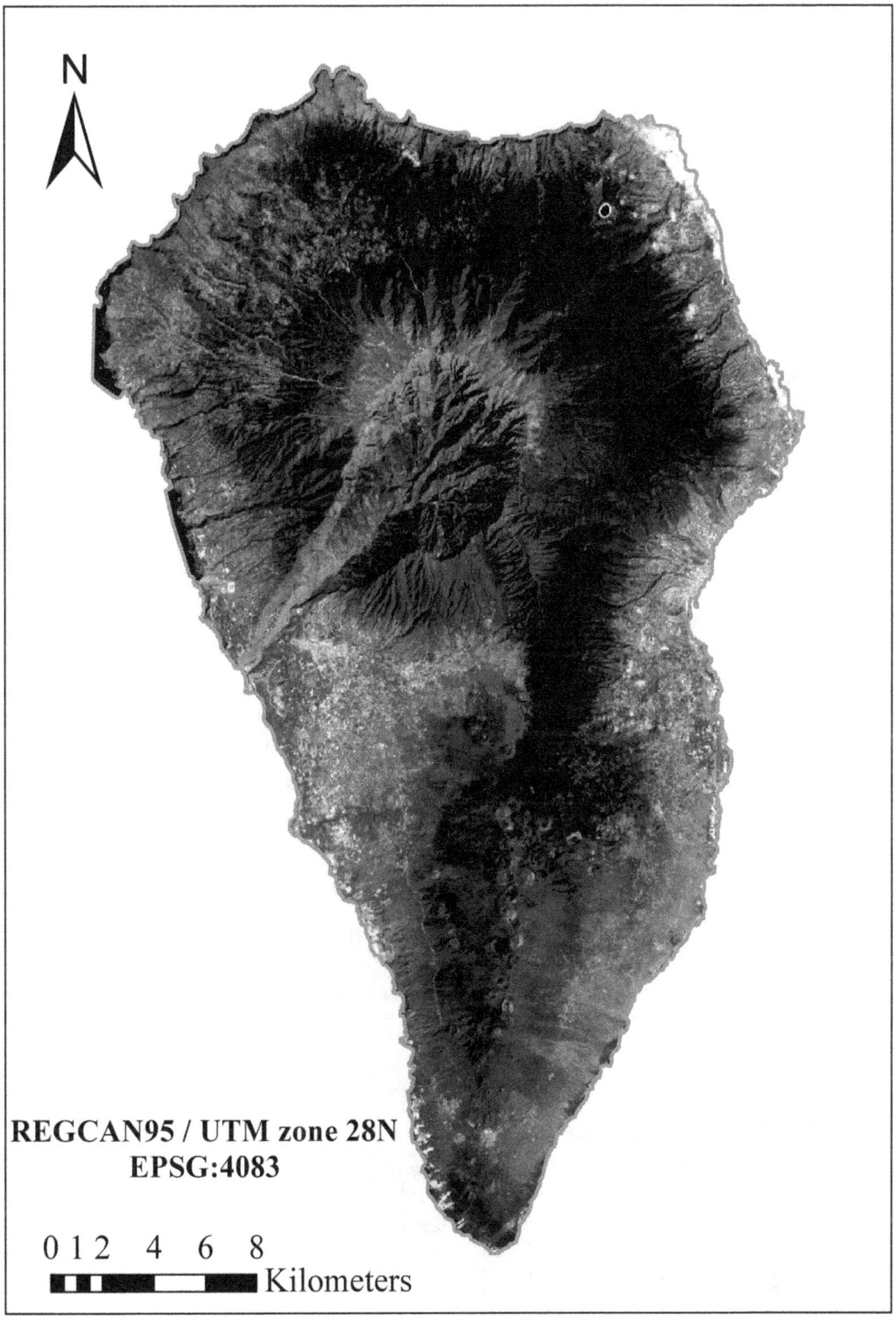

Figure 2. A colour infrared image of the island of La Palma
Source: Own elaboration from the Landsat 8 satellite image

Figure 3. A short-wave infrared image of the island of La Palma
Source: Own elaboration from the Landsat 8 satellite image

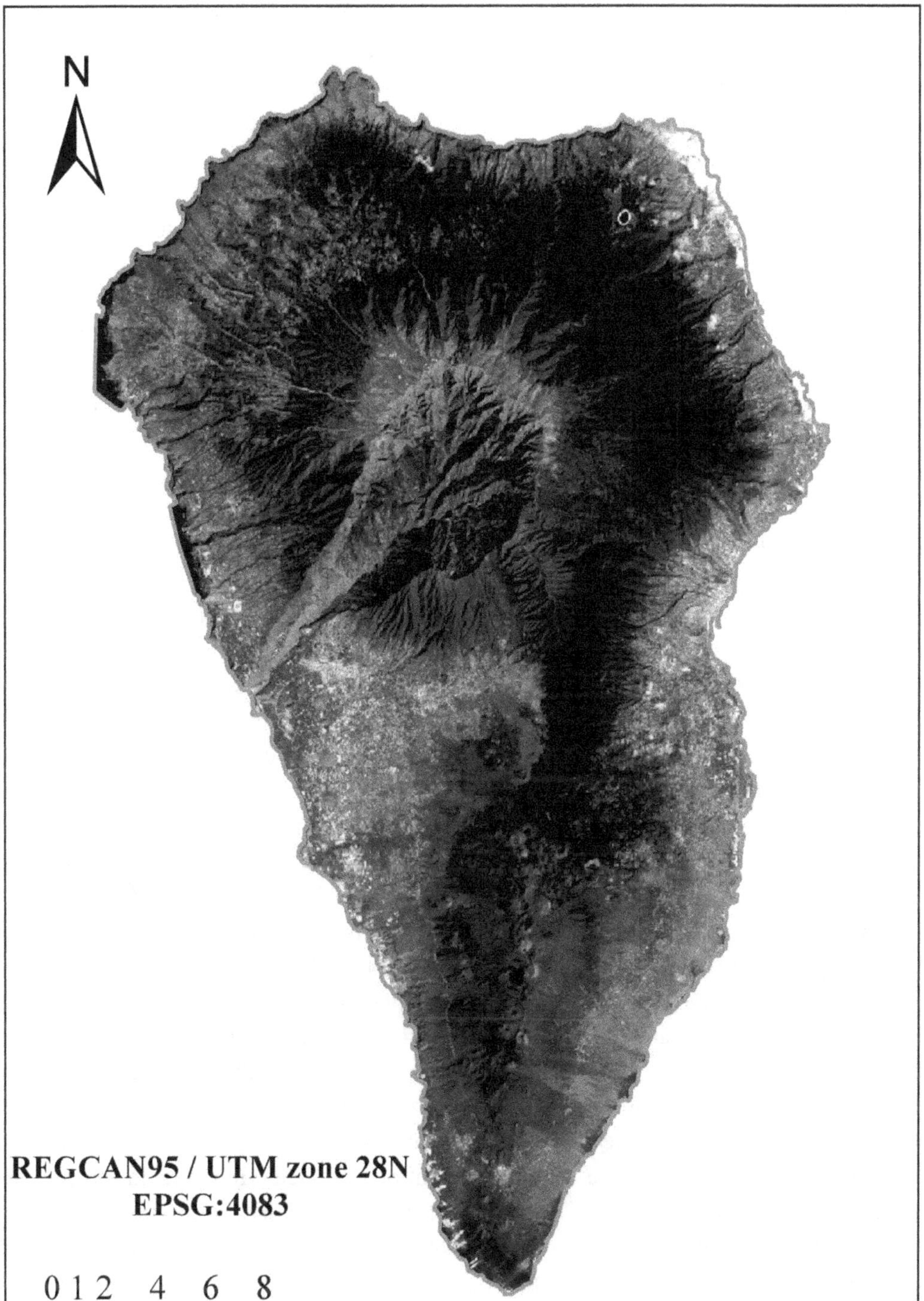

Figure 1 shows that there is an abrupt orography in the north and central parts of the island. Subsequently, the unevenness gradually descends until it becomes practically non-existent in certain regions of the island when it meets the coastline. Likewise, regarding the vegetation, the contrast of the intense dark green colour of the area that is between the steepest area of the island and the coastal regions can be observed enormously. It is precisely in these coastal areas that urban settlements can be seen.

Although the image obtained for real colour (Figure 1) is of great help, for vegetation analysis the use of the image obtained for infrared colour (Figure 2) is unquestionable. Because it helps to identify without any doubt where the greatest amount of vegetation is found. Simply by looking at Figure 2 it can be seen that the most intense and bright red colour is found in the same area previously identified in Figure 1. But now it is possible to discern where the vegetation is healthier. In this case, it is located in the north-western and central part of the island in the strip located between the steepest area and the coastline. Likewise, it is possible to notice the existence of vegetation in the western and eastern half and other areas along the coastline. Due to the existence of vegetation of agricultural origin.

As for the short-wave infrared image (Figure 3), it shows that there are areas with little or no vegetation in the steepest area of the island, and in some areas of the coastline. The healthiest vegetation is again found in the strip between the steepest area and the coastline. However, from this image (Figure 3) it can be seen that in the north-central area there are lighter shades of green, indicating the existence of lower quality vegetation. Urban areas located dispersedly in some areas of the island and more agglutinated in the areas of the coastlines can also be seen in cyan.

Regarding the analysis that allows us to carry out the image obtained for agriculture (Figure 4), we can observe that the healthiest vegetation with a darker green colour appears in the area that surrounds the highest areas of the island. However, this vegetation does not correspond to agricultural land, but native vegetation. However, in coastal areas, healthy vegetation for agricultural use does appear. Although it is true that in the southern part of the island in the eastern part there is no agricultural vegetation and there is in the western part. Precisely going up along this coastline, it is possible to observe more and more healthy agricultural vegetation in the central area, which subsequently decreases until it reaches the north. As for the eastern flank of the island, the behaviour is similar. Although, the beginning of the most intense agricultural vegetation appears further north than on the eastern flank. Likewise, bare vegetation represented by shades of magenta appears in different parts of the island, although predominantly in the south and surrounding the steepest part of the island.

The eruption of the Cumbre Vieja volcano on September 19, 2021, makes the image (Figure 5) of interest, which allows for analysing the geology of the island. Interestingly, it can be seen that the territory is steeped in the north. However, the volcano is located further south. Although, it is also true in an area with also steep orography. In another order of things, it can be seen how from the areas with higher altitudes to areas of lower altitude, the island seems to have been sculpted as if an intrusion of the earth had emerged from the sea, forming the island, as might be expected. However, in the central-western part of the Cumbre Vieja volcano by the year 2021, there were areas of crops and houses (Figures 3 and 4), since the slope was descending gently. This area has been modified since the eruption of the volcano as it has been the natural path of lava descent.

In addition to the previously obtained images, representative images of different indices were also obtained, such as NDVI (Figure 6), GNDVI (Figure 7), NDWI (Figure 8), and MSI (Figure 9).

Figure 4. An agriculture image of the island of La Palma
Source: Own elaboration from the Landsat 8 satellite image

Figure 5. A geology image of the island of La Palma
Source: Own elaboration from the Landsat 8 satellite image

The image for NDVI (Figure 6) shows values always above 0.1. A large part of the image is occupied by the highest values, that is, between 0.2 and 0.5 corresponding to vegetation with very good vigour. Therefore, there is good quality vegetation on the island. Although there are indeed other higher values and these are less present, they also exist on the island. Lastly, the values with less vigour seem to extend over the steepest part of the island and along the line parallel to the coastline located in the westernmost part.

Figure 7 shows the GNDVI showing negative values for areas of water or bare soil in the southern part of the island, in some coastal areas and in the steepest part of the island, and even in the central and southern part of the island in the dividing topographic line. As for the highest values corresponding to high-quality vegetation, they are found again in the zone that is intermediate between the northern zone, the coastal zone, and the steepest part of the island. Also, going down the central and steep part of the island to the south, the area that is between the central steep area and the coastline.

Regarding the moisture content of the plants and the soil shown in Figure 8, the values above 0 clearly show where there is a greater amount of water on the island. The southern area stands out again and the areas are located on the western coastline to the north of the island and central areas in the steepest part of the island.

Regarding the stress in the canopy and the possible productivity of the vegetation, Figure 9 shows the MSI. Thus, it is possible to determine that the lowest values that represent a higher soil moisture content corresponding to the areas located in the previously mentioned strip where there was higher quality vegetation located in the intermediate strip between the steepest area of the island and the coastline located in the easternmost part of the island. However, it stands out with magenta colour in Figure 9 as the maximum value obtained for punctual values located dispersed on the island, corresponding to the areas with the highest humidity. On the contrary, the greater predominance of values corresponds to intermediate values, which would show vegetation with little intermediate water stress in most of the island.

CONCLUSION

The different images obtained are shown as an effective methodology to analyse the vegetation on the island of La Palma, using images obtained using a satellite just before the eruption of the Cumbre Vieja volcano, which has irreversibly changed the existing vegetation affected both by the eruption zone and by the area occupied by the lava.

The disposition of the orography has determined the place where the different urban settlements have been produced. In this regard, the settlements are scattered in the interior of the island, and on the contrary, they are grouped in areas located on the coastlines. As for the amount of vegetation, the largest amount of it is concentrated in a strip delimited by the steepest part of the island and the coastline, located between these two areas. Regarding the quality of the vegetation, it is generally of medium or very good quality, since there is no great hydric stress caused by the plants. Because there is an amount of water considered as average that can nourish this vegetation. Likewise, the vegetation intended for agricultural use is concentrated mostly in coastal areas, and also has good vigour, which allows us to affirm that its productivity, provided that weather conditions and geological phenomena allow it, could also be optimal.

Figure 6. NDVI image of the island of La Palma
Source: Own elaboration from the Landsat 8 satellite image

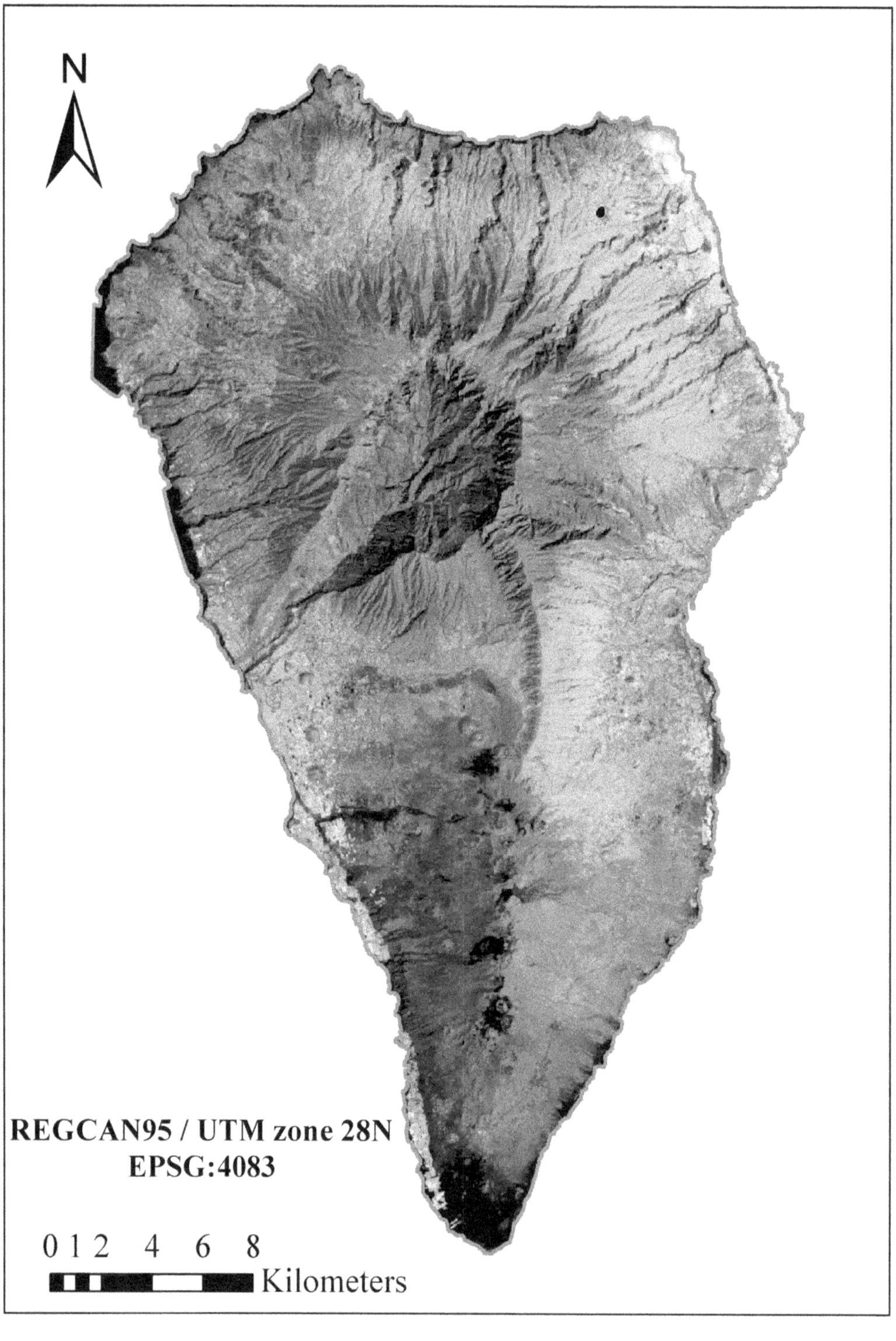

Figure 7. GNDVI image of the island of La Palma
Source: Own elaboration from the Landsat 8 satellite image

Figure 8. NDWI image of the island of La Palma
Source: Own elaboration from the Landsat 8 satellite image

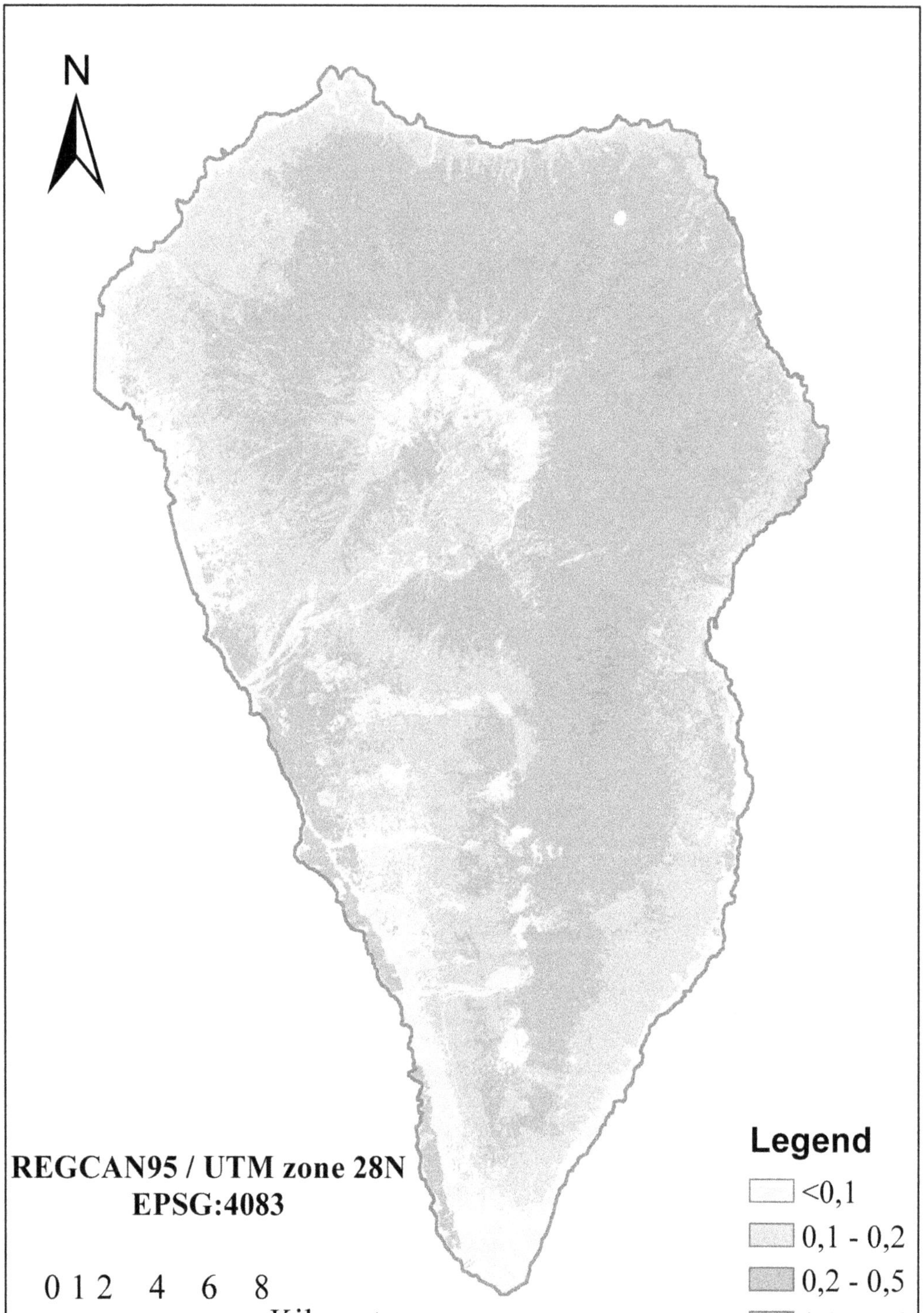

Figure 9. MSI image of the island of La Palma
Source: Own elaboration from the Landsat 8 satellite image

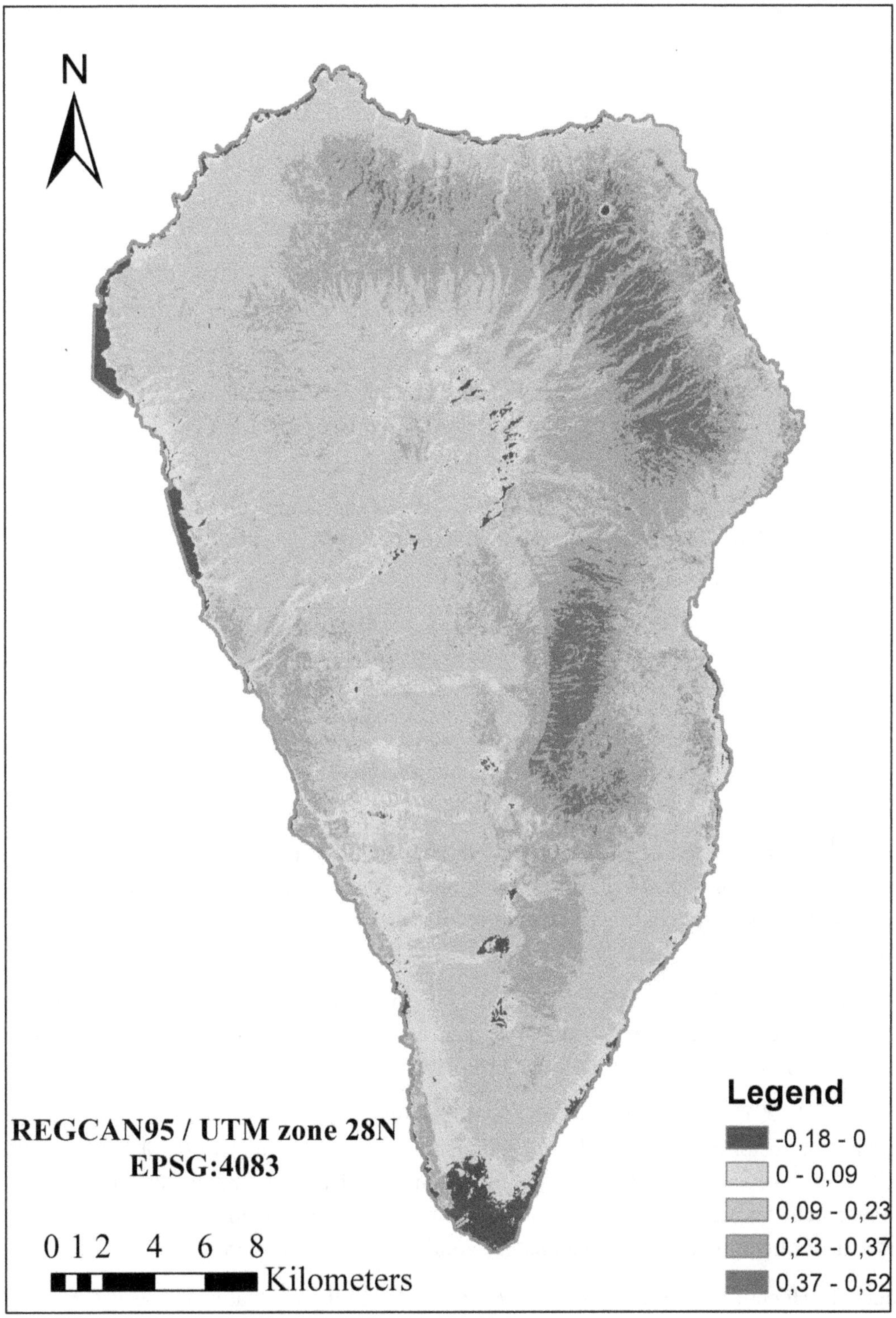

REFERENCES

Castanho, R. A., Lousada, S., Naranjo Gómez, J. M., Escórcio, P., Cabezas, J., Fernández-Pozo, L., & Loures, L. (2018). Dynamics of the Land Use Changes and the Associated Barriers and Opportunities for Sustainable Development on Peripheral and Insular Territories: The Madeira Island (Portugal). In Land Use: Assessing the Past, Envisioning the Future. doi:10.5772/intechopen.80827

Corbau, C., Zambello, E., Rodella, I., Utizi, K., Nardin, W., & Simeoni, U. (2019). Quantifying the impacts of human activities on the evolution of Po delta territory during the last 120 years. *Journal of Environmental Management*, *232*, 702–712. doi:10.1016/j.jenvman.2018.11.096 PMID:30529412

Diaz-Pacheco, J., & Gutiérrez, J. (2014). Exploring the Limitations of CORINE Land Cover for Monitoring Urban Land-Use Dynamics in Metropolitan Areas. *Journal of Land Use Science*, *9*(3), 243–259. doi:10.1080/1747423X.2012.761736

Feranec, J., Hazeu, G., Christensen, S., & Jaffrain, G. (2007). Corine land cover change detection in Europe (case studies of the Netherlands and Slovakia). *Land Use Policy*, *24*(1), 234–247. doi:10.1016/j.landusepol.2006.02.002

Feranec, J., Jaffrain, G., Soukup, T., & Hazeu, G. (2010). Determining changes and flows in European landscapes 1990–2000 using CORINE land cover data. *Applied Geography (Sevenoaks, England)*, *30*(1), 19–35. doi:10.1016/j.apgeog.2009.07.003

Jaraíz-Cabanillas, F. J., Mora-Aliseda, J., Jeong, J. S., & Garrido-Velarde, J. (2018). Methodological proposal to classify and delineate natural protected áreas. Study case: Region of Extremadura, Spain. *Land Use Policy*, *79*, 310–319. doi:10.1016/j.landusepol.2018.08.034

Kucsicsa, G., Popovici, E. A., Bălteanu, D., Grigorescu, I., Dumitraşcu, M., & Mitrică, B. (2019). Future land use/cover changes in Romania: Regional simulations based on CLUE-S model and CORINE land cover database. *Landscape and Ecological Engineering*, *15*(1), 75–90. doi:10.100711355-018-0362-1

Montero, M. J., López-Casares, S., García, L., & Hernández, J. (2005). Visual Impact on Wetlands: Consequence of Building Sprawls in Rural Areas of the West of Spain. *MODSIM Intnl Cong on Modelling and Simulation; Modelling and Simulation Society of Australia and New Zealand: Melbourne, Australia*, 170–176.

Mora, J., Garrido Velarde, J., & Mora, C. (2018). Gobernanza de los recursos hídricos transfronterizos: Una Propuesta [Governance of transboundary water resources: A Proposal]. *Vertentes do Direito*, *5*(2), 1–15. doi:10.20873/uft.2359-0106.2018.v5n2.p1-15

Mora Aliseda, J., & Castellano Álvarez, F. J. (2005). Evolución y redefinición del concepto de desarrollo rural: de la sectorización hacia el enfoque holístico. Junta de Extremadura. Mérida [Evolution and redefinition of the concept of rural development: from sectorization to the holistic approach. Regional Government of Extremadura]. In Políticas urbanas y territoriales en la Península Ibérica. Tomo II [Urban and territorial policies in the Iberian Peninsula. Volume II]. Academic Press.

Plieninger, T., Höchtl, F., & Spek, T. (2006). Traditional land-use and nature conservation in European rural landscape. *Environmental Science & Policy*, *9*(9), 317–321. doi:10.1016/j.envsci.2006.03.001

Romero González, J. (2005). El Gobierno del Territorio en España. Balance de iniciativas de coordinación y cooperación territorial [The Territorial Government in Spain. Balance of coordination and territorial cooperation initiatives]. *Boletín de la Asociación de Geógrafos Españoles*, (39), 59–86.

Sheppard, S. R. J. (1989). *Visual Simulation: A User's Guide for Architects, Engineers and Planners*. Van Nostrand Reinhold.

Zoido, F. (2000). Proteger y realzar el paisaje [Protect and enhance the landscape]. *Boletín de la Asociación de Geógrafos Profesionales de Andalucía*, (7), 7–15.

Zoido, F. (2001). La Convención Europea del Paisaje y su aplicación en España [The European Landscape Convention and its application in Spain]. *Ciudad y Territorio, Estudios Territoriales, 33*(128), 275-281.

Chapter 13

Waste Management and the COVID-19 Pandemic in Developed and Developing Countries: A Mexican Case Study

Sharon Reyes Gonzalez
WSB University, Poland

ABSTRACT

Most countries have considered waste management as a high priority since it is a problem of environmental pollution and resource depletion. The main goal of implementing environmental policies is to find or create ways to manage waste beneficial for social, environmental, and economic aspects. In the face of the COVID-19 pandemic, the amount of waste generated got out of control, and therefore, the traditional way of handling waste was insufficient. Contextually, this chapter aims to analyze the strategies, regulations, and measures imposed by the governments of Japan, the United Kingdom, Mexico, the United States, and China to compare between them their way of controlling waste and the effects that the pandemic has left regarding the increase in waste and their way of managing them.

INTRODUCTION

Environmental pollution constitutes one of the most critical problems in the world. In this regard, there is a need to raise awareness, searching for alternatives to resolve it. Reyes González & Castanho (2021) comment that government agendas are interested in environmental issues such as proper waste disposal, open and informal dumps, recycling promotion, and climate change.

DOI: 10.4018/978-1-6684-4548-8.ch013

Many of the products that we use today have a short life cycle. The production of goods, merchandise, and services are so high that the companies that manufacture and supply them need to release them to continue producing and earning quickly. So, this is why governments are trying to implement policies and processes to reduce waste and pollution problems.

Separating waste is a conscious and socially responsible way of caring for our environment, as we facilitate the recycling of its components that, otherwise, would end up in a sanitary landfill, polluting the soil, air, and water. Recycling is a good idea because we give the waste we generate a second chance. Avfall Sverige (2020) states that "recycling means that the waste will be used as replacement for another material".

There are other ways to treat waste. For example, reuse, where items are separated, washed, and sorted so they can be used again without further treatment. Thereby, this positively impacts the environment as energy and natural resources are saved by producing new products.

In biological treatment, waste is treated by anaerobic digestion or composting. In energy recovery, all waste that cannot be recovered or recycled in any way goes through a process that can generate electricity or heating, for example.

All waste that should not be recycled or sent to any of the aforementioned processes is sent to the landfill. They are stored for a long time, and neither fuels nor organic waste can be deposited there. The European Union count on a waste hierarchy. According to European Commission (2010) "this sets an order of priority, starting with the preferred option of waste prevention, followed by preparing waste for re-use, recycling and energy recovery, with disposal (such as landfill) as the last resort".

The issue of waste has affected the environment over the years. Nevertheless, the amount of waste generated by the current health emergency (COVID-19 pandemic) is an increasing problem since products with everyday use, such as gloves, masks, cleaning products, and other personal protective equipment, increased dramatically. In addition, the measures imposed, such as social distancing and quarantine, caused an increase in household waste generation and put at risk the correct waste separation.

CLASSIFICATION OF WASTE

Through time, different classifications of waste have been developed by diverse authors. We will focus on the one made by Nuestra Esfera (2014), who divides waste according to its origin, biodegradability, and composition.

According to its Origin

Household waste: It is the garbage generated in households due to the activities that are carried out daily (eating, cleaning, etc.). During the COVID-19 pandemic, this kind of waste increased, and their way of handling it was modified. Households with infected family members are not allowed to use the waste containers to separate organic and recyclable materials, which pushes to mix the waste creating a big problem and letting the use of single-use plastic bags become more prominent.

Municipal waste: Are those that result from the cleaning of the streets in daily life and/or special events (does not include household waste).

Industrial solid waste: These are the materials resulting from industrial or semi-industrial processes (including waste from administrative activities or food).

Hospital waste: These derive from any medical activity and can include infectious-type waste, surgical medical material, sharp items, remains of human tissues, and remains of drugs. Due to their characteristics, they must have special handling when discarded.

Before the pandemic started, the medical waste disposed of from hospitals and the daily domestic garbage (papers, documents, lunch boxes used by patients, etc.) were separated and received a different treatment; nevertheless, during the outbreak, both wastes were treated as hospital one in order to be appropriately handled and prevent the spread of the virus.

Liang, et al., (2021) mention that "an evaluation of waste generation rate data from all over the world shows that medical waste from hospitals is about 0.5 kg per bed per day and that [...] it is estimated that medical waste from healthcare facilities associated with COVID-19 is increasing by 3.4 kg/person/day". Due to the pandemic, this kind of waste increased rapidly at the hospitals and self-protective equipment to prevent the COVID-19 spread.

Construction waste: As it is the result of construction work, it is generated in large volumes. Therefore, its management and disposal are somehow tricky.

According to its Biodegradability

Organic waste: These are wastes that derive from edible products and are biodegradable; they can ferment and go through the decomposition process.

Inorganic waste: Are those wastes that, unlike those previously mentioned, are not biodegradable because they do not contain organic elements and take longer to disintegrate. Among these are cans, bottles, metals, plastics, and other everyday products of industrial origin.

According to its Composition

Papers and Cardboard: Trees play an essential role in the environment. They reduce CO2 emissions which is the most harmful greenhouse gas. The less we consume paper, the fewer trees will be cut down for production, representing other advantages for the environment such as energy and water saving, contribution to the preservation of animal and plant species, improvement of the air quality, and soil contamination reduction.

Glass: The most common things made of glass used on daily basis are bottles, food containers and windows. "Glass is a material that, in many of its forms, is reusable. This is particularly true of bottles and other glass vessels, which are still able to satisfy their role as a container long after their original use is over. This has led to manufacturers and governments recognizing that savings in terms of energy and resources could be achieved through the recovery and reuse or recycling of glass container (Worrell & Reuter, 2014). Unfortunately, not all kind of glass are recyclable at all, for example bulbs, lamps or some ceramic objects.

Scrap and Metal: Includes cans of food recycled often, pipes, copper gotten from the cords, and windows frames. The probability of metal recycling depends on its design and composition. Through proper and conscious design, metals and their compounds can contribute to sustainability and other products can also be recovered or recycled in large quantities, however, the limitations on the rate of recycling can be affected by a complex combination of factors. materials and metals (Worrell & Reuter, 2014).

Paints and Oils: The automotive industry disposes of the majority of these kinds of products, and as they are considered flammable, they should not be disposed of with other garbage.

Plastic: There are many types of plastic products, but bags, toys, and cleaning products containers are the most common ones. In Europe, the handling and recycling of this kind of waste is a widespread practice and has become a priority issue worldwide as it causes several damages to the ecosystem. According to Penga et al., (2021), we verify that "More than eight million tons of pandemic-associated plastic waste have been generated globally, with more than 25,000 tons entering the global ocean". It is essential to highlight that the items disposed of in significant quantities are personal protection equipment and online-shopping package material.

As the outbreak occurred unexpectedly, neither governments nor companies were prepared to face the challenges, the waste treatment capacity was less than the demand for plastic products. Unfortunately, not all kinds of plastics can be processed or recycled, so that they might end up in landfills or the ocean. Penga, et al. (2021) state that "(…) as of August 23, 2021, the total excess of mismanaged plastics waste generated during the pandemic is calculated as 4.4 to 15.1 million tons". This number reflects excessive consumption of plastics since the beginning of the pandemic.

As in all kinds of waste, the treatment can vary depending on the characteristics and nature of the material. For plastic, recycling is the best way of treatment as it is the most environmentally friendly. However, incineration has become the most common nowadays as it is the most effective to mitigate infections during waste handling.

PET Plastic Bottles: This kind of material and containers can be recycled and used to produce textiles and packaging. After the recycling process, they can be converted in "fiberfill, fibers, textiles, strapping, non-food bottles, and industrial paint. With chemical recycling can form into appliance handles, auto components, bath tubes and insulation" according to Copperthite (1989).

HDPE Plastic Bottles: This type of plastic is found mainly in detergent bottles, bleach, and milk containers. Some of the benefits of recycling this kind of plastic are that it reduces costs of production compared to the one of virgin plastics and, after recycled, HDPE can have other applications like: "pipe, toys, trash cans, lumber substitute, flower pots, tubing and non-food bottles" according to Copperthite (1989).

Textiles: Cotton and linen are usually reusable waste. Fabrics impregnated with contaminants such as paint, fuel, etc. are not recyclable. Many companies nowadays are implementing the use of recycled materials to produce clothes or shoes, however, one of the most challenging facts is that "they see the use of recycled fibres as a money saving opportunity but certainly not as a selling point to consumers." according to Watson et al., (2017).

Batteries: They are found in many electrical, mobile, and other appliances that contain materials like copper, aluminum, and lithium. Batteries contain toxic substances that, after use, must be disposed of properly. Each battery is different, and therefore its recycling procedure is different, so it is vital to classify batteries in the appropriate categories: Lead Acid, Nickel-cadmium, Nickel-metal-hydride, Primary Lithium, Lithium-ion0, and Alkaline.

COVID-19 AND ITS EFFECT IN WASTE GENERATION

During the pandemic, the generation of waste increase due two factors. According to Liang, et al., 2021, these wastes are known as:

COVID-19-related waste. Are all waste generated due confirmed cases of people at hospital (mainly) and the medical and healthcare material used for their treatment which includes gloves, masks, gowns and other protective equipment. After used, it should be collected and disposed of with additional precautions and care. Roy, P., et al., (2021) say that, "globally, every month, 129–210 billion disposable masks and 65 billion disposable gloves are used".

Indirect waste. Includes waste from social distancing and other preventive measures imposed by the governments to decrease the number of infected people. The lifestyle changed and it brought as result the increase of single-use plastics (masks, packaging and bags, for instance), also known as disposable plastics which are used one time before being disposed. When the pandemic began, people made panic purchases so demand of products like masks, gloves and food increased. Afterwards, due isolation, people started changing the way of purchases, ordering online products or food to deliver at home reflecting increase in transportation and packaging waste (carton, plastics, etc). Based on Liang et al., (2021) "Compared to the reference period in January and February 2020, the online trading across 20 different industries increased in June, and even though it decreased on September 6, online transactions still increased (16.6%) compared to the index period".

Recycling and Benefits

According to EPA (2020), recycling " (...) is a series of activities that includes collecting used, reused, or unused items that would otherwise be considered waste; sorting and processing the recyclable products into raw materials; and remanufacturing the recycled raw materials into new products". The common belief that recycling is simply reusing a thing or transforming it into another can be misleading. Since some waste products cannot undergo a cleaning process (which could make them a valuable source for other products), people must still seek new ways of recycling to reduce waste rather than increase it.

The process of recycling is an act of high importance for society since it involves the reuse of elements and objects of different types that would otherwise be discarded, contributing to the accumulation of more garbage. It is linked to ecology and the concept of sustainability, which implies that humans must be able to take advantage of nature's resources, but without abusing them and without damaging the environment.

Reyes González & Castanho (2021) mention that, " (...) when recycled or reused materials are used in a new production process, other non-renewable resources can also be saved and the consumption of others can be reduced, such as the amount of energy used in a process with non-recycled materials since this is much higher than the used with recycled/reused resources".

The essential benefits that recycling implementation in the countries reflect are:

Environmental Impact Reduction: The production process of new products made from recycled materials reduces the extraction of the raw ones. Also, less energy is needed for this process than the energy required in production with raw materials, which creates large amounts of valueless by-products or wastes with little/non-economic value and sometimes might contain harmful substances.

Zero Waste: The waste that all manufacturers and producers generate can be used as raw material for a second manufacturer, reducing the waste generated in each production process.

Costs: Economic recycling enables waste to become a resource. Hill (2003) says that: "When products are made from recycled materials it saves energy, air emissions and resources." It reduces the use of energy as certain virgin materials are more intensive than reprocessing of recycled materials.

Growing Demand for Recycled Materials: Nowadays, people are more conscious of the problem that waste represents for the environment and health. Therefore, the demand for recycled products is growing, so the companies increase the demand for recycled materials for their production.

Increase of Number in Waste Handling Activities: Reyes González & Castanho (2021) tell us that " (…) all activities aiming to reduce pollution and pro-environment generate new job offers giving the opportunity to develop a new economic sector".

Innovative Processes and Products: There appears to be a new tendency to make products from recycled material, such as clothes or shoes. Companies are spreading sustainable initiatives and are innovating processes with the reuse of materials to reduce emissions that represent cost reduction in their production or operation. In fact, according to Erdoğdu et al., (2016) "(…) in recent years, companies started to consider greening and sustainability issues in their supplier evaluation and selection processes".

Challenges

According to EPA (2020), " (…) although recycling saves resources and energy, cities often struggle to implement a successful recycling program" due some important factors as listed below:

Quality: Materials that can be recycled must have specific quality standards. Therefore, the sorting and treatment should be done carefully.

Contamination: Recyclable items can be contaminated if they are not correctly separated from the non-recyclable ones and can be considered contaminated, losing the opportunity to be used for future products.

Volatile Markets: The market can change from time to another, bringing price fluctuations up. When it happens, it affects the recycling facilities and their cost of operation, which might end in closure.

High Operating costs: The recycling process and all its activities can represent high costs of handling, transportation, and labor.

Financing for Capital Investments: As per the high amounts of investment that the recycling facilities usually need external financing. UN Environment (2018) states that "investment costs include all costs related to developing and constructing a project, such as preparation, planning, studies, permits, public consultations, designs and land costs, amongst others."

Lack of Processing Facilities: The main issue that leads to a lack of facilities is the required infrastructure. Many places do not have the necessary infrastructure to run this facility.

Lack of Appropriate Technology: The recycling process can be done in multiple ways depending on the kind of a waste to be handled. Some of them require advanced technology to be recycled and should not be processed through an inadequate machine as they can damage them.

Environmental and Health Concerns: Handling the waste since it is collected, sorted, transported, and treated requires natural resources to complete the process. However, at the same time brings some consequences like massive water use, air pollution. Implementing an accurate recycling culture and process, the sanitary condition and health problems can be improved and the raw materials and energy used for production can be saved (UN Environment, 2018).

Incorporation of the Informal Sector: The local governments face the challenge of incorporating the informal sector into recycling as there are entities that have already been in the business for a long time. Nevertheless, the budget they have destined for this purpose is not enough to incorporate people oi the informal sector into the formal one.

RECYCLING AND WASTE MANAGEMENT DURING COVID-19 PANDEMIC IN DEVELOPED AND DEVELOPING COUNTRIES

United Kingdom

The Department for Environment, Food, and Rural Affairs (2021) shows that the total household waste generation in the UK in 2019 was 26.4 million tonnes, from which 46.2% was recycled. The packaging waste generated (plastic, metal, glass, paper wood, and others) reached 11,836 thousand tonnes, and 68.2% was recovered/recycled. As in all countries, the COVID-19 pandemic affected the consumption of products and waste generation negatively. In England, at the end of May 2020, as per Liang et al. (2021), more than 90% of local governments reported an increase in waste collection and generation compared to the usually generated, and 58% mentioned that the increase was around 0%-20% more than expected.

Generally, the government has cared about waste problematic. They count on environmental policies or laws to regulate waste disposal and ensure that the process it receives is the most convenient for the citizens and the environment. As per the current situation, the UK has increased the capacity of its facilities to manage the waste generated in a pandemic.

As per Law and your Environment (2017), below we can find a summary of most important laws created to regulate the waste management in UK:

- Environmental Protection Act 1990 (EPA): Establishes the responsibilities of the authorities involved in pollution control of land, air and water and the rules to prevent the waste generation.
- Hazardous Waste (England and Wales) Regulations 2005: Basically, specifies how to get rid of the hazardous waste in a proper way and the requirements that the movement of this kind of waste should fulfil, for example Law and your Environment (2017) says that "collectors of hazardous waste must keep thorough records and provide the Environment Agency with quarterly returns".
- The Waste (Household Waste Duty of Care) and England & Wales Regulations 2005: This law makes some changes to the Environmental Protection Law of 1990. Its objective is to promote among its citizens the correct disposal of their domestic waste and to ensure that waste is collected by authorized carriers.
- Waste Electrical and Electronic Equipment (WEEE) Regulations 2013: The main goal of this legislation is to decrease the amount of waste that is destinated to landfill, therefore, they implement the extended responsibility in producers and distributors who are responsible for the collection and treatment of their own production. (Legislation.gov.uk, 2013). It also establishes the target levels for electrical and electronic equipment recovery and reuse.
- Producer Responsibility Obligations (Packaging Waste) Regulations 2007: This regulation is aimed at companies that exceed a turnover of more than 2 million pounds sterling and that use more than 50 tons of packaging per year.
- Waste (Wales) Measure 2010: Defines the objectives of local governments in terms of recycling, reuse and composting of waste. It also promotes the extra charge for the use of single-use bags and the reduction of landfill use.
- Waste (England and Wales) Regulations 2011: It imposes the rule on companies to apply the hierarchy of waste when disposing of it. It also establishes the selective collection of paper, metal, plastic and glass waste since 2015 and introduces a system where authorized waste transporters must be registered.

Japan

For Japan waste management and environment is a topic with high importance. One important legislation is "Law for the Promotion of Sorted Collection and Recycling of Containers and Packaging (Container and Packaging Recycling Law)," which was implemented in 1997 to reduce PET and glass bottles, paper cartons, steel, and aluminium cans but in 2000 plastic & paper containers and packages plus cardboard boxes were included.

Japanese recycling campaign is based on a philosophical concept: "The 4R campaign: Reduce, Reuse, Recycle and Respect". Together with it, the Ministry of the Environment (2018) estates that, Japan has implemented the following procedures in order to reach their recycling targets:

1. Businesses cut-off plastic bags so citizens have to bring their own bags to do shopping.
2. Collection and separation of plastic into 4 (Packaging, PET bottle, PS tray, Goods)
3. Four containers for different kind of plastic (Bottle, cap, PS tray, Egg pack)
4. Electronic Money System machines where people take their bottles to recycle and receive money back in cards according to the weight.
5. Use of reusable cups in cafeterias

According to the Ministry of Environment (2021), the largest type of waste generated is household and in 2019 the average waste produced by person every day was 918 grams, number that reflects a decrease from 919 grams in 2018. Also, the statistics show that the total amount of waste generated in 2019 was 42,740,000 tonnes (0.02% more compared to 2018). The landfill waste in 2018 was 3,840,000 tonnes and in 2019 was 3,800,000. The recycling rate decreased from 19.9% in 2018 to 19.6% in 2019. Both were down, nevertheless the recycling amounts were affected by the COVID-19 pandemic due the social distancing and isolation measures established by the government. According to Liang et al., (2021), in Tokyo, " (...) the commercial waste amount decreased by 57% in May, 2020, while the household waste amount increased by 110%".

United States

Although there is no national regulation established concerning recycling processes, the existing Environmental Protection Agency (EPA) (belonging to Resource Conservation and Recovery Act - RCRA) regulates this topic. Inside the US, there are associations such as The National Recycling Coalition (NRC), which consists of around 6000 recycling businesses in the country, and some others that are part of the international recycling industry like Bureau of International Recycling (BIR) and the Institute of Scrap Recycling Industries Inc. (ISRI). According to EPA (2020), in 2018 in US, over 292 million tons of MSW (Municipal Solid Waste) were generated and the treatment that they received is presented in Table 1:

Nevertheless, as per Liang et al. (2021), the pandemic also affected the demand in the United Stated for consumer-packaged goods growing by 9.5% as result of panic purchases and, it is estimated that waste generation during COVID-19 will increase to 2.5 million tons/month compared to the 5 million tons/year leading a bigger problematic as there were imposed restrictions to recycle the waste in order to avoid the infection spread.

Table 1. Management of MSW in the United States in 2018 (292 million tons)

Landfilling	50%
Recycling	23.60%
Composting	8.50%
Other Food Management Pathways	6.10%
Combustion with Energy Recovery	11.80%

China

China is an industrial city where many products are created every day. Due to the pandemic, the industrial facilities decreased or stopped their production, letting the emissions they generate be reduced positively. However, at the same time, it impacted them economically in a negative way.

The effects of the COVID-19 pandemic were visible immediately. Within the positive ones, we can mention that emissions dropped down due to the social distancing rules imposed by the government. Tourism was affected as borders closed, so the waste that tourists typically generate at the beaches or tourist places decreased.

For Liang et al., (2021), "MSW generation in Macao, a typical tourist city in China, decreased 17%–25% between February and May". Also, the use of cars decreased which helped the air to be cleaned up and the noise to reduced.

Regarding waste, for Roy et al., (2021), "(…) in Wuhan, the generation of medical waste increased from 40 to 240 tonne/day due to this pandemic, surpassing the city's maximum incineration capacity (49 tonne/day)".

Mexico

Mexico is a developing country; therefore, the waste management topic is different from the processes in the developed ones. Although in Mexico there are no legislations created for waste management, there are some laws that regulate from a general perspective the environment care from hazardous materials:

- The Ley Federal de Derechos Act of 1981
- Ley General del Equilibrio Ecológico y la Protección al Ambiente Act of 1988
- General para la Prevención y Gestión Integral de los Residuos Act of 2003

According to Secretaría de Medio Ambiente y Recursos Naturales (2020), 102,895 tons of waste are normally generated every day in Mexico. There are 1,060 Collection Centers for recyclable materials in 21 states of the country where there are received around 38,431 kg of waste every day and then, the recyclable materials are transported to separation plants. The total recovered quantity in 2017 was 277.835 tonnes a day.

The numbers above are measured according to the registered establishments, however, mostly of the country is handled by informal sector and in both cases, the waste that cannot be recycled or reused, is sent to the dumpsites or transfer facilities, which, as per EPA (2020) " (…) can include small, highly decentralized and un-mechanized facilities such as empty lots that serve as temporary disposal lots, where residents and commercial establishments may dispose of their waste or where primary collectors (e.g. collectors using handcarts and bicycles) deposit waste they have collected".

The quantity of waste generated is enormous daily. However, Martínez Arroyo et al. (2020) tell us that the total generation of waste due to the pandemic is estimated between 81,214 t/day - 92,338 t/day (from 3.3-16.5% additional to that generated under standard conditions) of medical and solid waste.

Jiménez Martínez and Nancy (2021) comments that, before COVID-19 pandemic, there were generated 1.5 kilos/person at the hospitals but in June of 2020, the total amount was 16,000 tonnes and by January 2021 there were generated 29,000 tonnes which represents 300% more than the last previous year without pandemic.

There is an association called ECOCE, a private company that promotes recycling and waste collection. It works together with big corporations to recover the waste of their productions and send those materials to recycling. They handle different materials and implement some programs and campaigns of waste separation in some states in Mexico. They also cooperate with other private entities involved in recycling and waste collection and have monthly meetings to evaluate the collection and international factors that can affect their collection or the prices of the materials they work with. Also, according to Sáenz Guzman (2021), an extensive recycling plant in Mexico city was inaugurated in July 2021, which seems to be the most modern plant in Latinoamerica and can receive 400,000 tonnes of waste every day and to process 1000t. From those 1000 tones, 400 will be organic waste sent to the nearest compost plant, 300 will be converted into fuel for cement kilns, and 60 will be recoverable and recyclable materials.

Nevertheless, the problem in Mexico is more severe as there are no imposed processes to manage the waste disposed of from households. The waste collection can be done by local governments or by informal collectors called "pepenadores." They collect the bags from households and put them into their pushcarts, bicycle carts, or vehicles. Inside the same carts, they separate the waste according to their type (plastic, glass, carton, for instance) and deposit the non-recyclables into a bin (including organic waste). In general, households do not separate their garbage, so most of the disposed waste is dirty or mixed with, for example, organic waste, which makes the separation more complicated.

According to Jiménez Martínez and Nancy (2021): "(...) 10% of the municipalities do not provide public garbage collection service, which means that 16% of the waste are not collected and they stay somewhere of the streets. The majority of the waste is taken to open dumps and only 3.7% of those places comply the required standards".

METHODOLOGY

This research aims to analyze how waste is handled in developed and developing countries and how it was affected during the COVID-19 pandemic. An investigation was made through various articles and books to get information and statistics on how developed countries control waste disposal.

A questionnaire to 106 Mexican citizens was done online during the COVID-19 pandemic. The survey was elaborated and conducted on April 2021 online using Google Forms, considering the distance and the numerous entities covered in the country. People were open to answering the questionnaire. Some participants provided feedback directly on how important the topic is nowadays for them as the pollution generated by waste disposal is a critical problem in Mexico that affects the environment and health.

Also, interviews were conducted with four people who work in waste collection and separation (three Mexican and one polish citizen who worked in the Netherlands) to know their personal experience as waste handling, collectors, or separation workers. No more interviews were conducted; the reason is that people who work in waste collection or as scavengers are not willing to share information. In fact, they can feel marginated or discriminated against when they are asked about their job.

RESULTS

The questions conducted were: Do you separate garbage at home? 59% responded that they separate their waste from the total participants, while 41% said no.

The participants who separate waste were asked which kind of waste they separate. The results are that the separated waste is organic with 43.30% while glass is less separated with 1.50%, as indicated in Table 2.

Table 2. Which kind of waste do you separate?

Glass	1.50%
Batteries	7.50%
Paper and cardboard	16.40%
Organic waste	43.30%
Plastic and metal	31.30%

In this survey, it was found that 80.20% of the participants dispose of their waste in the collector handcars or trucks. From this question, it derives that 77.40% mention that there are no containers destinated for waste sorting into paper, glass, plastic, organic (for instance) near their house and, the majority of them (54.70%) do not know any recycling point in their city or town, and 45.30% replied that could recognize at least one of these facilities.

The situation of waste disposal and impacts is visible in the whole country. One more important question was made to know which aspects they consider problematic to properly separate garbage (Table 3).

Table 3. Which aspects do you think make the waste separation difficult?

Reduced place for bags/containers per type of waste	35.80%
Considered as waste of time	5.70%
Lack of knowledge of waste sorting	30.20%
Lack of awareness of future benefits	28.30%

35.80% of the people commented that the main reason is that there is not enough place to dispose of waste according to their type in different bags or containers. The lack of knowledge on waste sorting was the second reason voted with 30.20%, and the lack of awareness of the future benefits that a correct waste separation can bring was a response of 28.30%. Just 5.70% of the participants said it could be considered a waste of time.

Regarding the interviews, some critical questions about recycling practices during the COVID-19 pandemic were done, such as: a)Do you recycle at home? b)How do you handle waste? c)How do you collect waste? d)What is the final destination? e)Which facilities of final disposal do you know? f)Do you have any support from the government? g)Do scavenging workers have benefits from the government? h)Do you think recycling is economically profitable?

Two of the fourth interviewed replied that they recycle at home (a polish citizen and one Mexican who owns a company in the waste separation). The other two people do not recycle and do not separate waste and commented that there is no culture on citizens. The separation of waste before throwing it is not optimal as all is mixed when being collected by the trucks or carts. Regarding how the waste is handled and collected, one of the people works for the government collecting the garbage. However, they have specific days to cover certain areas of the town, and all the waste disposed of by households is mixed in the collection truck.

According to the answer of Mexican participants, the majority of the waste is disposed of at the landfill. However, as two of them are involved in the waste collection/separation business, they comment that the materials like plastic, glass, metal, and carton are separated and delivered to some recycling facilities. The polish interviewed who worked in the Netherlands hired by a polish company in a metals recycling company states that they were transporting the not functional materials to Poland where they have facilities for destroying them. They were keeping the valuable parts of the metals. The risk of working in this company was high; he commented that the company did not offer personal protection equipment to them. The rest of the national employees in this company had, and the salary was not very well paid.

According to the comments of the interviewed people in Mexico, they do not receive any support from the government. The person who collects waste in the truck designated by the government mentions that the working conditions are unfavourable and the salary is low. The other two participants comment that as they run their business government is not involved directly, but, according to scavengers' comments, the government is trying to incentivize their work.

Nevertheless, the four participants confirm that waste is a good way of doing business. Nevertheless, not all people are motivated to work in this activity, as marginated. Also, the Mexican interviewed to confirm that the resources in their areas are not enough to develop a correct recycling system and that there is a lack of culture and environmental consciousness.

FINAL THOUGHTS

The generation of waste is a consequence of daily activities (from household activities to industrial processes). These must be treated according to the type of material, and all people must contribute to its proper handling and disposal. The amount of waste generation depends on the consumption dynamics of the population or the modification of social habits. From the current situation caused by the COVID-19 pandemic and the government measures, people have had to make changes in their lifestyles, like isolation. Staying and working from home has brought health and environmental repercussions related to the increase in the generation and management of waste.

Regarding the waste recovery activities during the pandemic, the governments of developed countries are taking extreme security measures to prevent the spread. Roy, P., et al., (2021) states that changes to MSW management have been implemented in both developed and developing countries because of the COVID-19 pandemic to ensure adequate protection against the transmission of the virus. Unfortunately, these measures require resources that developing countries do not have or cannot easily take since lifestyles differ. In countries like Mexico, where the working conditions for people dedicated to waste management are not so favorable, it is not easy to stay at home since they do not have a fixed salary or formal employment. Like them, many other citizens work in activities that they cannot carry out from home.

There are multiple studies and researches that analyse the situation in developed and developing countries. For example, "Waste Management Outlook for Latin America and the Caribbean" and, "Global Waste Management Outlook" which provide a panorama of finance and governance of waste management all around the world and give some proposals to contribute to a proper waste handling.

FUTURE RESEARCH LINES

Waste generation is a high priority topic for countries. Avfall Sverige says that for the EU there are some goals already posted for the upcoming years:

2024: The separation and collection of biowaste should become mandatory in the EU.
2025: Municipal waste might be recycled to new materials in at least 55%. From 1st of January, separation and collection of textiles will be obligatory. From 1st of April, collection of packaging waste and recyclable paper should take place from all residential properties and, the collection of bulk packaging will be required.
2030: Municipal waste might be recycled to new materials in at least 60%.
2035: Municipal waste might be recycled to new materials in at least 65%. And maximum 10% of waste might go to landfill.

Regarding Mexico, the topic is opened as local governments take some measures. The direction for future research includes the analysis of results obtained in the recycling plants established in Mexico City.

Governments took action during the COVID-19 pandemic regarding the waste management topic.

REFERENCES

Copperthite, K. (1989). *Rigid Container Recycling: Status and Impact on the Rigid Container Industry*. U.S. Department of Commerce, International Trade Administration.

Department for Environment, Food, and Rural Affairs. (2021). UK Statistics on Waste. Government Statistical Service.

ECOCE. (2019). *Plan de Manejo ECOCE.* Recovered from: https://www.ecoce.mx/plan_nacional

Environment, U. N. (2018). *Waste Management Outlook for Latin America and the Caribbean*. United Nations Environment Programme.

EPA. (2020). *Best Practices for Solid Waste Management: A Guide for Decision-Makers in Developing Countries*. United States Environmental Protection Agency.

Erdoğdu, M., Mermod, A., & Aşkun, Y. (2016). *Social and Economic Perspectives on Sustainability.* IJOPEC Publication.

European Commission. (2010). *Life Cycle Thinking and Assessment for Waste Management.* Publications Office of the European Union.

González, R. S., & Castanho, R. A. (2021). *Comparison in Recycling Processes and Methods in developed and developing countries: A Brief Preliminary Study through Literature Review-Based Research.* Online International Conference on Empirical Economics and Social Sciences (e-ICEESS'21), Turkey.

Guzmán, S. C. (2021). *GobCDMX inaugura planta recicladora de basura, la más moderna de América Latina.* Gobierno de la Ciudad de México. Recovered from: https://www.capital21.cdmx.gob.mx/noticias/?p=25324

Hill, T. (2003). *Reduce, Reuse, Recycle.* Ready-Ed Publications.

Law and your Environment. (2017). *Law Search.* Recovered from: http://www.environmentlaw.org.uk/rte.asp?id=88

Legislation.gov.uk. (2013). *The Waste Electrical and Electronic Equipment Regulations 201.* Recovered from https://www.legislation.gov.uk/uksi/2013/3113/contents/made

Liang, Y., Song, Q., Wu, N., Li, J., Zhong, Y., & Zeng, W. (2021). Repercussions of COVID-19 pandemic on solid waste generation and management strategies. *Frontiers of Environmental Science & Engineering, 15*, 115.

Martínez, J. N. (2021). *A la pandemia se suma otro problema... la generación de desperdicios.* Dirección General de Comunicación Social. Recovered from: https://www.dgcs.unam.mx/boletin/bdboletin/2021_080.html#:~:text=Con%20el%20confinamiento%20en%20M%C3%A9xico,Regional%20de%20Investigaciones%20Multidisciplinarias%20(CRIM

Martínez Arroyo, A., Ruíz Suárez, L. G., Gavilán García, A., Ramírez Muñoz, T., & Huerta Colosia, D. (2020). *Panorama de la generación y manejo de residuos sólidos y médicos durante la emergencia sanitaria por COVID-19.* Instituto Nacional de Ecología y Cambio Climático INECC.

Ministry of the Environment. (2018). *Japan's Resource Circulation Policy for Plastics.* Author.

Ministry of the Environment. (2021). *Municipal solid waste emissions and disposal in FY2019.* Author.

Nuestra esfera. (2014). *¿Cómo se clasifican los residuos?* Accessed at http://nuestraesfera.cl/zoom/como-se-clasifican-los-residuos/#:~:text=Los%20residuos%20pueden%20ser%20clasificados,cartones%2C%20vidrios%2C%20por%20ejemplo)

Penga, Y., Wua, P., Schartupb, A., & Zhanga, Y. (2021). Plastic waste release caused by COVID-19 and its fate in the global ocean. *Proceedings of the National Academy of Sciences of the United States of America.*

Roy, P. K., Mohanty, A., Wagner, A., Sharif, S., Khalil, H., & Misra, M. (2021). *Impacts of COVID-19 Outbreak on the Municipal Solid Waste Management: Now and beyond the Pandemic.* ACS Environmental.

Secretaría de Medio Ambiente y Recursos Naturales. (2020). *Diagnóstico Básico para la Gestión Integral de los Residuos*. Author.

Sverige, A. (2020). *Swedish Waste Management 2019*. Recovered from: https://www.avfallsverige.se/fileadmin/user_upload/Publikationer/SAH_2019_publ20_eng.pdf

UNEP. (2015). *Global Waste Management Outlook*. United Nations Environment Programme.

Watson, D., Elander, M., Gylling, A., Andersson, T., & Heikkilä, P. (2017). *Stimulating Textile-to-Textile Recycling*. Nordic Council of Ministers.

Worrell, E., & Reuter, M. A. (2014). *Handbook of Recycling*. Elsevier.

Chapter 14
Analysis of Productivity and Participation Relationship in Performance

Jabbarov Tural
WSB University, Poland

Nurdan Ozrecberoglu
European University of Lefke, Turkey

ABSTRACT

Performance is a multidimensional concept and only multiple indicators can evaluate it as a whole. The main purpose of the study is to discuss the variables that can evaluate the business performance as a whole with different aspects, the factors that can affect these variables, and the relationships between them. This study was carried out using quantitative and qualitative methods in order to find answers to questions such as how the performance of employees in enterprises is measured, which methods are used, who is employed as performance evaluators in enterprises, which mistakes are made due to performance evaluators, and the attitude of the employees whose performance is evaluated. The study is a descriptive and descriptive qualitative case study; the data were structured with a questionnaire form and provided with semi-structured face-to-face interviews and document review and analyzed with a descriptive analysis approach.

1. INTRODUCTION

In the global competitive environment, the importance of measuring the performance of businesses has increased even more. The reason for this is that businesses can be more successful by determining what results they have achieved during their activities in line with their strategies and goals, and acting appropriately and establishing an effective performance measurement system.

DOI: 10.4018/978-1-6684-4548-8.ch014

In both the private sector and the public administration, it is of great importance to know the extent to which employees contribute to the achievement of organizational goals, how much they adopt the assigned duties, their skills and their contribution to the effective and efficient operation of the organization. Performance evaluation, which is defined as all of the efforts to determine and improve the success of the employees of the organization, is one of the most important tools of modern human resources policy.

Evaluating the performance of businesses with modern methods is useful in adapting to developing market conditions. It is the main tool to evaluate the general or specific performance of the organization in every respect, to determine its ability to fulfill its own duties and to know how close it is to the mission, vision, goal and objectives set in the first place. Managers who have an important role here should evaluate the activities of the enterprise and its employees with an accurate, fair and logical performance measurement system.

In the subsections included in this study, performance development and definition, the importance of performance evaluation, performance measurement methods, the relationship between productivity and participation in the evaluation system, research questions on the relationship between job satisfaction and performance evaluation in Azerbaijan and Poland, analysis of the employees who constitute the research sample for the evaluation of employees' performance has been made. In the conclusion part, the results obtained from the study were evaluated and interpreted.

2. THEORETICAL FRAMEWORK OF PERFORMANCE AND PERFORMANCE EVALUATION

Determining to what extent the duties assigned to the employees are fulfilled by the employees or the ability of employees to work is one of the most important problems faced in organizations in our age. This problem has led to the rapidly gaining importance of the concept of performance in organizations.

There is no common definition of performance, which covers almost all areas of individual and social life. However, the Turkish Language and Historical Society qualify the term of performance as "my success". Performance is a qualitative (quality) and quantitative (quantity) expression of where an individual, a group or an enterprise doing a job can reach the intended goal, in other words, what it can achieve (Baş, 1990).

Since the contribution of the employee with low performance in the enterprises will not be at the desired level, working with such an employee will not be advantageous in terms of management. This situation may lead to employee turnover. It is clear that such a situation will be a cost increasing factor for businesses (Akın, 2002).

Performance is the amount of goods or services produced in a certain time unit and is expressed as a result of the interaction between the concepts of "efficiency", "productivity", "output" according to its function, as well as the individual's ability and motivation (Schermerhorn, 1987). Performance is a good, service or an opinion put forward in order to fulfill the task and to fulfill the purpose, in a way that meets the criteria determined in the framework of the task (Pugh, 1991).

Performance evaluation is extremely important in achieving organizational goals and determining what individual contributions are to the goals (Ludeman, 2000). Performance evaluation is a method of objectively measuring how the tasks are performed within a program in the performance of products, services or transactions (Demirkaya, 2000). Performance evaluation refers to the evaluation of how well the employee is doing his/her job (Önder, 2000).

Palmer (1993) defines performance evaluation as the process of a manager evaluating the employee's performance at work by comparing and measuring with predetermined standards.

As a result of the above understandings, what is expected from performance evaluation is to determine the needs of the employee as a whole, to meet their training, motivation and similar needs, to increase their creativity and to achieve the difference with incentives such as appreciation, leave, premium, wage increase, reward and bonus by performing the desired level of work and efficiency.

3. PERFORMANCE EVALUATION EFFICIENCY AND PARTICIPATION RELATIONSHIP

Some current developments also play a role in the prominence of the concept of productivity. The most important places of these can be listed as follows. Although there is no definite judgment as to what should and should not be done in order to get high efficiency from the employees, there are some points to be considered. First of all, it should be ensured that employees clearly understand the desired goals and develop a belief about how important these goals are. It is also important for each individual to know what to do and to understand how they should work together to achieve goals. However, job satisfaction alone is not a sufficient factor. At the same time, the individuals who make up the team must be competent in their work. In short, every individual should have the technical skills and abilities to achieve the desired goals, and also have the characteristics required for teamwork to work in the best possible way. Mutual trust should also be promoted.

There is a direct relationship between employee performance and productivity. Performance will increase the efficiency of the work done in the enterprise. Total productivity in the enterprise is related to how effective the employees are individually. Employees' job satisfaction, working with high performance will enable the enterprise to reach its goals and increase its productivity. When employees have a say in matters related to their work and in making decisions that will concern them, their trust in the business and their commitment to work increases.

In an enterprise, the level of employee involvement determines the nature of performance management practices. If the participation is at a low level and extremely centralized in the enterprise, the target setting, performance evaluation and reward systems are formalized and administered by the management. In cases where participation is high, it is necessary to ensure that both management and employees participate in setting performance targets and in performance appraisal.

Employees who do the job may think that they have the expertise that should be taken into account in making the decision, or they may think that the decision to be made will significantly affect the work, and therefore taking their opinions on this issue means respecting them.

4. REVIEW OF PERFORMANCE EVALUATION METHODS

There are various approaches and methods for performance appraisal. Among these methods, businesses; they make performance evaluations by choosing according to their goals, the qualifications and structures of their employees, and sometimes using several methods together.

A number of methods are used in the performance evaluation of employees, and these methods should be the most appropriate studies that are important for the enterprise. It is in 2 groups, being classical and modern (Aktan, 2002). Some of them are explained as followings:

Simple Sorting Method: It is the simplest of the methods. Certain qualifications or workers in general are ranked. In this method, where the workers are ranked according to their success, there are differences in the technique.

Binary Comparison Method: Today, it is a limited method used. The result is reached by comparing the employees with each other. Not only two people, but a few people can be compared. This method has many drawbacks: It is not correct to use these results for pay as the specific and obvious differences between performance levels are not fully explained.

Table 1. Binary comparison method https://aydancag.com/geleneksel-performans-degerlendirme-yontemleri/

Comparison	A	B	C	D	E		A	B	C	D	E
A		+	+	-	-			-	-	-	-
B	-		-	-	-		+		-	+	+
C	-	+		+	-		+	+		-	+
D	+	+	-		+		+	-	+		-
E	+	+	+	-			+	--	+		
Evaluation related to work + Better - Worse											

Mandatory Distribution Method (Forced Distribution Method): Managers can determine that there is nothing different about the success of their workers or that there is no improvement in their performance and they can turn to important information in the evaluations. Accurate determination of success between people is the main process. This method provides several limitations to the evaluators. Evaluators should be placed on the five-scale format, as the method applies, to the normal distribution curve.

Table 2. Mandatory distribution method (Akın, 2002)

Highest	high	medium	low	Very Low
10%	20%	40%	20%	10%

Drawbacks: This method cannot be done in an organization where few workers work. There may be errors in the rating of the employees (Doğan, 2014).

Figure 1.

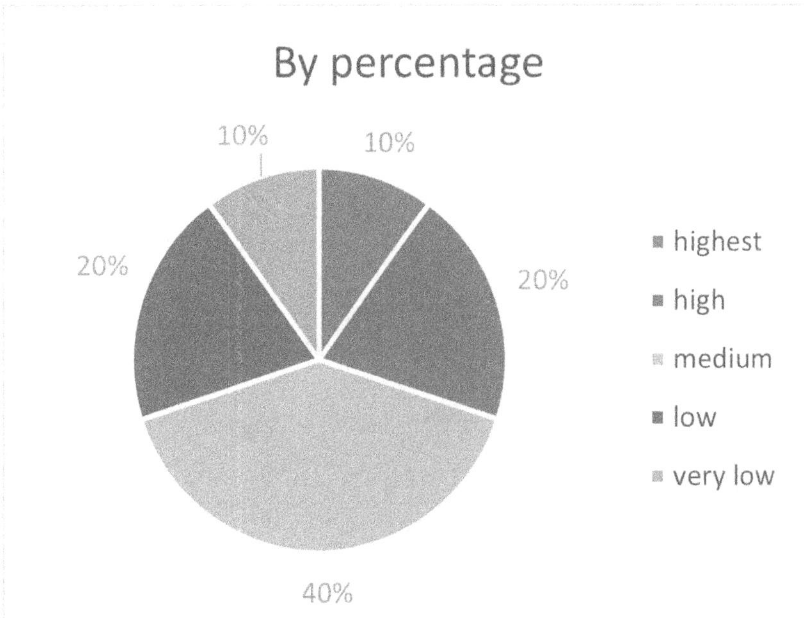

Critical Events Method: It is the method that reveals the positive or negative behavior of the evaluators when they do their work. The reason why this method is called critical events is that it explains the behavior of the observed, the success or failure of the workers in the right direction. The feature of this method is beneficial in that it provides reliable feedback by evaluating the strengths and weaknesses of the worker. The evaluators are asked to record the events for this form.

Alternative sort method: It is one of the easiest, inexpensive, less time consuming methods. They rank by the managers according to the success of the employees. Managers divide workers who have differces, by dividing them into sub-groups under the groups and sorts them from good to bad (Ceylan, 2007).

Graphic Rating or Standard Scoring Method: Graphical grading method is named as traditional assessment scales in some places; it is the most used method in businesses. Conducting multilateral analysis by applying a good arrangement is appropriate in this method. It is organized by grouping according to the specific success levels of each worker. For example, with 5 selections, collective workers would be an analyzed with particular feature. The important thing is the type of factor to be measured; such as personality traits, amount of work, quality. Before the managers evaluate, the limits of the factors are determined and explained. Sometimes correct results are not known, because it is boring to evaluate 30-40 people on other factors at 5 degrees` scale. The new points have been added to fix the shortcomings in the system: The items evaluated should be clear and opened, and the items should not be similar, the same performance should not be evaluated with many items (Akalin & Koç, 2014).

Table 3. Critical Situation (event) evaluation form https://aydancag.com/geleneksel-performans-degerlendirme-yontemleri/

Worker's sensitivity to the problems of the institution					
a. he couldn't see the problems b. he didn't care about the cause of the problems c. he did not look for solutions to problems			a. he anticipated that there might be problems. b. he emphasized on the causes of the problems. c. he solved the source of the problems.		
Date	Option	event	Date	Option	event
12.07.2017	C	He/she caused the special announcement to be announced that to be delayed.	23.08.2017	C	With his/her personal efforts, he/she prevented the burn from coming off.
		Explanation: He/she did not report it to the concerned in time.			Explanation: He/she prevented the burn from coming out by putting himself/herself at risk.

Table 4. Alternative sort method (Schermerhorn, 2017)

Chapter date	Valuer
Names of workers to be listed	1. the most successful
Fatih	2. the most successful
Fikrat	3. the most successful
Neslihan	4. the most successful
Erol	5. the most successful
Ali	5. the most unsuccessful
Asli	4. the most unsuccessful
Ahmet	3. the most unsuccessful
	2. the most unsuccessful
	1. the most unsuccessful

Figure 2.

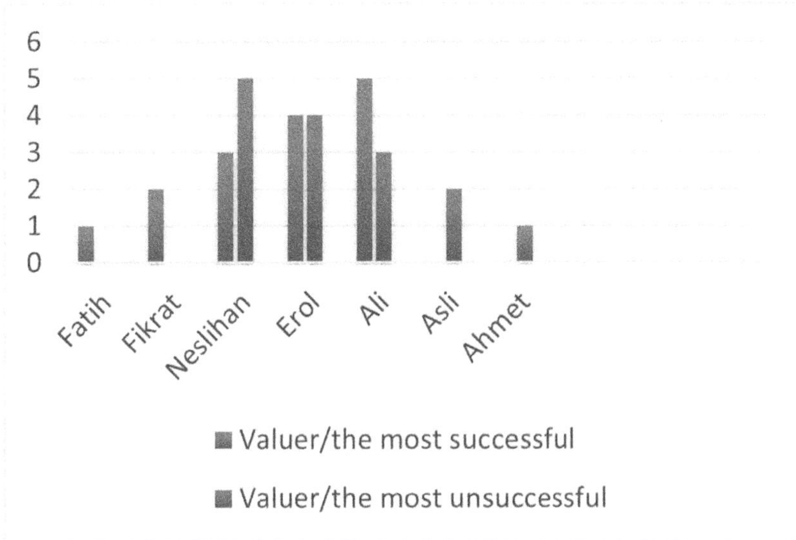

Table 5. Graphic ratting or standart scoring method (Sabuncuoğlu, 2013)

Evaluation factor	rate
The quantity of the work: It reaches the size of the work in the desired amount	3
The nature of the work: It makes its work full and complete in terms of quality.	2
Job information: He/she knows the job requirements and task dimensions.	3
Collaboration: He/she is willing to cooperate with others and their duties.	1
Commitment: He/she is meticulous and careful in her participation and completion of the Study.	2
Willingness to work: He/she is willing to present his ideas and increase his duties.	2
3 = excellent 2 = satisfactory 1 = unsatisfactory	

Assessment method according to aims and objectives: This method, which is one of the contemporary methods, was developed to reduce the problems of other evaluation methods, to help the development of the organization, and to emphasize that it is important in evaluating the success of workers. Stages: targets are determined for subordinates, top and subordinate determine this stage together, the ideas of both parties are heard and the common point is reached and the necessary plans are prepared for reaching the goals and how to determine the goals by coming together again at the end, feedback is provided for the works by mutual discussion and the performance of the worker is revealed from this application.

360 degree assessment method: The 360 degree performance evaluation method is the evaluation of the performance of the employee by everyone in the working environment.

The skill areas are graded with the questionnaire form. A report is then prepared and feedback is given to the worker. The main goal is to reveal strengths and weaknesses. The feature of the method: It is to obtain more accurate results by reaching more resources by evaluating the worker by more than one person.

Figure 3. 360 degree performance evaluation (Mathis, 2015)

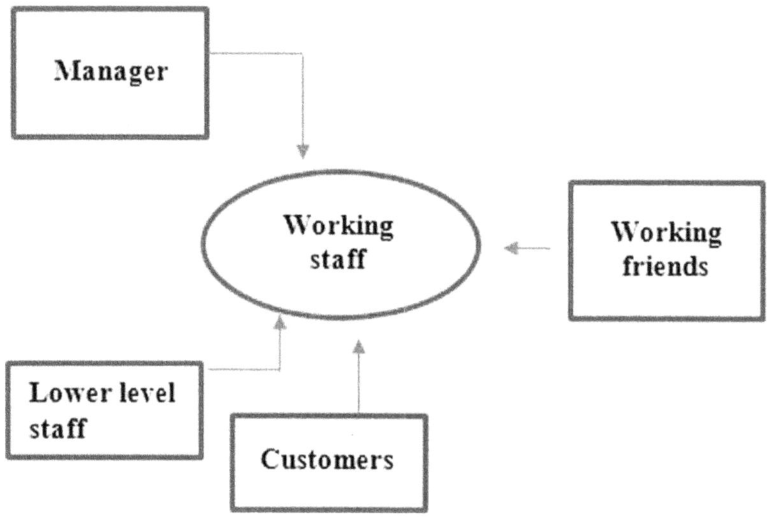

Table 6. Comparing performance evaluation (Feldman & Arnold, 1996)

	Graphic grading method	Sorting method	Evaluation according to goals and objectives
The accuracy of the evaluations	low	low	high
Contribution of staff to their decisions	medium	medium	medium
Contribution to the determination of awards	low	high	high
Contribution to identify need for improvement	very low	very low	medium
Time and money required to develop the method	low	very low	medium high
The power to motivate those assessed	low	low	High
The money required for the execution of the method	low	low	medium high
Accepted by person who evaluated	low medium	low medium	high
Accepted by people who evaluated	low medium	low medium	high
Skills required in assessors	low	low medium	high

Benefits: The contribution of the customer in the evaluation by strengthening the communication, always taking into account the customer ideas and focusing in that direction and at the same time, the managers' evaluation by the workers also helps the managers to see themselves.

Figure 4.

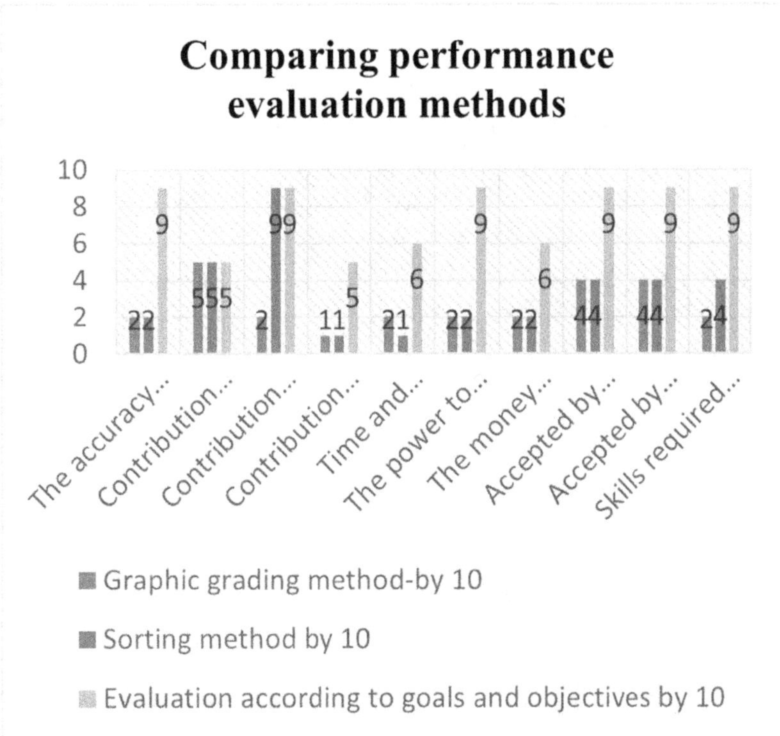

5. ANALYSIS OF THE RELATIONSHIP BETWEEN JOB SATISFACTION AND PERFORMANCE EVALUATION IN AZERBAIJAN AND POLAND

A survey was conducted to investigate the relationship between productivity and participation in performance appraisal. 241 people took part in the survey. First of all, the age, gender, country of residence and education level of the respondents were studied. Thus, 50% of the 241 respondents in the online survey were aged 18-25, and 36% were aged 26-30. It can be concluded that the survey was of more interest to people under 30 years of age. As for the gender category, it should be noted that 60% of respondents were men and 41% were women.

An interesting aspect of the survey results is that Azerbaijani and Polish citizens participated in the survey with interest. Thus, about 58% of respondents are from Poland and 42% from Azerbaijan.

In addition, the level of education of the participants was studied. Based on the results, we can say that 63% of the participants were masters, 18% were bachelors and 15% were PhD graduates. This section on the level of education is considered satisfactory. Thus, the level of education of the participants allows them to answer the questions in section 2 of the questionnaire more accurately.

Figure 5. Country division

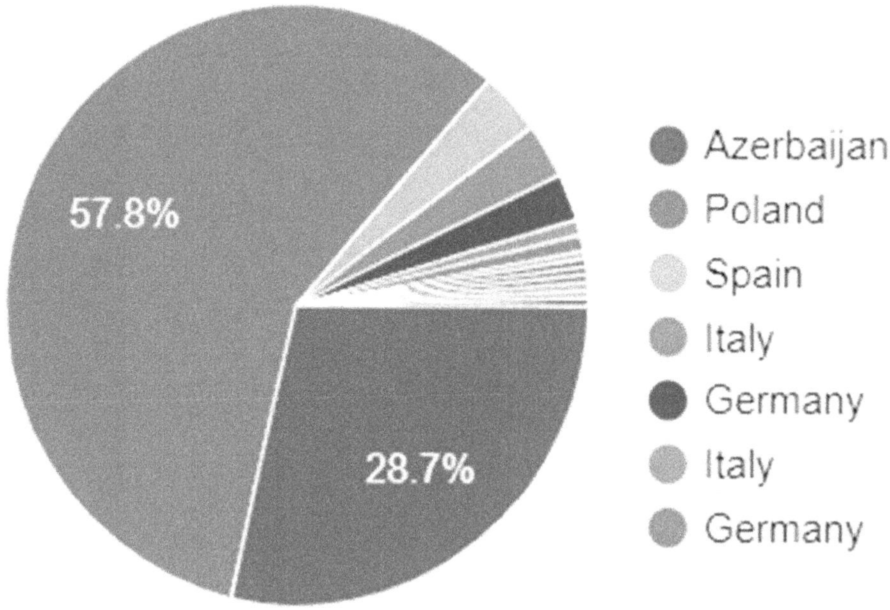

Figure 6. Level of study

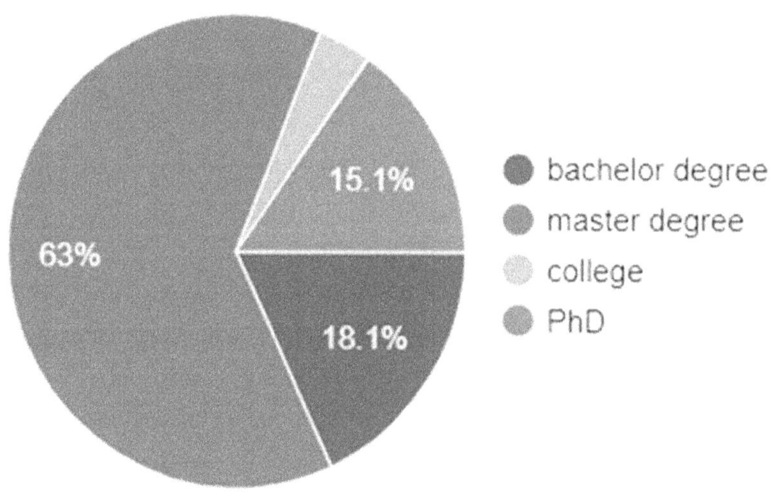

It should be noted that while 27% of respondents are students, 71% work in various organizations. Undoubtedly, working people can clearly see the role of productivity in performance appraisal in their enterprises.

In order for the analysis to lead to more accurate results, the questionnaires included a question about the current situation of the employees. Based on the answers to the question, it can be concluded that 24% of respondents are managers, 21% are assistants, 24% are employees and about 13% are technical assistants.

Figure 7. Current status of person

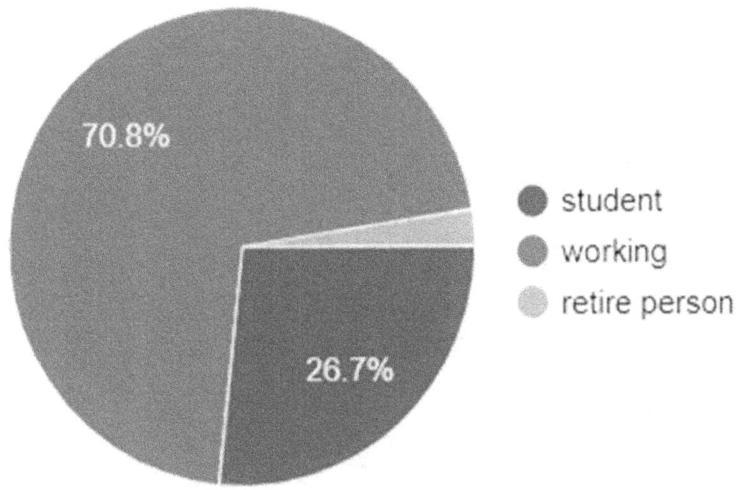

Figure 8. If you work, what is your current position?

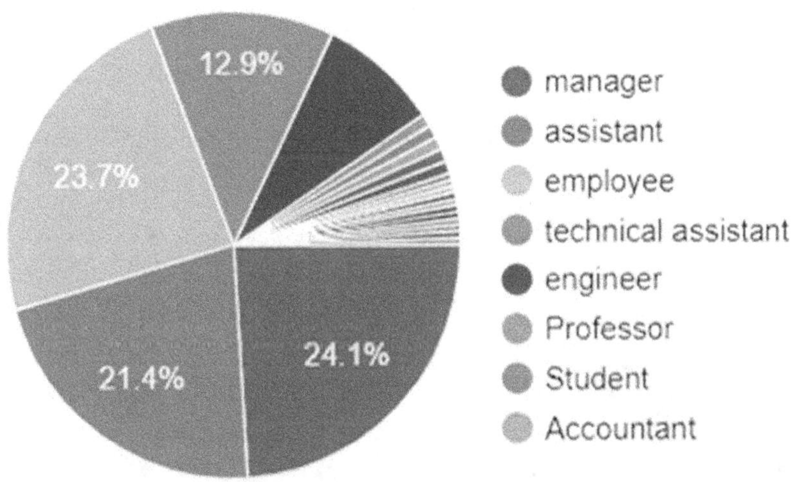

It is no secret that the long-term stay of an employee in the same company does not mean an effective result. By remaining in the same position, an employee can only become passive. In this sense, it is very important to work in the same organization for several years. In response to this question, 39% of respondents stated that they remained in the same position for 1 year, and 45% for 3 years. However, 10% of respondents said they had stayed in the same organization for more than five years.

If we recall the first question in the main part, it should be noted that the majority of survey respondents are bachelor's and master's students. Therefore, when asked how many years of work experience they had, 61% said they had 1-3 years of work experience and 30% had 5-7 years of work experience.

Figure 9. How many years did you stay in the same job position at most?

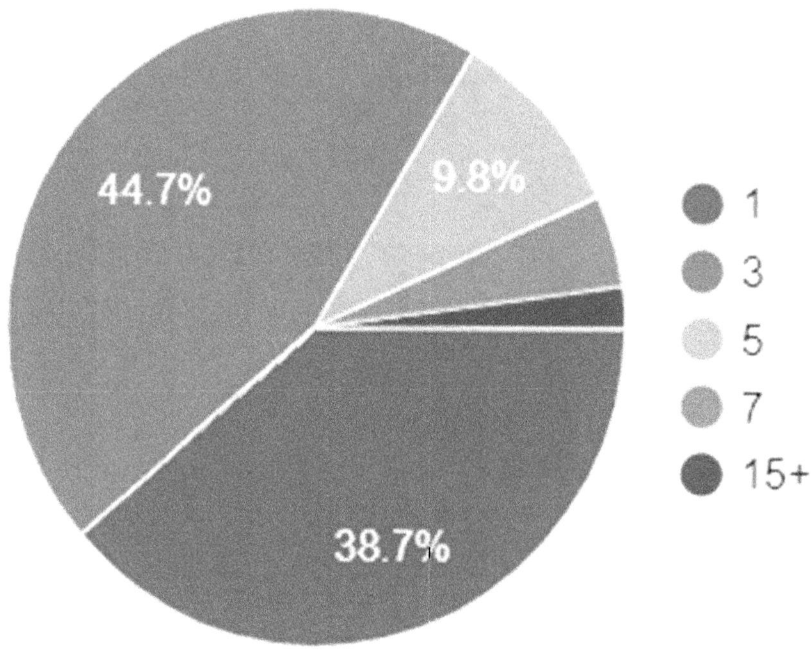

Figure 10. How many years working experiences do you have?

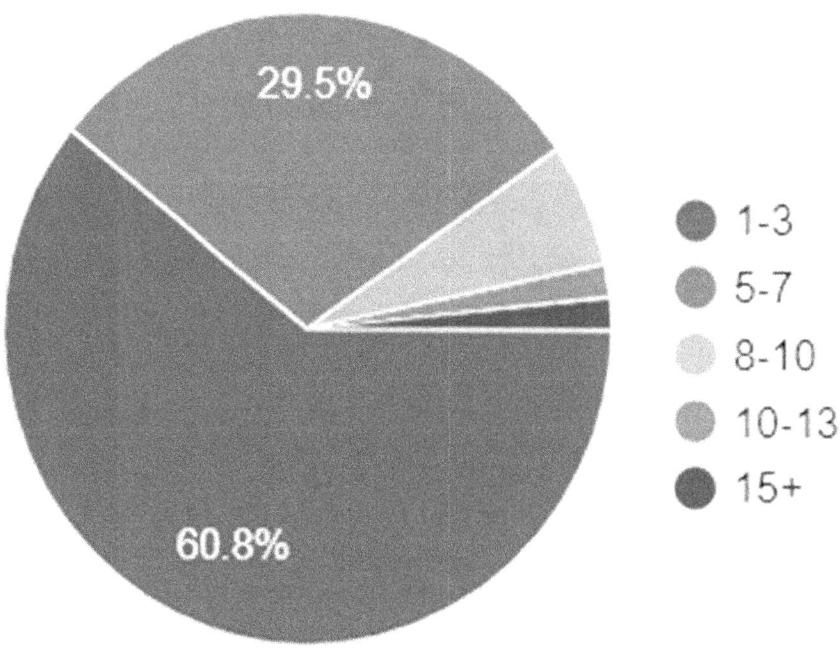

Since one of the main purposes of the presented article is the factors affecting productivity, the survey tried to get information about it from the respondents. The analysis of the answers shows that about 49% of respondents based their responsibility on the main factor of productivity, and 33% on honesty. Only 12% of respondents pointed out that the factor that affects productivity is creativity.

It is important to note that productivity is measured not only by the factors that affect it, but also by the conditions created by the company for employees through motivational measures and benefits. In order to come to a definite conclusion, this question was also brought to the fore during the survey, and the answers mentioned are very attractive. Thus, 39% of respondents think that there is a high salary, 33% - experienced employees, and 18% - motivation.

One of the most controversial aspects of human resource management is the issue of performance evaluation in the workplace. So while some see this as a positive development, others are wary of subjective approaches. The survey also asked that 25% of respondents say that performance should be evaluated according to work experience, 16% said that a degree should be taken as a basis, and 40% said that employees should be determined by the benefits they bring to the company. I think this approach is correct. An employee's performance evaluation should be compared to his or her productive work.

Respondents rate their performance in education was asked whether it affected the assessment. About 64% of respondents think that educated employees should always get high results in the assessment. 31% of participants said that education is not important, it is the attitude to the main work. In order to achieve a high result in the evaluation of an employee's performance, not only education, but also the results of the work should be taken into account.

In the end, respondents were asked how to evaluate the performance of employees within the same company, 56% of respondents said that the process is correct by managers, and 23% said that it should be determined by its impact on productivity.

In addition to the above survey results, the following frequency analysis was conducted among 113 individuals to achieve deeper results in the study.

Figure 11. What are the factors that affect productivity?

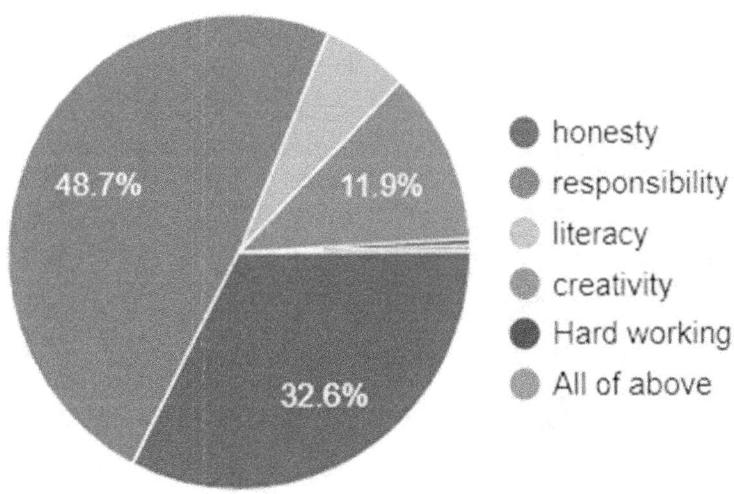

Figure 12. What conditions of the company do you think have a positive effect on work productivity?

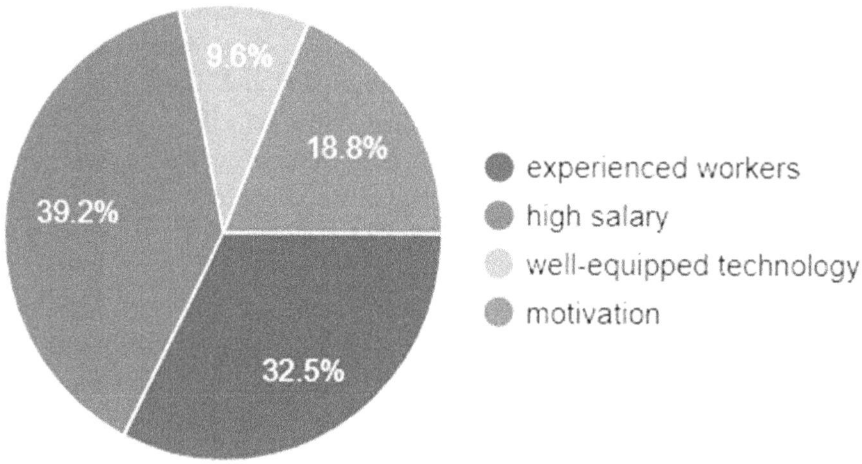

Figure 13. How do you think performance should be evaluated in the workplace?

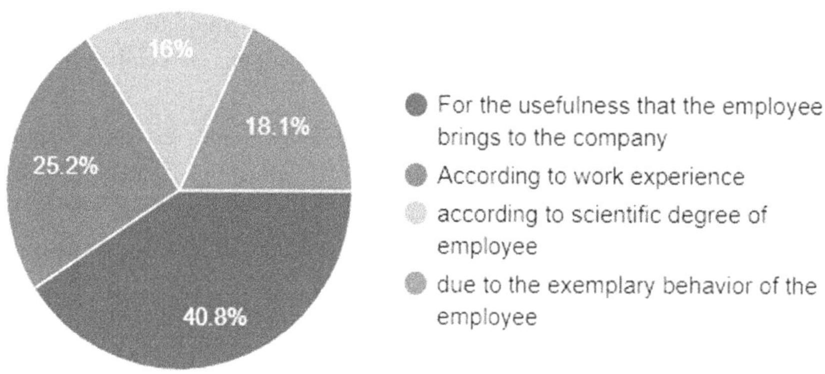

Figure 14. Do you think it is correct to evaluate an employee's performance based on his / her educational status?

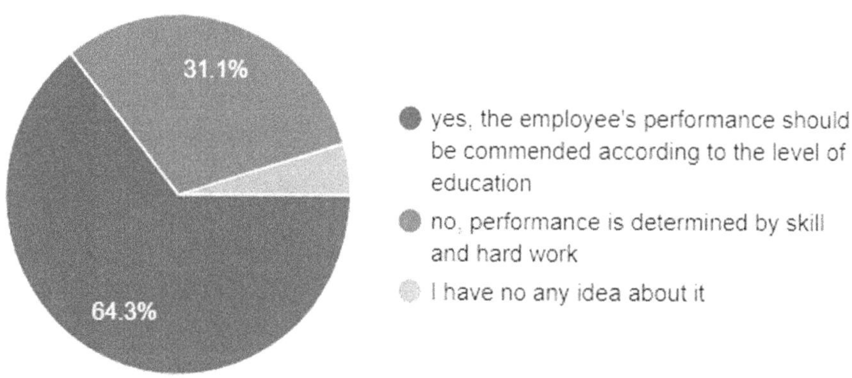

Figure 15. How do you think performance appraisals should be conducted for employees within the same company?

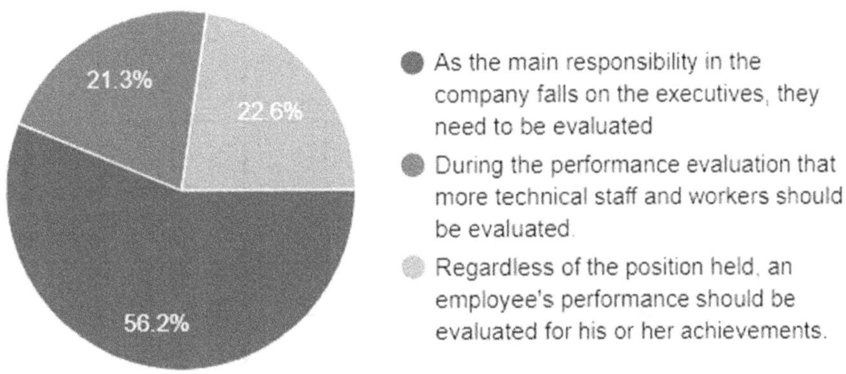

Frequency Analysis

In the frequency analysis phase of the research, the questions in the questionnaire are grouped and explained in tables. There are a total of 24 questions related to job satisfaction and these questions are divided into 7 separate subheadings as follows. From thus, there are a total of 4 questions in the survey regarding the wage satisfaction of the employees. The frequency table of these questions and questions is given below.

Question 1: "I find the wages I get sufficient for my work."
Question 2: "When I compare it with my colleagues, I do not feel that I am being injustice with my salary."
Question 3: "When I compare myself with employees doing similar jobs in other institutions, I see that my wages are not low." Question 4: "I am satisfied with the wage increases in my job."

When the frequency values related to Question 1 are examined, 45% of the employees do not find the wages sufficient. 24 out of 113 employees who filled out the questionnaire could not decide on the adequacy of the wage. Those who think their wages are sufficient is only 38 people out of 113 people.

When the frequency values of Question 2 are examined, it is seen that 27 people are undecided about the fairness of wages. 46% of the employees do not think that they have suffered injustice when compared to their colleagues regarding wages.

The rate of those who think that their wages are not low when they compare themselves with employees doing similar jobs in other institutions is 30% (Question 3), while the rate of those who answer this proposition as "I strongly agree" is 14%, and the rate of those who respond as undecided is 23%.

When the frequency values regarding wage increases are examined in question 4, the rate of those who are satisfied with the wage increases in their jobs is 21.2% (those who say I agree and strongly agree) in total. The rate of those who are not satisfied with the wage increases is 22%, and the ratio of those who are satisfied and not satisfied are very close to each other. It is the ratio of indecisive people that plays an important role here. A very high rate of 26% belongs to the undecided.

Table 7. Frequency analysis of fees-related questions

Questions about wage	I strongly disagree		I don't agree		undecided		I agree		Absolutely I agree		Total	
	Fr.	%	Fr.	%	Fr.	%	Fr.	%	Fr.	%	Fr.	%
S1	13	11,5	38	33,6	24	21,2	34	30,1	4	3,5	113	100
S2	11	9,7	17	15,0	27	23,9	48	42,5	4	3,5	113	100
S3	8	7,1	29	25,7	26	23,0	34	30,1	16	14,2	113	100
S4	14	12,4	46	40,7	29	25,7	20	17,7	4	3,5	113	100

In the questionnaire, there are a total of four questions about the relations of the employees with their colleagues. The frequency table of these questions and questions is given below.

Question5: "As colleagues, we get along well with each other, my colleagues are compatible."
Question6: "I am happy to be able to form close friendships with the people I work with."
Question7: "We work in cooperation with my colleagues, we do not hesitate to help."
Question8: "We get on well with my colleagues."

It is seen that the harmony and relations of the employees in the company with each other are quite good. The rate of those who think their colleagues are not compatible is only 0.9%. According to the frequency values for question 6, 89 out of 113 people stated that they were happy to be able to establish close friendships with the people they worked with. However, the rate of those who say they are indecisive (15%) is not less at all.

In question 7, the number of employees who stated that they were working in cooperation and helping each other was 87. This group of 87 people constitutes a high rate of 77% among the total employees. The rate of those who answered this question as "disagree" and "strongly disagree" is 14% in total.

According to the frequency values for Question 8, there is no employee who stated that they could not get along with their colleagues. Although 91 people out of 113 stated that they got along well with their colleagues, the rate of those who answered "I do not agree" (7%) and "I am undecided" (12%) is also noteworthy.

Table 8. Frequency analysis of questions regarding employes with their colleagues

Questions about the relations of the employees with their colleagues	I strongly disagree		I don't agree		undecided		I agree		Absolutely I agree		Total	
	Fr.	%	Fr.	%	Fr.	%	Fr.	%	Fr.	%	Fr.	%
S5	1	,9	7	6,2	14	12,4	58	51,3	33	29,2	113	100
S6	1	,9	6	5,3	17	15,0	57	50,4	32	28,3	113	100
S7	1	,9	15	13,3	10	8,8	58	51,3	29	25,7	113	100
S8	0	0	8	7,1	14	12,4	59	52,2	32	28,3	113	100

There are a total of four questions regarding the knowledge and skill levels of the employees in the questionnaire. The frequency table of these questions and questions is given below.

Question 9: "I can do the activitics rcquircd by my job in the best way without difficulty."
Question 10: "I think I have enough of the information I need while doing my job."
Question11: "I am doing the job that best suits my abilities."
Question12: "I think I am really successful in my job."

There is no employee among the employees who stated that they do their job with great difficulty (Question 9). However, the rate of those who responded "disagree" is 4%. The rate of those who agree (57%) and strongly agree (32%) is quite high with 89% in total.

When the frequency distribution of question 10 is examined, it is seen that the rate of those who think they have the necessary knowledge while doing their job in the enterprise is 83%. Those who think they do not have the necessary knowledge are only 6 people.

A total of 94 people think that they are working in a job suitable for their abilities (Question 11). The rate of those who state that they work in a job that is not suitable for their abilities is 20%. According to the results of the analysis, those who stated that they work in a job that is not suitable for their abilities or that they are indecisive on this subject represent almost half of the total employees (41 people).

When the frequency values of Question 12 are examined, it is concluded that most of the employees (94 people) who work in the enterprise think that they are successful in their job. Those who think that they are absolutely not successful in their job are very few with a ratio of 0.88%.

Table 9. Frequency analysis of questions related to knowledge and skill

Questions about the knowledge and skill levels of the employees	I strongly disagree		I don't agree		undecided		I agree		Absolutely I agree		Total	
	Fr.	%	Fr.	%	Fr.	%	Fr.	%	Fr.	%	Fr.	%
S9	0	0	4	3,5	9	9,0	64	56,6	36	31,9	113	100
S10	1	,9	5	4,4	13	11,5	58	51,3	36	31,9	113	100
S11	9	8,0	13	11,5	19	16,8	46	40,7	26	23,0	113	100
S12	0	,9	0	0	18	15,9	58	51,3	36	31,9	113	100

There are four questions in the questionnaire regarding the perception of the employees towards the company they work for. The frequency table of these questions and questions is given below.

Question13: "Working in this organization gives me dignity in the society."
Question14: "It is sufficient to inform its employees about the policies and targets of the enterprise I work for."
Question15: "Working in this organization offers me a safe future."
Question16: "My organization always fulfills its promises to its employees."

When the frequency values of Question 13 are examined, it is seen that the rate of those who do not think that they gain prestige by working in the enterprise is quite low (6%). The rate of undecided is 22%. The ratio of those who think that the institution has gained prestige is quite high with a total of 72%.

More than half of the employees (52%) state that they have been informed about the policies and targets by the company (Question 14). However, employees representing a ratio of 28% think that they are not informed by the company. It can be said that the rate of those who stated that they were indecisive on this issue (20%) was higher.

The rate of those who think that a secure future is not provided by the enterprise is very low with 6% (Question 15). The rate of those answering question 15 as "strongly agree" and "agree" is 62%. In short, more than half of the employees think that their businesses provide them with a secure future. However, it can be said that the number of those who stated that they were indecisive on this matter (39 people) had a high share in the total.

Table 10. Frequency analysis of questions related to the perceptions of the organization

Questions about the perception of the employees towards the company they work for	I strongly disagree		I don't agree		undecided		I agree		Absolutely I agree		Total	
	Fr.	%	Fr.	%	Fr.	%	Fr.	%	Fr.	%	Fr.	%
S13	1	,9	6	5,3	25	22,1	57	50,4	24	21,2	113	100
S14	9	8,0	22	19,5	23	20,4	47	41,6	12	10,6	113	100
S15	7	6,2	10	8,8	39	34,5	42	37,2	15	13,3	113	100
S16	6	5,3	20	17,7	27	23,9	43	38,1	17	15,0	113	100

The frequency values of Question 16 think that 24% of the employees do not fulfill the promises of the enterprise. Although the rate of undecided people is the same as those who expressed negative opinions on this issue (24%), 53% in total are of the opinion that the enterprise fulfills its promises to its employees.

In the questionnaire, there are four questions about the relations of the employees with their supervisors. The frequency table of these questions and questions is given below.

Question17: “My supervisors are helpful when I have a problem with my job.”
Question18: “My superiors have a positive relationship with their subordinates.”
Question19: “My supervisors take my complaints into consideration.”
Question20: “My superiors display a supportive attitude in the relations of the subordinates with the top management.”

The rate of those who stated that they received help from superiors in question 17 is 56%. The rate of those who state that they definitely get help is 24%. In short, it can be concluded that the supervisors help their employees in this business. However, the rate of those who expressed their opinions as “disagree” and “strongly disagree” is 12% and it should be taken into consideration.

When the frequency values of question 18 are examined, the rate of those who evaluate the relationship of superiors with their subordinates positively is 68% in total, and the rate of those who evaluate it negatively is 12%. Although he stated that the rate of undecided is high, at 20%, it is important that the majority of employees stated that their relationship with their supervisors was positive.

Table 11. Frequency analysis of about relations of employees with supervisions

Questions about the relations of the employees with their supervisors	I strongly disagree		I don't agree		undecided		I agree		Absolutely I agree		Total	
	Fr.	%	Fr.	%	Fr.	%	Fr.	%	Fr.	%	Fr.	%
S17	2	1,8	11	9,7	10	8,8	63	55,8	27	23,9	113	100
S18	3	2,7	10	8,8	23	20,4	52	46,0	25	22,1	113	100
S19	4	3,5	15	13,3	28	24,8	46	40,7	20	17,7	113	100
S20	10	8,8	17	15,0	34	30,1	37	32,7	15	13,3	113	100

According to question 19, the rate of those who think "absolutely" that their superiors take their complaints into consideration is 18%, and the rate of those who think they absolutely do not consider it is 4%. More than half of the employees (58%) stated that their supervisors took their complaints into consideration.

When the frequency distribution of Question 20 is examined, it is seen that the number of those who stated that superiors did not show supporting attitudes towards their subordinates was 27 (24%). The total rate of employees who answered this question as "I agree" and "I absolutely agree" is 46%. The rate of undecided is quite high with 30%.

There are four questions in the questionnaire regarding the working conditions of the employees. The frequency table of these questions and questions is given below.

Question21: "I am satisfied with my working conditions."
Question22: "The environment in which I am doing my job has the physical facilities to be found."
Question23: "I have the opportunity to work freely alone."
Question24: "Measures taken for occupational health and safety are at a satisfactory level."

When the frequency values of Question 21 are examined, the rate of those who are generally satisfied with the working conditions (89%) is quite high. The rate of those who responded as "undecided" (3%) is quite low.

Table 12. Frequency analysis of questions on working conditions

Questions about the working conditions of the employees	I strongly disagree		I don`t agree		undecided		I agree		Absolutely I agree		Total	
	Fr.	%	Fr.	%	Fr.	%	Fr.	%	Fr.	%	Fr.	%
S21	0	0	10	8,8	3	2,7	53	46,9	47	41,6	113	100
S22	0	0	11	9,7	13	11,5	56	49,6	33	29,2	113	100
S23	5	4,4	18	15,9	9	8,0	56	49,6	25	22,1	113	100
S24	2	1,8	10	8,8	17	15,0	49	43,4	35	31,0	113	100

The rate of those who answered "absolutely agree" and "agree" to question 22 is 79% in total. There is no employee who states that they do not find the working environment appropriate. In short, it can be concluded that most of the employees are satisfied with the physical facilities related to the working environment.

According to question 23, the rate of those who stated that they do not have the opportunity to work freely is 4%. In general, it can be thought that the enterprise provides this freedom to its employees.

When the frequency values of Question 24 are examined, it is clearly seen that the employees are satisfied with the job security. While the rate of those who say that the precautions are absolutely sufficient is 31%, the rate of those who say they are definitely not sufficient is only 2%. 84 out of 113 people stated that the measures taken for occupational health and safety are satisfactory.

The cross table (Table 13) and explanation of the marital status of the employees and their satisfaction with their relations with their superiors is given below.

As a result of the cross-table and chi-square statistics between marital status and superiors helping their subordinates, the p-value being lower than the alpha level ($p = 0.045 < 0.05$) indicates that there is a relationship between the marital status of the employees and the variable expressing the attitudes of their superiors.

As a result of the analysis, 50 of 68 married people were satisfied with the behaviors of their superiors towards their subordinates and the support they received with the senior management (73%), while 26 out of 35 single people (74%) seem satisfied with their superiors' attitudes. With a very small difference (1%), it can be said that single employees are more satisfied with their supervisors than married ones.

The cross table (Table 14) and explanation of the relationship between the employee's seniority and job satisfaction is given below.

Table 13. Cross table on material status and employess's relationship with supervisors

		Supervisors			Total
		I don't agree	undecided	I agree	
Marital status	married	5	13	50	68
	single	5	4	26	35
	divorce	3	3	4	10
Total		13	20	80	113

Chi-square tests			
	Value	Degree of freedom (df)	Asymptotic Significance (2 Sided)
Pearson Chi-square	7,409(a)	4	,116
Likelihood ratio	6,833	4	,145 ,045
Linear-by-Linear relationship (M-H)	4,026	1	
Number of Valid data	113		

Table 14. Cross Table for relationship between employee seniority and job satisfaction

		Job satisfaction			Total
		I don't agree	undecided	I agree	
Seniorit y	0-5 years	5	6	27	38
	6-11 years	2	5	26	33
	12-17 years	1	4	21	26
	+18 years	0	1	15	16
Total		8	16	89	113

Chi-square tests			
	Value	Degree of freedom (df)	Asymptotic Significance (2 Sided)
Pearson Chi-square	5.133(a)	6	,527
Likelihood ratio	6,109	6	,411 ,038
Linear-by-Linear relationship (M-H)	4,215	1	
Number of Valid data	113		

Cross-table and chi-square statistics were obtained between the seniority status of the employees and the variable of job satisfaction. The fact that the p-value as a result of the test is lower than the alpha level, which is the determined error level ($p = 0.038 < 0.05$), indicates that there is a relationship between the seniority status of the employees and the job satisfaction variable.

As a result of the analysis, most of the employees (94%) for 18 years or more have reached job satisfaction, while 27 (71%) of 38 people with a seniority of 0-5 years stated their job satisfaction negatively.

6. CONCLUSION

Job satisfaction refers to the attitudes of employees towards their jobs in general and is related to whether they like their job or not. Job satisfaction has become an important issue for today's businesses, as it also affects job productivity and the stress that job dissatisfaction can cause in-group non-compliance and similar negative effects. Businesses that want to determine the needs and satisfaction levels of their employees make measurements with different methods and benefit from the results they obtain at the stage of making managerial decisions.

According to the findings obtained from the data collected by the survey method in order to investigate the relationship between job satisfaction and performance evaluation, a positive relationship was found between job satisfaction and performance evaluation. Other findings obtained as a result of the research are as follows:

The higher the seniority and education levels of the employees, the higher their job satisfaction. Considering that elementary, middle school, high school and associate degree graduates working in the business are in the blue-collar staff, it is concluded that the satisfaction of the white-collar employees is higher than the blue-collar employees.

As the education level of the employees increases, their satisfaction with working conditions increases. From this, it is concluded that, compared to blue-collar employees, white-collar workers are more satisfied with their working conditions.

As the seniority of the employee increases, the rate of seeing himself as more competent and well-informed also increases.

There is a positive relationship between performance appraisal practices in the company and job satisfaction of the employees.

There is a relationship between job satisfaction and wage satisfaction. However, the most important factor affecting job satisfaction is not wage, other variables (relations with supervisors, corporate perception, skills of the employee, friendship relations) come to the fore. This point reached also supports the theoretical knowledge in the first part of the study that today wages alone cannot provide a satisfaction-enhancing effect.

As a result, employees achieve job satisfaction as long as their requests and expectations are met. Businesses that want to increase the job satisfaction of their employees must first be able to accurately determine the needs of their employees. Employees, who see that their gains increase in return for their performance, should increase both their performance and job satisfaction in this way. That is, the total performance of an enterprise with a high performance and job satisfaction group will also increase.

REFERENCES

Akalin & Koç. (2014). Satış personeline yönelik performans değerlendirme ölçeğinin geliştirilmesi ve psikometrik özelliklerinin incelenmesi. *Eğitim Bilimleri Araştırmaları Dergisi*, *4*(2), 227-241.

Akgemci, T., & Güleş, H. K. (2009). *İşletmelerde Stratejik Yönetim*. Gazi Kitabevi.

Akın, A. (2002). İşletmelerde insan kaynakları performansını değerleme sürecinde coaching (Özel rehberlik). *C.Ü. İktisadi ve İdari Bilimler Dergisi*, *3*(1).

Aktan, C. C. (2002). Performans yönetimi: Organizasyonlarda performans değerlendirme ve ölçme. *Organizasyon ve Yönetim Bilimleri Dergisi*, *1*(1), 25–33.

Baş, İ. M., & Atar, A. (1990). *İşletmelerde Verimlilik Denetimi: Ölçme ve Değerlendirme Modelleri*. MPM.

Ceylan, R. (2007). Genel işletme. Anadolu Üniversitesi Yayını.

Demirkaya, H. (2000). *Performans ölçüm rehberi*. Sayıştay Yayın İşleri Müdürlüğü.

Doğan, H. (2014). *Örgütsel Adalet Algısı İle İş Performansı Arasındaki İlişki: Afyonkarahisar'da Beş Yıldızlı Termal Otel İşletmelerinde Bir Araştırma* (Master's thesis). Afyon Kocatepe Üniversitesi, Sosyal Bilimler Enstitüsü.

Feldman, C. D., & Arnold, J. H. (1996). *Managing Individual and Group Behavior in Organization*. Mc.Graw-Hill International Book Company.

Ludeman, K. (2000). How to conduct self-directed 360. *Training & Development*, *54*(7), 44–44.

Mathis, R. L. (2015). *Human resource management: Essential perspectives*. Cengage Learning.

Önder, B. (2000). Kamu Yayın Kurumlarında Personel Yönetimi. TRT.

Pugh, S. D. (1991). *Organizational Behaviour*. Prentice Hall.

Sabuncuoğlu, Z. (2013). İnsan kaynakları yönetimi (uygulamalı). *BETA Bulletin of Experimental Treatments for AIDS*.

Schermerhorn, J. R., Jr. (1989). *Management for productivity*. https://aydancag.com/geleneksel-performans-degerlendirme-yontemleri/

Chapter 15
Low Density Tourism in the Global South:
Second Home Tourism in South Africa as a Form of Visiting Friends and Relatives

Unathi Sonwabile Henama
https://orcid.org/0000-0002-1111-0729
Central University of Technology, Free State, South Africa

Lwazi Apleni
https://orcid.org/0000-0001-8521-6174
University of Zululand, South Africa

Jankie Kgalabi Phale
Tshwane University of Technology, South Africa

ABSTRACT

Tourism has grown since the first democratic elections in 1994 in South Africa, which led to the election of Nelson Mandela as President. The high levels of concentration of tourism in major urban centres has limited the developmental potential of tourism. The first type of second home tourism is located in high amenity areas and is dominated by the upper- and middle-class South Africans. The high amenity nature of these localities has led to the emergence of a strong leisure and business component alongside second home tourism. The second home tourism market in South Africa is dominated by working-class South Africans who work in urban centres and have homes in former apartheid-created homelands, where family and extended family reside on ancestral land. These working-class travellers dominate domestic tourism trips and the visiting friends and relatives market in South Africa.

DOI: 10.4018/978-1-6684-4548-8.ch015

INTRODUCTION

Tourism has emerged as an economic diversification avenue for African economies, as the tourism product offering is produced and consumed at the destination area, which allows for the majority of the value adding to occur at the destination area. Tourism to Africa is disproportionally important for African economies, even though Africa is yet to reach its peak as a tourism destination. “Africa’s tourism industry remains relatively underdeveloped. Of the 1.2 billion people traveling internationally in 2016, only 58 million arrived in Africa-roughly 5% of the world’s inbound tourism” Signe (2018: 3). Africa receives less than 10% of global tourism receipts, is as much as tourism has become a major export sector for many African countries. As noted by Sharpley & Telfer (2004) international tourism is still largely dominated by the industrialised world, with the major tourism flows being primarily between the more developed nations. Africa has grown in leaps and bounds in attracting tourism arrivals and seeking to benefit from the perennial nature of tourism growth, making it the largest industry in the world.

Signe (2018) noted that during the global financial crisis in 2007-2008, Africa was the only region to show continued growth in the tourism industry, and international tourist arrivals on the whole continent have increased by nearly 36 million between 2000 and 2017. “Africa has the highest growth rates in visitors over the decade spanning 2005-2014” Daly & Gereffi (2017: 2). “Tourism to Africa is critical for growth and yet international arrivals are focused mainly on South Africa, Egypt, Morocco, Tunisia and Mauritius, who collectively obtain nearly three quarters of Africa’s tourism receipts” Sifolo (2016:1). The outbreak of contagious diseases such as Ebola negatively impacted on tourism arrivals in Africa. Maphanga & Henama (2019) noted that the 2014 and 2015 Ebola outbreak in West Africa presented many challenges for the hospitality sector. “The potential for tourism growth in Sub-Saharan Africa is significant. The region has abundant assets, with expansive beaches, plentiful wildlife, extensive natural and cultural attractions, and adventure opportunities” Christie et al. (2013: 3). Tourism is a transport intensive industry, and the competiveness of African destinations depends on making it easy, cost-effective and pleasurable to travel to and from their home destinations to Africa for tourism consumption.

The pursuit of tourism has been the enemy of many tourism destinations that have succeeded in attracting tourists, exceeding their carrying capacity which in turn created conflict with the local community. Increased traffic congestion, increased property prices, too many tourists, gentrification, and the loss of local culture and uniqueness. Tourism can compete with locals and destination communities for scarce resources, by overloading infrastructure, damaging nature, which led to the emergence of a phenomenon of overtourism. Overtourism had caught the attention of tourism academia as the most pressing issue before the Covid19 pandemic. Tourism consumption in low density area is imperative as a means for diversifying the economic activities which may be dominated by declining fortunes in agriculture and mining. As other economic sectors are mechanising and shedding jobs, tourism is labour-intensive and can use labour shed from other economic sector, after reskilling. The gaze that the chapter seeks to highlight is on area of low tourism consumption, where tourism uses existing local infrastructure, immersed into the local culture and conceptualised by the local community, representing relevant cases of low density tourism from the Global South. “In South Africa, the promotion of tourism has been identified as a keys strategy that can led to economic growth, community development and poverty alleviation” Ramukumba et al., (2012: 39). Tourism more than any other economic sector has the potential to drive and address the stubborn structural challenges of poverty, unemployment and inequality that continue to haunt South Africa.

The tourism economy in South Africa is dominated by the tourism triangle of Cape Town-Durban-Johannesburg, and major tourism corridors such as the Garden Route (Cape Town to Port Elizabeth), Kruger National Park and Mpumalanga for the bush experience. The geographic spread of the developmental benefits of tourism is major initiative of the state.

As many local authorities observed the decline in their economic fortunes, tourism emerged as a relatively easy economic sector that could be used to restore economic prosperity. In as much as the vast majority of South Africans reside in urban areas, rural hinterlands are important, and restoring their economic independence is important. "South Africa has 278 local municipalities (LMs), the vast majority of which fall into the category of small towns and rural areas" Rogerson (2016: 6). As noted by Donaldson (2018) many small centres are experiencing continued poverty, the loss of job opportunities in the formal sector, the outmigration of skilled workers and the dependence of much of the towns' population of state welfare. Tourism in such low density areas, located in rural areas, far from the urban fringe can have a positive impact in diversifying rural economies. Cyclical movements remains a feature of life in South Africa between the urban centres of employment and commerce and the rural areas. What is encouraging is that there are established migratory patterns which can be used to further strengthen pro-poor rural tourism. Rural tourism is one of the very few avenues to ensure that the developmental benefits of tourism are geographically spread within the Global South.

THEORETICAL FRAMEWORK: LOW-DENSITY TOURISM

The global discourse in the recent years has been the emergence of overtourism as a major factor that impacts negatively on destinations in the world, as destinations became successful in attracting tourists, they experienced overtourism. Overcrowding is by no means new-but it seems to be coming to a head in popular destinations across the world" McKinsey &World Travel & Tourism Council (2017: 4). This can be attributed to two things, the first, the fact that all countries have jumped on the tourism bandwagon to become tourism destination, and the institutionalization of the habit of holidays, which has meant that as the middle class grows, it grows with tourism consumption as part and parcel of normal human life. "2017 was one of the strongest years of GDP growth in a decade with robust consumer spending worldwide. This global growth transferred again into Travel & Tourism with the sector's direct growth of 4.6% outpacing the global economy for the seventh successive year" Oxford Economics (2017: 1). OECD (2014) noted that international tourists arrivals are projected to surpass 1.8 billion by 2030, on the back of rapid growth in emerging tourism economies.

Promoting less visited places is important is ensuring that overtourism is avoided, and ensuring the geographic spread of tourists and their expenditure. McKinsey &World Travel & Tourism Council (2017) noted that spreading visitors involved doing the following:

- Promoting less visited attractions: Some countries and cities are shifting the focus on promotions away from the most-visited attractions. In addition to altering their own materials, tourism authorities can partner with tour operators, travel agencies and travel journalists to alter the messaging and imagery they use to promote less-visited attractions,
- Developing new routes and attractions: Some destinations are developing new routes and attractions to get people to travel to less-congested sites. The success of this approach requires close

collaboration between the public and private sectors, as well as creativity in developing related infrastructure. Some development are done with specific tourist segments in mind.

Spreading the developmental benefits of tourism is important because both in area of high tourism demand and spreading geographically to areas of low tourism density. This is important to spread revenue within destinations, and sustain jobs and fragile economies. The tourism industry is the world's largest industry and spreading its developmental benefits to areas of low density is an opportunity to correct the mistakes of lack of planning in the destinations today that are facing overtourism. The success of low density tourism within the South African context may be in promoting route-tourism, linking several towns and cities for tourism consumption. One of the established route is the Garden Route, that begins in Cape Town (Western Cape Province) and ends in Port Elizabeth (Eastern Cape Province). As noted by Donaldson (2018) present-day South Africa has a shortage of tourism routes. This can only be achieved if community-based tourism is implemented that is pro-poor in nature, with the intention to share the developmental benefits of tourism.

POST-PRODUCTIVIST RURAL HINTERLANDS

According to Visser & Hoogendoorn (2015) the post-productivist countryside is closely associated with the decline in the number of farmers in addition to broader processes of agricultural and rural change such as recession, technological advancement, saturation of markets, agricultural decline, freezing of assets, and disinvestment. "On-farm diversification that incorporates tourism activities has become one of the main reasons by which all of the towns involved aimed to maintain or expand their economic base". Visser & Hoogendoorn (2015: 116). If second home tourism are carefully planned to be pro-poor, it can benefit existing economic supply chains such as agriculture, manufacturing and construction. It can increase the demands for goods and services in a locality and would increase tax coffers through rates and taxes for rural municipalities with limited tax bases. The purchase of second homes was motivated my lifestyle needs, seeking a break from intense urbanisation over weekend and holidays. They bring in "new" money into the fragile rural economy and can infuse their entrepreneurial capital into the local economy, which can in turn create more job opportunities and benefit the local economy.

Since the end of apartheid in 1994, the rural hinterlands and countrysides have undergone many changes. The highly subsidised farming industry ceased to exists, and this led to intense consolidation of farms. Urbanisation has increased in line with national spatial planning initiatives that sought to consolidate South Africans into urban centres, to concentrate development in those urban centres. Irvine et al. (2016) noted that post-productivism has seen a shift away from traditional activities in rural areas such as agricultural productiction. Rural areas have evolved to become places of consumption-related economic activities that increasingly include tourism and leisure spaces for play. "Outside the metropolitan areas of South Africa, tourism has assumed an equally pivotal role in the search for economic diversification in secondary centres and the shift to a post-productivist countryside accompanying rural restructuring" Rogerson & Rogerson (2014a: 95). This has seen a fusion of agriculture and tourism, into special interest tourism such as farm tourism, rural tourism, agritourism and adventure tourism. The "move back to the rurals" has emerged as a counter to urbanisation, and technology in the last decade has facilitated the professionals that work from home, and those that commute for work between areas rural and urban areas. The second home owners would but houses and farmlands and invoke entrepreneurial creativity that would seek to supplement their incomes, and this has seen the emergence of coffee shops, art galleries and places of entertainment due to their entrepreneurial capital.

SECOND HOME OWNERSHIP IN AFRICA: FROM PLEASURE PERIPHERY TO MIGRANT LABOUR VISITING FRIENDS AND RELATIVES (VFRS)

Visiting Friends and Relatives represents a sizable share of domestic tourism in South Africa. Department of Tourism (2018) in the *State of Tourism Report 2017/18* noted that the majority of international tourists travelled for leisure, recreation and holiday purposes (56%), followed by those who travelled to visit friends and relatives (VFR). The majority of VFR stay in unpaid accommodation and this may lead to an under-representation of the economic importance of VFR tourism. "The critical significance of VFR travel expenditures for destinations was recognised particularly in terms of spending on transport, food and gifts" Rogerson (2017: 6).

Rogerson (2017) noted that VFR dominates tourism development in small towns and distressed underdeveloped regions of the country. Donaldson (2018) is a vital component of the tourism industry in South Africa, which is often overlooked. Second home ownership are associated with consumption-led migration in amenity rich areas, where the purchase of a second home allows for occasional travels on weekends and holiday periods away from the primary home. The primary home would in all probability be located within the urban areas, which dominate economic activity in South Africa. Increased urbanisation, led to the emergence of several towns and cities on the urban fringe, that became targets for purchasing second homes tourism, located in areas of low-economic density, a form of rural tourism. VFR is closely associated with second home tourism within the context of South Africa in the Global South. Rogerson & Rogerson (2014) identified the accommodation sector as the base for tourism-led local economic development in South Africa.

"Entrepreneurship plays a prominent role in the early development stages of tourism in small and rural towns where international hotel chains and multi-national corporations are most likely to be attracted to do business and provide potential investments" Mtsiliza (2017:2). Rural entrepreneurship using tourism has much promise to be more sustainable, with fewer leakages and be owned by locals. Strengthening tourism-agricultural linkages are can achieve immediate positive economic beneficiation for the destination. Within the South African case, local government is prescribed by legislation to be developmental in nature, and as such tourism-led local economic development have been adopted in many localities. Hoogendoorn & Visser (2010) noted that the existence of second homes created an opportunity for leisure activities such as horse riding and golf to be developed for consumption by second home tourists and their guests. Second homes are bought primarily for recreational place utility as noted by Long & Hoogendoorn (2013). Second homes serve as holiday homes, and often with investment returns in mind in the South African context as noted by Hoogendoorn et al. (2005).

Second homes tourism can also drive up entrepreneurship in a locality. Hoogendoorn & Visser (2004) noted that the completion of the completion of the Lesotho Highlands Water Project in Clarens, the town was plunged into recession and a resourceful entrepreneurs proceeded to buy large number of properties and market the town as a unique leisure hide away.

Hoogendoorn et al. (2005) noted that second homes development benefitted the local community. The purchase of second homes, benefits the local economy as improvements are made on the second homes. "Second homes are thought to bring about the inflow of capital to local areas, but their impact varies depending on the local context" Irvine et al. (2016: 392).

In research conducted in Clarens by Hoogendoorn & Visser (2004) the second home owners used local contractors when building or renovating their second homes. Second home tourism increased the economic base of towns and cities by creating jobs and business opportunities that would have never existed in the absence of second home tourism. Second homes can also attract investment into a destination. In research about Clarens by Hoogendoorn & Visser (2004), there were plans to small shopping centre complex. "All the participants employed a gardener and domestic worker, in addition to the obligation of paying the monthly rates, electricity, refuse removal, water and security bills" Hoogendoorn & Visser (2004: 112).Hoogendoorn et al. (2005) noted that a large number of towns and villages along the South African coastline, and increasingly in the interior, have grown significantly as a result of consumption-led migration. "A majority of owner's primary residence was within one hour travelling distance from the second home. This would suggest that these second homes probably act as weekend homes" Hoogendoorn et al. (2005: 126). The investment outcome in amenity rich areas may be for retirement outcomes. In research conducted by Hoogendoorn et al. (2005) 97% of respondents bought their second homes with the intention of retiring in Clarens at some time in the future.

"Internationally, the most widely used term to describe the phenomenon of people owning or utilizing some sort of dwelling for recreational and secondary purposes is a second home" Marjavaara (2008: 7). "Second home tourism comprises an important component of the discourse in tourism geographies, since its re-emergence in the literature in the late 1990S. Second homes are often situated in amenity-rich areas within pristine natural environments along coastal areas, river banks and lakes" Hoogendoorn & Fitchett (2018: 1). Several towns in South Africa are associated with second-home ownership, such as Franschhoek (Van Laar, 2011), Clarens (Hoogendoorn & Visser, 2004), George (Ramukumba et al., 2012), Pringle Bay (Pienaar & Visser, 2009), Hermanus (Pienaar & Visser, 2009), Dullstroom (Hoogendoorn et al., 2005), Greytown (Visser & Hoogendoorn, 2015), Rosendal (Hay & Visser, 2014), Nieu Bethesda (Hoogendoorn & Visser, 2010), and Hartbeespoort (Long & Hoogendoorn, 2013). Van Laar (2011) noted that second home owners are economically active and they are categorized as upper-level income earners, indicating a high purchasing power. Rogerson & Rogerson (2014a) noted the vast majority of second home owners were identified through their rate-based address listings on the municipality database. Rogerson (2016) identified that local municipalities that have developed into second home destinations such as Hermanus, Plettenburg Bay, Graaff Reinet, Beaufort West and Dullstroom exhibited higher levels of both leisure and business travel than those located in the former homelands.

"Second home owners in these towns are predominantly white, middle-aged, male, wealthy, highly-educated, highly mobile and able to migrate fairly regularly for leisure purposes. The overwhelming majority of second home owners in these towns come from the main metropolitan regions of the country (Johannesburg, Durban and Cape Town) and some secondary cities such as Bloemfontein and Port-Elizabeth" Hoogendoorn (2011: 41). South Africa has two types of second homes, firstly, consumption-led second homes owned primarily by the white, and wealth South Africans and secondly, residency led second homes located in rural hinterlands associated with the homelands systems created during apartheid, primarily owned by Black South Africans. "In the case of South Africa, a number of debates have emerged; however, second home research has focused on white, wealthy, mobile and highly educated second home owners, mirroring many of the debated of the Global North" Rogerson & Hoogendoorn (2014: 168). There are positive and negative impacts associated with second homes. Hoogendoorn & Visser (2015) noted that in South Africa, second homes may push up property prices, which may take away housing equity, but the presence of second homes owners does allow for more services to be established, which is the core problem in main areas where service delivery is severely lacking and tax bases small. Residents of Rosendal in research conducted by Hay & Visser (2014) noted that second home owners and guests quad-bike in the streets and tend to host rowdy parties during their visits, disrupting the tranquility of the town.

Careful planning is therefore necessary to ensure that second homes do not create the conflicts associated with mass tourism. This is dependent on having a capable and competent public service, especially at local government level to undertake this responsibility, as the level of government, closest to the public. In research conducted by Hoogendoorn et al. (2005), not all addresses point to predictable higher-income areas, with some lower income areas including and surprising even former "black group area" neigbourhouds. In research conducted by Visser & Hoogendoorn (2015) noted that during the past 10 years, it become quite clear that in the case of South Africa, second home tourism is not exclusively to the White, wealthy, mobile and educated. Hoogendoorn (2011) noted that second homes usage in the developing world, for especially low-income earners, is almost exclusively used for purely tourists or leisure purposes. According to Rogerson & Hoogendoorn (2014) VFR travel in South Africa is the segment of travel that is massively dominated by "ordinary" or working-class travellers, and 78% of VFR travellers are black population. These "black group areas" included the bantustands (artificial independent black republics) and there acted as labour reservoirs, and influx control laws kept the "unemployed" in the rural areas, and they served as the primary homes. "From 1948 onwards Black South Africans effectively were stripped of their citizenship making them legally citizens of one of the ten ethnically-based and nominally self-governing Bantustants or tribal homelands" Rogerson (2014: 4).

Black South Africans were considered to be 'temporary' residents in urban areas, which were the areas of work and commerce. Rogerson & Hoogendoorn (2014) noted that the majority of VFR travel is mainly towards the rural municipalities in Limpopo, Mpumalanga and KwaZulu-Natal provinces. During apartheid in South Africa, the tourism industry catered exclusively for the Whites, due to the repressive racial policies. This led to the development of thriving VFR amongst Black South Africans. "With the growth of black urbanisation from the 1930s it is likely that the phenomenon of VFR travel involving return visits to rural areas which already was an established form of mobility, albeit uncounted in official tourism statistics" Rogerson (2015:145). Rogerson & Hoogendoorn (2014) noted that in the Global South, there working class migrate between first and second homes through circular migration as a consequence of labour migration. Rogerson & Hoogendoorn (2014) noted that for low-income earners, the "second home" is not used for leisure purposes. Actually, it can be regarded as the first and primary home, as it's located possibly on ancestral land, and would be occupied by family and extended family. According to Sake24 (2010) a third of black South Africans have second homes, located in former townships, rural properties, informal accommodation and farms. Hoogendoorn (2011) noted that second home ownership tourism forms an important part of labour migration and a midway point between Visiting Friends and Relatives (VFR) tourism and economic survival. Rogerson (2015) noted that most migrants maintain intensive contact with relatives in rural area, and visit occasionally during school holiday periods, for celebrations and ceremonies (funerals, weddings and circumcisions) as well as for traditional healing. The working class citizens that commute between the rural homelands and the places of work, may not consider their activities as a form of tourism, specifically VFR tourism.

At the urban centres, a place of dwelling would be acquired but upon retirement, the ancestral land in the rural hinterlands located in the homelands, would have preference as the primary dwelling place, and therefore its utility would change from being a second home, to becoming the primary residence, whilst the urban home becomes a second home. This unique South African reality means that the rural areas must be supported with good infrastructure such as roads, electricity and telecommunications as a means of opening up the areas for the easy movement of people. The rural hinterlands and the second homes of the wealthy therefore experienced peak demands during the school holidays, long weekends and for important events, when movement is achieved from the urban centres. "Peripheral spaces of national tourism economies are important regions for research by tourism scholars" Rogerson (2017:14). This is imperative in the context of South Africa tourism, which wishes to ensure geographical dispersal of the tourists, to ensure that there is shared growth in tourism. Tourism consumption in rural areas is important to diversify the rural economies, and inject new money in the rural economy that may be dependent on a single industry such as agriculture and mining.

The rural areas disproportionally depend on local government for the delivery of services such as water, electricity, basic services and maintaining the road infrastructure. Unfortunately, in the context of South Africa, local government has become a perennial under-performer and has therefore become a stumbling block for the growth of tourism in South Africa. VFR can contribute toward leisure consumption in economically distressed localities should local authorities working with the tourism industry package and sell leisure experiences. Marketing collateral and efforts must be also in vernacular languages, to ensure that deeply held views that "tourism is a white man's thing" are dispelled, so that VFR can contribute to increased leisure consumption. Such acts of tourism publicity could allow for entrepreneurship by locals within the former homelands to also offer authentic tourism experiences, which are off-the-beaten-track and participate as product owners within the tourism industry. The sharing economy through companies such as Airbnb can offer homestays and tailor-made tours using Airbnb Experiences that have changed the supply chain dynamics within the tourism economy and unearthed destinations and experiences that were off-the-beaten track. The established second home destination with high amenity value would achieve the objective of urban-to-rural retirement as the owners would retire to the rural hinterlands and occupy what was once a second home, on a full-time basis. As the high amenity areas which were once second homes, become primary residence, it would stimulate VFR tourists from family and friends. In all honesty, the high amenity second homes would exhibit the second homes experiences of the majority of working class South Africa, becoming their primary residence, primarily located in the rural hinterland, outside the urban fringe.

RHODES VILLAGE, EASTERN CAPE PROVINCE

Du Preez & Hosking (2011: 268) "Rhodes village is located at the foot of the southern Drakensburg Mountains in the North-Eastern Cape, South Africa. Rhodes lies deep in the southern Drakensberg, in the quiet of the highest mountains in the Eastern Cape, where the purest water springs from the earth. Here in Rhodes lies a retreat from civilisation. Tiffindell Ski and Alpine Resort was established in 1993 on the slopes of the Ben McDhui peak of 3001 metres, and is the only ski resort in sub-Saharan Africa. In the winter months from June to August, over 5 000 visitors from all over South Africa take part in the snow fun activities on the slopes. The natural snow is supplemented by a snowmaking system that has guaranteed on average 95 days of skiing a year. The demand for snow results in an overflow of visitors into the nearby hamlet of Rhodes according to the Drakensberg Mountains (2018). Commercial activities in the Rhodes region comprise wool and meat farming and tourism-related businesses". "Rhodes village is 1940m above sea level and 16km south of Lesotho and more than 600km from major city" Hoogendoorn et al. (2009: 77). " Rhodes became a popular venue for permanent families in the late 1960s and 1970s. Rhodes was "rediscovered" shortly thereafter by a group of people seeking an alternative lifestyle. This period in local parlance was referred to as the "Hippie era". The advent of the hippie-era signalled the beginning of the current post-productivist/ second home and tourism phase" Hoogendoorn et al. (2009: 78). According to Hoogendoorn & Visser (2010) since the 1980's Rhodes has gradually become better known as a tourist destination for professionals from South Africa's main metropolitan region, primarily Johannesburg and Pretoria. Rhodes is not the only snow tourism destination in Southern Africa, in neighbouring and landlocked Lesotho, Afriski, is a winter and ski resort.

This created the perfect conditions for establishing ski's recreation using the perennial snow supply that falls during the winter months from May to August at the Tiffendell Skii Resort. Tiffindell (2019) noted that Tiffindell is at an altitude of 2720m, nestling on the slope of Ben McDhui (3001m), the highest peak in the whole Cape Province, making it the highest resort in South Africa. The entire town of Rhodes, situated in the valley 20km away, is steeped in history, and is a national monument, boasting a 100-year history starting in Victorian times. Tiffindell (2019) noted that Tiffindell came into existence in 1993, born by avid skiers and run by passionate snow lovers. Since then the resort has grown to become a true Alpine resort for all seasons. The area is one of South Africa's premier fly-fishing destination with well stocked streams throughout the year. Du Preez & Hosking (2011) noted that trout fishing around Rhodes provides: firstly, a food product and secondly, a recreational angling resource, which is one of the fastest growing tourism attractions in South Africa. "The South African trout industry is well established and most, importantly, generates both income and jobs in some of the poorest and most underdeveloped rural areas of South Africa" Du Preez & Lee (2010: 242).Being one of the few areas that have enough snow for recreational skiing, led to the outsiders buying housing stock in Rhodes for amongst others the snow experience, the abundance of trout and the rural lifestyle of the town. Jopson (2018) noted that Alpine Swift Trails is another big reason people come here – for trail running, hiking and mountain biking over an impressive facility built around a 25-metre swimming pond fed by springs and cleaned by water lilies and koi. This attracts sports tourism and adventure tourism enthusiasts which helps to diversify the base tourism in Rhodes from snow tourists and second home tourism.

As a result of the snow opportunity, Rhodes has attracted second home tourists that have bought property and this has gradually changed the character of the town. Hoogendoorn & Visser (2010) noted that second home owners spend on average R744 at local restaurants, R694 at art galleries and R819 on petrol at the local petrol stations on previous trips. Hoogendoorn & Visser (2011) uncovered in research that second home in Rhodes stated an average of 22 days a year in their properties in Rhodes. Hoogendoorn et al. (2009) noted that 48% of the second home owners reside far away in Gauteng. Du Preez & Lee (2010) the large majority of the population earned less than R800 per month, the population amounting to 600, and the employment opportunities being limited to farming and tourism. "Second home owners and guests who rent second homes in Rhodes generally take part in a variety of leisure-based activities such as fly fishing, game drives, 4x4 trails, snow-skiing and horse-riding. These activities have proven to have a significant economic impact on the surrounding rural economy" Hoogendoorn & Visser (2011:285). Burakowski & Magnusson (2012: 5) noted that in 2009/2010, more than 23 million people participated in winter sporting activities as measured through visits to downhill ski resorts and snowmobiling, adding $12.2 billion in economic value to the U.S. economy. The negative impacts of climate change are impacting on climate dependent destination, and in this case on snow destinations. Enough snow must fall on a destination, to make it safe and secure for the consumption of the snow experience. As a consequence, poor snow conditions can have a negative impact on a snow destinations' ability to attract and retain tourists. "Colorado, around 98% of resorts are reliant on what can be created by artificial snow machines" World Travel Guide (2015:1).

This means that destinations must adapt and reflect agility, in planning for the immediate and omnipresent challenge of less snow coverage by using artificial snow to ensure that the guest experience to consume a snow destination is not disturbed. Extending the length of the ski season which had been traditionally determined by climate and geography. However, technology has allowed the extension of the ski season, through the use of artificial snow to ensure client satisfaction and business sustainability. Snowmaking machines have been relied upon to mitigate the decrease of snow cover in many destination, as the impacts of climate change are felt. Climate change will have a sizable impact on the future success of snow tourism around Rhodes. Hoogendoorn & Fitchett (2018) noted that the increase in temperatures and reduction in humidity resulting from the mean annual precipitation reduction hamper the process of artificial snow production increase the rates of evaporation and sublimation of the snow, and thus are likely to reduce the length of the ski season. This would destroy the major drawcard of the region, which is snow and hamper efforts for promoting rural and low-density tourism in areas with limited economic activities outside of tourism and agrcuture. Africa remains an unexplored winter tourism destination specifically for ski tourism, which is an opportunity that needs to be exploited, through marketing outlays so that it becomes a bucket-list activity within the winter months of the South Africa, which is tourism's low season. South Africa is primarily sold as a sand, sea, sun and wildlife destination, and snow tourism is a niche that can be sold to diversify the product offering, especially for the rural economy of Rhodes in rural Eastern Cape. Tourism LED local economic development in Rhodes has seen locals acquire economic gain from meeting the needs of the second home owners and tourists that visit Rhodes. A plethora of locally owned enterprises have emerged such as offering hiking, mountain biking, trail running, bird watching, horse riding, 4x4 trails, winter skiing, and discovering Khoi-San rock art paintings which has given the residents of the town, another revenue stream.

CONCLUSION

These rurally located cities and towns will continue to derive economic benefit from the "new" and cyclical money and investments that second homes bring into the fragile rural economies. These investigation shed light on a neglected gaze in local government, the developmental impacts associated with second homes. Second homes can become a form of tourism-led local economic development for a locality that is rich is amenities. Furthermore, the second home literature has identified that the former homelands created by apartheid are also second homes, and they dominate VFR tourism between urban centres and these rural homelands. Economic remittances are transferred monthly from the urban areas, as the family and extended family continue to reside in the former homelands, which are ancestral lands. The second home tourism economy must complement the local economy, and allow it grow so that it can increase the capacity of meeting the needs of the second home tourists, whilst reducing opportunities for leakage. Community based tourism has emerged as the best means to achieve the aims of rural tourism as should allow for the sharing of benefits using pro-poor principles, whilst protecting the nature resource base upon which tourism is dependent upon.

The ability to empower the community depends on local communities having an asset base (such as land or scarce skills) which could be used to secure a string of benefits to local communities.

Rural localities that are better positioned to benefit from second home development, are those that are closely located to major urban areas. This would increase the frequency of travel. Johannesburg and the Gauteng Province remain the biggest losers from the these urban to rural migration, as residing in Gauteng after retirement is not attractive for those that have investment properties in high amenities areas and for the millions of working class South Africans that retire back to the homelands, where family and extended family resides. Rural and small towns are not homogenous in nature and therefore cannot present homogenous tourism outcomes, as this depends on the resource base, which is primarily natural, to develop tourism. There is need for a dedicated VFR strategy developed that would increase the frequency and the economic contribution of VFR tourism, especially within the former homelands, which dominate VFR travel, and therefore domestic tourism. Furthermore, both provincial and national government must investigate what infrastructure developments such as roads, airports, telecommunication and electricity need to installed to promote the development of VFR to the homelands, and catalyse the development in those homelands so that they can attract investment and become economically independent.

Rural tourism can only be successful if there is close collaboration between the business community and local government. It is therefore imperataive that there are tourism business chambers established in each municipality to ensure that the tourism industry has direct access to municipal leadership and the political leadershop of the municipality. This is important considering the fact that tourism is multi-sectoral and multi-dimensional and requires a 'Whole-of-Government-Approach to Tourism' to succeed. Municipal authorities and city planners must not ignore second homes, and should actively investigate their number, size, and locality for appropriate planning to be undertaken. This is important to investigate the phenomenon, and also to mitigate the possible negative consequences associated with second homes. Ensuring shared growth in South Africa's tourism growth is top of the agenda, to ensure that the developmental benefits of tourism are shared. In addition, South Africa must try and improve its competitiveness as a tourism destination. However shared growth and benefit from tourism remains elusive as the economically depressed areas benefit less from tourism. "Like most industries, tourism is still dominated by the developed countries. Since they are blessed with the highest incomes, developed countries correspondingly contribute the most international tourists to the world economy" Brakke (2004: 7). This means efforts must be made to attract foreign tourists to visit the destinations where second home tourism dominate, through proper packaging and route tourism. Second-home tourism-led local economic development will continue to dominate the geography of several small and rurally dispersed towns and cities in South Africa.

REFERENCES

Altman, M. (2006). Identifying employment-creating sectors in South Africa: The role of service industries. *Development Southern Africa*, *25*(5), 627–647. doi:10.1080/03768350601021871

Apleni, L. (2017). Residents' perceptions on urban tourism as a catalyst for economic development: A case study of Buffalo City, South Africa. *African Journal of Hospitality, Tourism and Leisure*, *6*(3), 1–14.

Apleni, L., & Henama, U. S. (2020). The impact of events in boosting local economic development: A case study of Port St Johns, South Africa. *African Journal of Hospitality, Tourism and Leisure*, (1), 1–12.

Apleni, L., Vallabh, D., & Henama, U. S. (2017). Motivation for tourists' participation in religious tourism in Eastern Cape: A case study of Buffalo City, South Africa. *African Journal of Hospitality, Tourism and Leisure*, *6*(2), 1–20.

Apleni, Mxunyelwa, & Vallabh. (2017). The implications of the proposed carbon tax in South Africa's tourism industry. *African Journal of Hospitality, Tourism and Leisure*, (1), 1–15.

Ashley, C., De Brine, P., Lehr, A., & Wilde, H. (2007). *The role of the tourism sector in expanding economic opportunities: Opportunity, Corporate Social Responsibility Initiative*. Report No.23. Cambridge, MA: Kennedy School of Government, Harvard University.

Brakke, M. (2004). International tourism, demand and GDP implications: A background and empiricial analysis. *Underground Economic Review*, *1*(2), 1–38.

Brelik, A. (2018). Globalization in Tourism. *Proceedings of the 2018 International Conference "Economic Science for Rural Development"*, 9-11. 10.22616/ESRD.2018.004

Brown, D. O. (1998). In search of an appropriate form of tourism for Africa: Lessons from the past and suggestions for the future. *Tourism Management*, *19*(3), 237–245. doi:10.1016/S0261-5177(98)00016-8

Celik, A. K., Ozcan, S., Topcuoglu, A., & Yildirim, K. E. (2013). Effects of the tourism industry on the balance of payment deficits. *Anatolia*, *24*(1), 86–90. doi:10.1080/13032917.2013.772529

Cerovic, S., Vukadinovic, P. & Knezevic, M. (2015). The influence of globalization of tourism and impact on tourism on other activities with an emphasis on greenfield investments in tourism. *Sitcon*, 47-52.

Christie, I. T., & Crompton, D. E. (2001). *Tourism in Africa*. The World Bank. Available at: https://www.worldbank.org/afr/wps/index.htm

Christie, I. T., Fernandes, E., Messerli, H., & Twining-Ward, L. (2013). *Tourism in Africa: Harness Tourism for Growth and Improved Livelihoods*. World Bank.

Cyberscope. (2019). *Maletsunyane Falls in Semonkong, Lesotho: the highest single-drop waterfalls in Africa?* Available at: https://cycloscope.net/maletsunyane-falls-semonkong

Daly, J., & Gereffi, G. (2017). *Tourism global value chains and Africa*. WIDER Working Paper 2017/17. Available at: https://wider.unu.edu

Department of Tourism. (2018). *State of Tourism Report 2017/18*. National Department of Tourism.

Divisekera, S. (2010). Economics of tourist's consumption behavior: Some evidence from Australia. *Tourism Management*, *31*(5), 629–636. doi:10.1016/j.tourman.2009.07.001

Donaldson, R. (2018). *Small Town Tourism in South Africa*. Springer. doi:10.1007/978-3-319-68088-0

Du Preez, M., & Hosking, S. G. (2011). The value of the trout fishery at Rhodes, North Eastern Cape, South Africa: A travel cost analysis using count data models. *Journal of Environmental Planning and Management*, *54*(2), 267–282. doi:10.1080/09640568.2010.505837

Du Preez, M., & Lee, D. E. (2010). The contribution of trout fly fishing to the economy of Rhodes, North Eastern Cape, South Africa. *Development Southern Africa*, *27*(2), 241–253. doi:10.1080/03768351003740654

Economics, O. (2017). *Travel & Tourism Power and Performance Report.* Available at: https:www.oxfordeconomics.com

Grundlingh, A. (2011). Mutating memories and the making of a myth: Remembering The SS Mendi Disaster, 1917-2007. *South African Historical Journal*, *63*(1), 20–37. doi:10.1080/02582473.2011.549372

Hay, A., & Visser, G. (2014). Socio-cultural and socio-economic features of second homes in Rosendal, South Africa. *Bulletin of Geography. Socio-Economic Series*, *26*(26), 157–166. doi:10.2478/bog-2014-0051

Heath, E. (2001). Globalisation of the Tourism Industry: Future Trends and Challenges for South Africa. *SAJEMS*, *4*(3), 542–596. doi:10.4102ajems.v4i3.2662

Henama, U. S. (2013). Attracting Indian Outbound Tourists to South Africa: A BRICS Perspective. *India Quarterly*, *69*(3), 229–247. doi:10.1177/0974928413489466

Henama, U. S. (2017). International Migration and National Tourism Industry: The Case of South Africa. *The EUrASEANs: The Journal of Global Socio-Economic Dynamics*, *2*(3), 72–82.

Honkanen, A., & Mustonen, P. (2007). Tourism consumption revisited-an empirical study of Finnish consumers. *Research on Finnish Society*, *1*, 43–58.

Hoogendoorn, G. (2011). Low-income earners as second home tourists in South Africa? *Tourism Review International*, *15*(1), 37–50. doi:10.3727/154427211X13139345020219

Hoogendoorn, G., & Fitchett, J. (2018). Perspectives on Second Homes, Climate Change and Tourism in South Africa. *African Journal of Hospitality, Tourism and Leisure*, *7*(2), 1–18.

Hoogendoorn, G., Mellett, R., & Visser, G. (2005). Second Home Tourism in Africa: Reflections on the South African Experience. *Urban Forum*, *16*(2-3), 112-154.

Hoogendoorn, G., & Visser, G. (2004). Second home and small-town (re) development: The case of Clarens. *Journal of Family Ecology and Consumer Sciences*, *31*, 105–115.

Hoogendoorn, G., & Visser, G. (2010). The role of second homes in local economic development in five small South African towns. *Development Southern Africa*, *27*(4), 547–562. doi:10.1080/0376835X.2010.508585

Hoogendoorn, G., & Visser, G. (2011). Economic Development through Second Home Development: Evidence from South Africa. *Tijdschrift voor Economische en Sociale Geografie*, *102*(3), 275–289. doi:10.1111/j.1467-9663.2011.00663.x

Hoogendoorn, G., & Visser, G. (2015). Focusing on the 'blessing' and not the 'curse' of second homes: notes from South Africa. *Urban Forum*, *47*(2), 179-184.

Hoogendoorn, G., Visser, G., & Marais, L. (2009). Changing Countrysides, Changing Villages: Second Homes in Rhodes, South Africa. *The South African Geographical Journal*, *91*(2), 75–83. doi:10.1080/03736245.2009.9725334

Irvine, P. M., Kepe, D. T., & Hamunime, N. P. (2016). Whose Mecca? Divergent experiences of post-productisvism and tourism in Nieu Bethesda, South Africa. *The South African Geographical Journal*, *98*(2), 386–401. doi:10.1080/03736245.2015.1052843

Jopson, T. (2018). *Rhodes Village-up where you belong*. Available at: https://www.getaway.co.za/travel-ideas/destinations-travel-ideas/rhodes-village/

Long, D., & Hoogendoorn, G. (2014). Second home owner perceptions of their environmental impacts: The case of Hartebeespoort. *Urban Forum*, *25*, 517-530.

Maphanga, P. M., & Henama, U. S. (2019). The Tourism Impact of Ebola in Africa: Lessons on Crisis Management. *African Journal of Hospitality, Tourism and Leisure*, *8*(3), 1–11.

McKinsey & World Travel & Tourism Council. (2017). *Coping with Success: Managing Overcrowding in Tourism Destinations*. WTTC.

Mogale, P. T., & Odeku, K. O. O. (2019). Perspectives on transformative legislative and policy framework promoting tourism for poverty alleviation in South Africa. *African Journal of Hospitality, Tourism and Leisure*, *8*(2), 1–14.

Mtsiliza, N. S. (2017). Seeking strategies for sustainability in Tourism Entrepreneurship in South Africa. *African Journal of Hospitality, Tourism and Leisure*, *6*(4), 1–10.

NovemberB. (2020). *About us*. Available at: http://www.berlinnovember.co.za/index.php/about-berlin-november

OECD. (2014). *OECD Tourism Trends and Policies, 2014*. OECD Publishing. doi:10.1787/tour-2014-

OECD. (2017). *Analysing megatrends to better shape the future of tourism: Issue Paper.* OECD Publishing.

OECD. (2018). Analysing megatrends to better shape the future of tourism. OECD Tourism Papers, No. 2018/02. OECD Publishing.

Pienaar, J. J., & Visser, G. (2009). The Thorny Issue of Identifying Second Homes in South Africa. *Urban Forum*, *20*, 455-469.

Ramukumba, T., Mmbengwa, M., Mwamayi, K. A., & Groenewald, J. A. (2012). Analysis of the socio-economic impacts of tourism for emerging tourism entrepreneurs: The case of George municipality in the Western Cape Province, South Africa. *Journal of Hospitality Management and Tourism*, *3*(3), 39–45. doi:10.5897/JHMT11.038

Rogerson, C. M. (2014). The Uneven Geography of Tourism in South Africa. *African Journal of Hospitality, Tourism and Leisure*, *3*(1), 1–14.

Rogerson, C. M. (2015). Revisiting VFR tourism in South Africa. *The South African Geographical Journal*, *97*(2), 139–157. doi:10.1080/03736245.2015.1028981

Rogerson, C. M. (2016). Outside the cities: Tourism pathways in South Africa's small towns and rural areas. *African Journal of Hospitality, Tourism and Leisure*, *6*(3), 1–10.

Rogerson, C. M. (2017). Visiting friends and relatives travel matters for sub-Saharan Africa. *African Journal of Hospitality, Tourism and Leisure*, *5*(3), 1–16.

Rogerson, C. M., & Hoogendoorn, G. (2014). VFR Travel and Second Home Tourism: The Missing Link: The Case of South Africa. *Tourism Review International*, *18*(3), 167–178. doi:10.3727/154427214X14101901317156

Rogerson, C. M. I, & Kiambo, W. (2007). The growth and promotion of regional tourism in the developing world: The South African experience. *Development Southern Africa*, *24*(3), 505–521. doi:10.1080/03768350701445608

Rogerson, J., & Rogerson, C. M. (2014a). Agritourism and local economic development in South Africa. *Bulletin of Geography. Socio-Economic Series*, *26*(26), 93–106. doi:10.2478/bog-2014-0047

Rogerson, J., & Rogerson, C. M. (2014b). Maximising the local development potential of Nature Tourism accommodation establishments in South Africa. *African Journal of Hospitality, Tourism and Leisure*, *3*(1), 1–20.

Saayman, A. & Saayman, M. (2010). Forecasting tourist arrivals in South Africa. *Acta Commercii*, 281-293.

Sharpley, R., & Telfer, D. J. (2004). *Tourism and Development: Concepts and Issues*. Channel View Publishers.

Sifolo, P. P. S. (2016). A reflection on tourism and related security implications on the African continent. *African Journal of Hospitality, Tourism and Leisure*, *5*(3), 1–10.

Signe, L. (2018). *Africa's tourism potential: Trends, drivers, opportunities and strategies. African Growth Initiative*. Brookings Institute.

Tech, B. (2019). *These 6 towns in South Africa are attracting a new wave of semigrants.* Available at: https://businesstech.co.za/news/property/344518/these-6-towns-in-south-africa-are-attracting-a-new-wave-of-semigrants/

United Nations Conference on Trade and Development. (2017). *Economic Development in Africa Report 2017. Tourism for Transformative and Inclusive Growth*. United Nations.

Visser, G. (2004). Second homes and local development: Issues arising from Cape Town's De Waterkant. *GeoJournal*, *60*(3), 259–271. doi:10.1023/B:GEJO.0000034733.80648.88

Visser, G., & Hoogendoorn, G. (2015). A decade of second home tourism in South Africa: Research prospects for the developing world? *The South African Geographical Journal*, *97*(2), 111–222. doi:10.1080/03736245.2015.1028976

World Economic Forum. (2018). *Travel and Tourism Competitiveness Report, 2018*. WEF.

KEY TERMS AND DEFINITIONS

LED: Local economic development.
VFR: Visiting friends and relatives.
WEF: World Economic Forum.

Chapter 16
Sustainable Accommodation in Low Density Areas:
The Case of Glamping Tourism in the Northern Lights

Stylianos A. Bouzis
https://orcid.org/0000-0001-5528-0108
International Hellenic University, Greece

Alexandros Filiopoulos
Uneversity of the Aegean, Greece

ABSTRACT

The main criterion for writing this research is the spatial dimension that tourism takes in some cases. In particular, tourism activity in areas exposed to the Northern Lights, which already exists and tends to develop further through sustainable glamping accommodations in low-density areas, passes into the sphere of interest for this chapter. The aim is to document and make known these attractive sustainable businesses, as well as the common characteristics they share on the supply axis. The geographic restriction concerns six countries in the Arctic Circle that exhibit these characteristics. In a second year, conclusions are drawn from the comparative case study; the research questions designed are answered, and future perspectives are presented that are fully in line with the new external changes that have formed in recent years.

INTRODUCTION

Taking as a given the spatial dimension of tourism, the research field of this research concerns the tourist activity in glamping accommodations, but with a geographical limitation, which is defined by the parallel of latitude (66°-33´-38´´´) north of the Equator and the main criterion of low density of geographical areas.

Including, however, the COVID-19 epidemiological crisis, which has brought the global community and its entire range of activities an uninterrupted recession, it is difficult to measure its impact (OECD,

DOI: 10.4018/978-1-6684-4548-8.ch016

2020). The health crisis caused enormous damage to the global economy, and this clearly had an impact on the broader tourism sector, which was negatively affected, and its hitherto upward trajectory was halted. However, given that it is a highly dynamic and complex phenomenon with many facets, one of its characteristics is its adaptability. The recovery must keep pace with the trends of the post-COVID 19 era and both the development strategy and new tourism products must be in line with them (Skryl & Gregoric,2022). Adopting the philosophy of sustainable tourism development combined with the promotion of alternativity and following the logic of adaptation, another variable, that of low density, has helped and continues to help the emergence of new ways of expressing tourism activities, in the light of now social distancing, as it results from the sequence of events. But can tourism become established and contribute to the development of low-density areas? Do tourism flows have the power to influence the sustainable development of tourism in these areas? These are two of the key research questions, the answers to which emerge effortlessly with the help of the case study research results.

By conducting the case study, we examine whether and to what extent tourist destinations, in low-density areas, can not only remain unscathed, but also evolve and emerge as fully sustainable tourism resources. What is the role of alternative forms of tourism and what is the role of glamping sites in this location? Questions also arise about the contribution of glamping accommodation to tourism recovery and the extent to which social factors influence the glamping - Northern Lights dichotomy.

The focus of this study is on six (6) countries, which are shown to have the best views of the Northern Lights and at the same time have low territorial and therefore population density. These countries are Norway, Sweden, Iceland, Canada, Lapland, and Greenland. The reference to these countries comes from a cross-check of independent and personalized online travel guides. The case of glamping sites in areas where the Northern Lights are visible is an important example with a strong sustainability element. Tourist sentiment has changed and this change, triggered by COVID-19, is expected to shrink both tourist preference and behavior. This will in turn increase the number of tourists travelling to destinations with low population and spatial density (Everingham & Chassagne, 2020). Thus, glamping accommodations, which seem to be inextricably linked to low-density areas, can gain a comparative advantage as they provide the sense of safety, social distance, open spaces, and lower density that COVID-19 has established (Duro et al., 2021).

With the restart of tourism and the identification of the future recovery phase by contemporary scholars, it is possible to transform the current production processes (Prideaux et al., 2020) and try to promote tourism flows in low-density areas, where the feeling of safety is more prevalent. At the same time, there are fears of an unrealistic approach, as governments will move towards models that serve job creation and reduce the high levels caused by the epidemiological crisis (Hall et al., 2020).

REFLECTIONS OF THEORIES AND CONCEPTS

Low Density Areas

The numerical density of a country can be calculated by dividing the country's population by its area. This combination is expressed in inhabitants per square kilometer (km2). It is often assumed that high densities are inherently bad and that low densities are inherently good (Forsyth, 2003). However, the link between density, aspects of tourism and sustainability remains a challenge for both theory and spatial planning practice.

The interactions of the concepts are inextricably linked and designed for new age tourism. In areas of low density and small economic scale, i.e., at the local level, alternative forms of tourism can be developed (Clarke, 1997). As tourism is a complex phenomenon with many facets and is highly dynamic, external changes constantly affect it, as mentioned above. Low-density areas, on the other hand, are areas that possess weak demographic and economic assets and involve multi-level concerns such as institutional thickness, endogenous growth potential, and cultural, psychological, and physical distance from more central areas. In addition, they show a general downward trend in their population (O'Donnell, 1997; Ramos & Fernandes, 2016). However, this trend is about to be changed and reshaped by glamping accommodations, which represent the golden mean of sustainability and tourism development.

Sustainability and Tourism Development

For many years now, when talking about development, an identification has been made with the sustainable development that the planet needs and the multiple impacts it reflects in different sectors, while sustainable development in turn can contribute to the achievement of sustainable tourism and vice versa (Sharpley, 2020). The need for this identification of these concepts is imperative in contemporary global society as much as in the global tourism arena. The UN's Agenda for Sustainability sets out a series of hierarchical goals (SDGs) for the well-being of the planet by 2030.

Sustainable tourism and its economic spectrum are at the center of studies in terms of their contribution to the sustainable development of tourism and the planet (Saarinen, Rogerson, 2014). In essence, when talking about sustainable tourism, the World Tourism Organization's formulation (UNWTO & UNDP, 2017) that sustainability is ensured when tourism resources are managed in such a way that social and economic needs are met is prevalent. At the same time, it requires the protection of resources for both conservation, ecological methods, and cultural integrity. Tourism is described either as a business or as a dynamic process that is distinguished by its great diversity and its ability to integrate and harmonize different dimensions. Consequently, the orientation of its development must be sustainable, so that it can become resilient over time.

The need for research and the incorporation of new policies has gradually become necessary as the massification of tourism has led to expectations of quick and easy enrichment, which has ultimately led to undesirable effects on society and the environment. Violent tourism development and the sacrifice of resources on the altar of profit led to the cultivation of new patterns of tourism travel that emerged after 1970 and in turn led to new consumer and social patterns. Then, and gradually, came the demystification of the mass model, with the integration of tourism development into another, different pattern. A model clearly more organized and accessible to the needs of today and, above all, tomorrow.

In this way, better use is made of resources, and income is channeled into productive investment and other sectors of the local economy, infrastructure, and environmental protection projects. According to Charta (1996), there are four types of tourism development. Development based on mass and organized tourism, development that is integrated into the economic and social context of the region, development in enclaves and finally, development based on the sustainable use of resources related to specific and alternative forms of tourism.

Over the years, and in line with current trends towards the latter type of development, mass tourism has been sidelined and the baton has been passed to experience tourism, i.e., experiential tourism based on the principles of sustainability. High-impact forms of tourism such as green tourism, ecotourism and responsible tourism have come to the fore and have become established (Dichter & Manzo, 2017). In line with the broader trends mentioned above, the concept of 'sustainability' has gradually broadened beyond the economic dimension to include social, community, environmental criteria, and concerns (Hardy, et al., 2002). This logic was also shaped by various factors that characterize modern tourism.

The importance of tourism development for each country, or region, is directly related to the type of development that followed. Concerns about the contribution of tourism development to sustainability also resonate at the local scale, following a sequence of negative reactions in destinations such as Barcelona, Iceland, and Venice (Hall, 2019). One of the most important objectives of sustainable development that is also related to the tourism sector is quality of life, in terms of service provision to tourists, as well as respecting the quality of life of residents in tourist destinations (Jeon et al., 2016). Therefore, by mentioning the term quality in tourism, the glamping phenomenon comes to prove that luxury and comfort in nature are essential elements for tourists' well-being and giving a quality touch to their holidays.

Glamping as a Part of Sustainable Accommodations

Talking about glamping accommodation is also talking about sustainable destinations. Reducing the environmental impact of tourist accommodation can lead to a certain economic improvement without necessarily increasing production/construction costs (Gurlek & Tuna, 2018). However, to achieve the above scenario, businesses need to plan their development according to the conditions of sustainability and with respect to the limits of the environment (Dias, 2009). A positive omen is the assumption that environmental actions have become entrenched in administrative, commercial, and economic functions (De Souza, 2002). In this context, tourists, using the services and activities provided, realize that their consumer behavior matters and influences the business practice in question, either positively or negatively towards nature and the environment. Conscious, therefore, consumer behavior considers all possible economic, social, and environmental impacts (Zimmer et al., 2019) with the socio-environmental characteristics of accommodation facilities being largely perceived and appreciated by most visitors (Song et al., 2020).

Promotional marketing strategies of accommodation facilities, in turn, perceive tourists' preferences and start making decisions considering the attractiveness of sustainable accommodation and its services (Ribas et al., 2017). Thus, the consumer starts to adopt this type of behavior and applies a sustainable filter in choosing not only the tourist destination, but more so the accommodation in which to stay (Zimmer et al., 2019).

In some regions, however, sustainable attractiveness is observed to be higher or lower, depending on the profile of the destination. This profile is shaped, among other things, by a combination of the demographic characteristics of the residents, the tourists, and the tourism offerings provided (Islam & Chandrasekaran, 2016). In this context of analyzing tourist behavior and considering the interest in sustainability in tourist accommodation, staying in glamping accommodation seems to be the ideal choice for conscious travelers.

It is now widespread that the choice of a glamping accommodation stems from the tourist's love of nature and the need to spend their holidays with a high level of comfort and relaxation. It is of course no coincidence that the supply of this type of accommodation has experienced tremendous growth in recent years, including the years when tourism was significantly reduced by the COVID-19 pandemic. As the years go by, it is estimated that the important elements of glamping tourism will be added to the important elements of glamping tourism and the possibility of social detachment offered to glampers, which will increase the number of interested tourists, given the individual responsibility that citizens must have.

In addition to social detachment, a strong element of glamping tourism is the escape from everyday life, as it seems to be chosen by tourists who usually wish to escape from the hectic pace of the city and relax in nature with the least possible impact on the environment (Bouzis & Poulaki, 2022).

Glamping and COVID-19

In view of the new normal and the social distancing aimed at many people due to the COVID 19 virus, it is recommended that a tourist site should only be shared by a group of tourists travelling together. More specifically, cleaning services must be provided, as in the past, and the handover of the key and entry to the accommodation can be done in an impersonal way, while the F&B area, which until now was provided together with the accommodation, is not a compulsory element. It is therefore a simple model of accommodation, which is more suitable for young tourists, but not restricted to other age groups. It is an excellent alternative accommodation solution in densely populated areas with the security of personal hygiene nowhere compromised.

Undoubtedly, the pandemic COVID-19 phenomenon has caused an extremely significant weakening of much of the global economic system. Taking a brief review and starting with the outbreak of the virus, gradually all economic and social sectors are being affected by this pandemic. As such, a new phase of innovation gradually emerges, which was to keep up with the current standards (Lew, Cheer... & Salazar, 2020). As a natural evolution of this, natural social distance can be applied. At this point, then, the question arises: "How will the new everyday life, including safety distances, continue?"

Focusing, on the most prevalent socio-economic phenomenon, tourism, several answers to the above question emerge, as well as opportunities for combining travel and personal protection. Although at first sight it sounds like an oxymoron, glamping tourism comes to overturn the status quo. It is, as mentioned above, a new alternative emerging trend in tourism, which refers to people who seek not only comfort, but also isolation on their holidays. Thus, by choosing glamping, the tourist lives in individual accommodation in nature without direct contact with other tourists. On the other hand, there are clearly some organized glamping complexes, which also host many diverse activities.

CASE STUDY

Destinations Where the Northern Lights are Visible

The global phenomenon with lights appearing in both the Northern and Southern hemispheres attracts thousands of tourists. The Northern Lights, more specifically, are one of the most spectacular sights that nature has to offer. The spectacular colors dancing in the night sky of the polar regions are the best excuse for an unforgettable journey, which can be described as an experience of a lifetime.

The Northern Lights, also known as the Aurora Borealis, occur more frequently over northern Scandinavia and really appeal to a small number of visitors. The lights also appear near the magnetic poles of other planets, Jupiter and Saturn, whose magnetic fields are stronger than those of Earth.

The aurora borealis is most prominent in areas of northern Scandinavia, in a zone extending between 66° and 69°, called the auroral belt. This zone may extend when geomagnetic activity is high. Therefore, the Northern Lights are often visible even from the northern parts of the United Kingdom. Essentially, this is because of the solar winds, as their ferocity comes from the sun. It is, clearly, likely to decrease, which means that the reaction now with the gases in the atmosphere will be less intense. This will result in the coming years in the Northern Lights being visible only from the Aurora Belt.

As solar activity slows down, moving further and further north will be a one-way street to the opportunity to witness a magical light show. For the best viewing of the phenomenon, visitors should make sure they are at a location within the Auroral Zone that is well away from any other light pollution. Artificial light from large cities and ski resorts can dazzle the Aurora to a great extent, so it is recommended that accommodation be located away from them.

But in which countries and destinations can the Northern Lights be enjoyed in winter? The Aurora complex stretches across Finland, Sweden, Norway, Iceland, Greenland, and Canada.

Finland

Finland often takes a back seat to its neighbors Sweden and Norway, but it is just as likely that the tourist will get a glimpse of the Northern Lights, as clearly as anywhere else. Indeed, the low density of the human population, where reindeer are more common than people, means that Finland's beautiful lake, forest and majestic landscapes are virtually free of any light pollution creating the perfect Aurora viewing conditions.

Finland has so many places from which the tourist can seek out the Northern Lights. In the northeast destinations such as Nellim, Muotka, Saariselkä, Menesjärvi and Inari are all extremely popular with experienced tourists seeking the Northern Lights. In the west, Harriniva, Jeris, Torassieppi and Kilpisjärvi offer comfortable locations in remote locations with the main requirement - a clear dark sky.

The ideal period is between September and March. The Northern Lights are visible in southern Finland only 10-20 nights a year. In contrast, in Finnish Lapland, this extraordinary celestial spectacle is seen around 200 nights a year. In Rovaniemi, the magnificent, starry show is visible from as early as mid-August until the beginning of April. In the capital of Finnish Lapland, tourists don't have to walk very far from the city center to find a good spot to enjoy the Northern Lights.

According to Forbes, Torassieppi Jerisjärvi (Figure 1), a futuristic glamping site nestled beneath the Northern Lights, is the ideal destination for Northern Lights viewing, with its surreal domes attracting thousands of visitors. This area is full of winter fun, including activities such as endurance skiing, dog sledding, snowmobiling, ice fishing and plenty of reindeer spotting. Close by is Saariselkä, Europe's northernmost ski resort, with 15 ski slopes.

Sweden

The average population of Sweden is about 23 people per km^2. Compared to the European average, this makes it the most sparsely populated country in Northern Europe. The wild nature of Swedish Lapland has a major influence on this figure.

It is one of the areas that is low in human population and extremely rich in natural wonders, being also one of the ideal locations for Aurora viewing. From the ice meadows of the Luleå archipelago, to the sweeping Torne River and the mountains of Abisko, Sweden can offer some simply staggering sights.

Figure 1. Glamping in Finland

Aurora hunting holidays in Sweden take place between December and late March and are designed to coincide with the arrival of the Northern Lights snow. Temperatures generally range from -6°C to -20°C but can drop further during the night.

Arctic night also means that during the depths of winter, in December and January, there are periods when the sun does not rise above the horizon. During this period the days are obviously short, but the short period around midday offers a unique polar light, casting a blue dusk over the entire landscape.

North of the Arctic Circle is Abisko, a very small town that has become a priority destination for Aurora hunters. The reason is simple, as there are very clear skies. Abisko is surrounded by a huge cluster of mountains, which manages to disperse the clouds and make the Northern Lights more visible. Finally, Sweden is home to the world's first Icehotel (Figure 2), created in 1989 from the frozen Torne River in Jukkasjärvi.

Figure 2. Glamping in Sweden

Norway

The coolest northern city, Tromso, has become synonymous with the Northern Lights and is no doubt a huge starting point for a Norwegian Aurora hunting spot. However, all the publicity of the Northern Lights around Tromso may be slightly misleading, given the artificial light pollution that distorts the Northern Lights and causes discontent among tourists.

Therefore, when talking about an Aurora Zone in Norway, there should certainly be a direction opposite of cities, islands, and mountains. There only the visitor will find dark skies that are absolutely necessary for viewing the Northern Lights.

The country is famous for its dramatic fjords, but there is much more to these magnificent waterways formed thousands of years ago by glacial retreat.

Islands like Senja and Sommaroy are sparsely populated and where they are inhabited, the colorful towns and fishing villages are achingly beautiful. They are also an ideal destination for Aurora hunting expeditions due to the lack of light pollution.

From September to March, lights regularly appear all the way north from the islands to the Russian border near Kirkenes. Norway's northern coastline is warmer than inland, which means that autumn is positively mild compared to most areas north of the Arctic Circle. Visiting places like Tromso or Sommaroy during this time offers excellent Aurora viewing opportunities.

Winter, on the other hand, brings snow of course, but that means there are even more activities, such as dog sledding. However, it's not just the Northern Lights that attracts visitors to this northern region. The Kirkenes Snow Hotel (Figure 3) offers travelers the chance to stay overnight in one of its themed rooms, but perhaps the most memorable experience is not that, as the stay includes fishing for king crab from the ocean, which is enjoyed by the professional chef as they look out over the Northern Lights.

Figure 3. Glamping in Norway

Iceland

Iceland has long been known for its stunning geology, but in recent years, the Northern Lights have also become a major attraction. The most famous destination in Iceland is Reykjavik and it is a tourist attraction especially during the Christmas holidays. Of course, the rule that one must get out of the city lights to see the Northern Lights applies in this case as well. In Iceland, however, the spectacle accompanies a night-time jeep ride.

That country has a surprisingly temperate climate, as the Gulf Stream helps this island stay much warmer than its name suggests, with average temperatures in Reykjavik in winter being just below freezing.

Holidays to Iceland classically follow the Aurora season and so start in late September or early October and are available until the end of March. Iceland is not the destination for tourists looking for a snowy, winter wonderland of sledding, dogs, and reindeer. Leave those attractions to Finland, Sweden, and Norway. Iceland is home to some of nature's greatest spectacles, and since none of them depend on snow cover, tourists can see the early aurora borealis without missing other highlights.

Beyond the capital, Hotel Ranga (Figure 4) on the South Coast has earned an almost legendary place on the list of accommodations from which the Northern Lights are visible. Its remote location offers great views of the pitch-black sky and the staff at the property checks for Auroras every 15 minutes facilitating and informing tourists. It has rightly become one of the most in-demand hotels in Iceland.

Figure 4. Glamping in Iceland

Greenland

Greenland has never been high on the tourist map but, as travelers become more adventurous, it is a destination where the number of visitors is growing. It is a remote, wild, and sparsely populated place where life can be difficult, but there are some benefits. Just as important as these benefits is the Aurora

Borealis that regularly blazes across the night sky and when seen from such an unusual destination, it can be considered an incredible experience.

Adventurous travelers who choose Greenland for a winter getaway, ignoring the extreme weather conditions there, will undoubtedly be rewarded for their courageous choice with an incredible spectacle of colorful light figures in the sky (Figure 5). The Northern Lights appear in the Greenland night sky from early autumn and according to a local legend of the Eskimo Inuit of Greenland, when the Northern Lights "dance" in the sky, it means that the dead are playing football with the skull of a walrus. Today, more than a few people believe that if a child is captured under the majestic glow of the Northern Lights, he or she will be particularly intelligent.

Finally, it is worth mentioning that Greenland enjoys around 300 days a year of clear skies. This makes it an ideal destination for Northern Lights hunters.

Figure 5. Glamping in Greenland

Canada

Finally, a worth mentioning destination is Canada, which has a reputation as the most expensive destination to enjoy the Northern Lights. But it may also be the most guaranteed for viewing it. The reason? In Northern Canada there are several areas that guarantee Northern Lights viewing, including Yukon, North Saskatchewan, the Northwest Territories and Newfoundland. This is explained by the fact that the Northern Lights appear in two rings (auroral ovals) around the north and south poles, i.e., near the Arctic Circle, thus making Canada super lucky.

Canada's Yukon region is home to tens of thousands of residents, caribou (reindeer), grizzly bears and other wildlife. Stunning, whitewashed mountain ranges, rushing rivers and beautiful lakes complete the scenery, which on icy winter nights is illuminated by the colored ribbons of Northern Lights swirling in the sky. A short distance from the town of Yellowknife in Canada's Northwest Territories is Aurora Village (Figure 6). The 'village' consists of teepee tents.

Figure 6. Glamping in Canada

FUTURE RESEARCH DIRECTIONS

Prospects and Trends

Summarizing the supply data and considering the trends that constantly emerge and govern the international and not only the tourism scene - always in connection with glamping tourism in the Northern Lights - the prospects for further development are listed, which at the same time constitute proposals for research.

The boom of social media and the need for social distancing created by COVID-19 are absolutely topical variables, with a major role in the development of this type of alternative tourism in all locations of interest.

During the health crisis of COVID-19, aspects that had long been in obscurity came to the surface, thus creating a new situation, completely unprecedented, which not only caused upheavals in the world economy but also in people's health, both physical and psychological.

Terms that have been studied in the past have come back to the fore, and one of the most prevalent is that of social distancing, which is related to social distance theory and concerns the understanding of inter-group relations. (Nyaupane, Timothy, & Poudel, 2015). When talking about social distancing under the shadow of COVID-19, actions such as limiting social interactions and avoiding crowded places and areas are meant.

An outcome of this attitude of this society ended up being the restriction of tourist travel (Sigala, 2020). The health crisis cultivated the major feelings of insecurity, anxiety and fear to such an extent that tourists now evaluate both their trips and their activities on them by other criteria (Chakrawarty,et al, 2020; Sigala,2020). Outdoor activities are preferred as the feeling of safety is greater, but questions arise about staying in low density areas.

Is it possible for glamping accommodations, combined with special and alternative forms of tourism to establish a new situation eventually? Furthermore, can these areas turn low density into an advantage, creating new tourism prospects? A notable variable in the above reflection, always in connection with the present comparative case study, is the Northern Lights, which is a rare natural phenomenon and can independently become an attraction. But is it enough to stimulate the triptych: glamping tourism - social detachment - safety, by rocking the tourist waters? Current and future research, focusing on the effects and residues left on tourists by this situation, especially at a psychological level, shows how much their temperament has been affected and, based on the trends that will emerge in the future from their attitudes, the next tourism strategies, which must be highly sustainable, will be drawn up.

On the other hand, the wide spread of social media in the global community and the daily friction with people is something that is taken for granted. Their role in travelers' decision making is also undeniable, as well as in tourism businesses, and they have been very much engaged in tourism and hospitality research (Alonso-Almeida et al., 2019).

According to Sigala et al. (2012), social media are characterized as suppliers, as they create conditions for participation in both production and consumption. They can even influence tourism practices and have a high degree of correlation with destinations (Jacobsen & Munar, 2012). Supplier-travelers may discover new places that are not yet popular and turn them into fashionable places through the social media network, a point that also carries risks of creating tourism flows that may even result in over-tourism (Alonso-Almeida et al., 2019). At the same time, however, they could highlight previously unknown and different aspects of destinations. (Stepchenkova & Zhan, 2013).

The case study follows the above reasoning, as the demand for alternative low-density, socially distant destinations can rapidly transform them into popular destinations, as the sense of security prevails, as mentioned above.

Of course, social networks may enhance the image of each destination, but they can just as easily cause readjustments and lead to the formation of a spectrum of negative consequences. In any case, glamping accommodations, governed by the principles of sustainability and identified with the need for new, more meaningful adventures, gain an opportunity for consolidation. The aesthetics of the accommodation (wooden houses, domes) and the desire of millennials to promote their new experiences on social media can be mentioned as additive factors. Can social media finally strengthen the position of glamping accommodations in the international tourism scene?

SOLUTIONS AND RECOMMENDATIONS

Northern Lights and Dark Tourism

In this context, darkness plays an important role, not least because without darkness there would be no lights, which may be at odds with the way Icelandic nature was created to be perceived in daylight as a picturesque yet sublime natural area. However, the darkness was never eliminated and with the tourism of the Northern Lights it became more visible, and so nature needs to be rearranged and managed to acquire aesthetic qualities.

This simultaneously changes the perception of nature by tourism promoters, who have now begun to add shades of darkness to tourism promotional materials. In this context, darkness becomes attractive and is complemented by the mysterious phenomenon of the Northern Lights, creating a special attraction. However, darkness, in a similar way to the flickering Northern Lights, "cannot be disciplined" (Bender 2002). In the darkness what light obscures emerges, meaning that the sense of time and space experienced in daylight changes, and enters into a space, where the forms and features of nature take on mobile and shimmering qualities

It is a space where culture and nature intertwine through narratives in which human and non-human beings appear and interact, which evokes an atmosphere that stimulates an aesthetic and sublime quality (Jóhannesdóttir 2010 & Edensor 2010). It is a nature that needs to be experienced, not just viewed, and therefore the task of the tourism industry is to manage it and turn it into an enticing option. That is, to put it on the stage for the imaginative and affluent tourist seeking to get in touch with nature in its purest form - hence green nature.

In an increasingly mobile and modern world, the tourist has come close to the exotic subarctic nature of not only Iceland, but also other tourist areas where the Northern Lights are visible in the winter season and beyond. The tourist is like stepping into the comfort of an airplane, named after a volcano, and decorated with images of the Northern Lights, to head into the darkness. This is the same exotic, dark nature to which scientists and explorers of the Romantic era travelled to discover and illuminate the pristine nature of lands of similar beauty to that of Iceland.

Today, however, tourists embrace the darkness and are embraced by it as they are transported in various types of vehicles into it, and into nature, to encounter the Northern Lights. In a nature enveloped in darkness, the "greenness" of sustainable destinations is nuanced through the projection of both tour guides and tour operators, commanding nature and emphasizing its sublime qualities in their efforts to make it welcoming to the tourists who visit it at any given time.

CONLUSION

Enhancing competitiveness requires improving the quality of supply of all types of accommodation, including campsites. Progress in development refers to new construction and improving the quality of existing supply, with glamping opening opportunities to achieve the primary strategic objectives.

Glamping, a new and creative type of accommodation, is directly linked to the creation of an innovative service that contributes to the conversion of campsites to glamping and represents the symbiosis of the hotel and camping industries. Glamping has been the focus of increasingly intensive research in recent years. An analysis of the existing literature reveals that glamping is associated with terms such as luxury, comfort, high quality services, adventurous activities, escape, relaxation, environmentally friendly, wildlife, extravagance, peace, innovative type of accommodation, unique experience, and intimacy. What is common to all definitions of glamping are the words 'luxury' and 'comfort'.

Although glamping represents an innovative outdoor accommodation, its roots can be found throughout history. Glamping accommodations gained the comparative advantage due to a health crisis.

The above factors, i.e., social media and social distancing, work positively and only as benefits and strong characteristics of the development of glamping tourism in low-density tourist areas. Firstly, because of the mass presentation of the spectacle to the public and secondly because of the location of the accommodation in these areas.

Moreover, low-density areas can undoubtedly create new conditions for alternative tourism and more specifically for glamping accommodations. However, the activities and services provided do not seem to be affected at all. Of course, with more and more passive activities such as yoga, sauna, jacuzzi and gastronomic tourism, glamping tourism in the Northern Lights can only grow spectacularly. In other words, even the pandemic crisis does not seem to have any impact on the tourist demand of the regions by sensitive and quality tourists.

Lastly, special, and alternative forms of tourism and low density are together the ions whose resulting tourism product is fully in tune with the future trends of the times and constitutes, especially at a psychological level, a protective wall against the external threat posed by COVID-19.

ACKNOWLEDGMENT

This research received no specific grant from any funding agency in the public, commercial, or not-for-profit sectors.

REFERENCES

Almeida Ramos, G. M. D., & Fernandes, J. L. J. (2016). Tourism territories in low density areas: The case of Naturtejo geopark in Portugal. *Journal of Tourism, Heritage & Services Marketing*, *2*(1), 14–21.

Alonso-Almeida, M. D. M., Borrajo-Millán, F., & Yi, L. (2019). Are social media data pushing overtourism? The case of Barcelona and Chinese tourists. *Sustainability*, *11*(12), 3356. doi:10.3390u11123356

Bouzis, S., & Poulaki, P. (2022). Cultural Routes as a Motivation for Choosing Glamping. The case of India. In Sustainable Development, Culture, Traditions (pp. 20-31). SDCT-Journal. doi:10.26341/issn.2241-4002-2021-1b-3

Chakrawarty, A., Ranjan, P., Klanidhi, K. B., Kaur, D., Sarkar, S., Sahu, A., ... Wig, N. (2021). Psychosocial and behavioral impact of COVID-19 on middle aged and elderly individuals: A qualitative study. *Journal of Education and Health Promotion*, 10. PMID:34485566

Clarke, J. (1997). A Framework of Approaches to Sustainable Tourism. *Journal of Sustainable Tourism*, *5*(3), 224–233. doi:10.1080/09669589708667287

De Souza, R. S. (2002). Evolução e condicionantes da gestão ambiental nas empresas. *READ - Revista Eletrônica de Administração, 8*(6), 1-22.

Dias, R. (2009). *Gestão ambiental: responsabilidade social e sustentabilidade* (2nd ed.). Saraiva.

Dichter, A., & Manzo, G. G. (2017). *Coping with success: Managing overcrowding in tourism destinations*. World Travel and Tourism Council.

Duro, J. A., Perez-Laborda, A., Turrion-Prats, J., & Fernández-Fernández, M. (2021). Covid-19 and tourism vulnerability. *Tourism Management Perspectives*, *38*, 100819. doi:10.1016/j.tmp.2021.100819 PMID:34873568

Edensor, T. 2010. Aurora Landscapes: Affective Atmospheres of Light and Dark. Conversations with Landscape, 227–240.

Everingham, P., & Chassagne, N. (2020). Post COVID-19 ecological and social reset: Moving away from capitalist growth models towards tourism as Buen Vivir. *Tourism Geographies*, *22*(3), 555–566. doi:10.1080/14616688.2020.1762119

Forsyth, A. (2003). Measuring density: working definitions for residential density and building intensity. *Design Brief, 9*(1), 2-8.

Gürlek, M., & Tuna, M. (2018). Reinforcing competitive advantage through green organizational culture and green innovation. *Service Industries Journal*, *38*(7-8), 467–491. doi:10.1080/02642069.2017.1402889

Hall, C. M. (2019). Constructing sustainable tourism development: The 2030 agenda and the managerial ecology of sustainable tourism. *Journal of Sustainable Tourism*, *27*(7), 1044–1060. doi:10.1080/09669582.2018.1560456

Hall, C. M., Scott, D., & Gössling, S. (2020). Pandemics, transformations, and tourism be careful what you wish for. *Tourism Geographies*, *22*(3), 577–598. doi:10.1080/14616688.2020.1759131

Hardy, A., Beeton, R. J., & Pearson, L. (2002). Sustainable tourism: An overview of the concept and its position in relation to conceptualisations of tourism. *Journal of Sustainable Tourism*, *10*(6), 475–496. doi:10.1080/09669580208667183

Islam, T., & Chandrasekaran, U. (2016). Religiosity and ecologically conscious consumption behaviour. *Asian Journal of Business Research*, *5*(1), 1–18.

Jacobsen, J. K. S., & Munar, A. M. (2012). Tourist information search and destination choice in a digital age. *Tourism Management Perspectives*, *1*, 39–47. doi:10.1016/j.tmp.2011.12.005

Jeon, M. M., Kang, M. M., & Desmarais, E. (2016). Residents' perceived quality of life in a cultural-heritage tourism destination. *Applied Research in Quality of Life*, *11*(1), 105–123. doi:10.100711482-014-9357-8

Jóhannesdóttir, G. R. (2010). Landscape and Aesthetic Values: Not Only in the Eye of the Beholder. In K. Benediktsson & K. Lund (Eds.), *Conversations with Landscape* (pp. 107–123). Ashgate.

Kokkossis, H., Tsartas, P., & Grimba, E. (2011). *Special and alternative forms of tourism*. Kritiki.

Lew, A. A., Cheer, J. M., Haywood, M., Brouder, P., & Salazar, N. B. (2020). Visions of travel and tourism after the global COVID-19 transformation of 2020. *Tourism Geographies*, *22*(3), 455–466. doi:10.1080/14616688.2020.1770326

Lund, K. A. (2016). Chasing the lights: Darkness, tourism and the northern lights. In *Green Ice* (pp. 49–71). Palgrave Macmillan. doi:10.1057/978-1-137-58736-7_3

Milohnić, I., Bonifačić, J. C., & Licul, I. (2019). Transformation of camping into glamping–trends and perspectives. *Tourism in Southeast Europe.*, *5*, 457–473. doi:10.20867/tosee.05.30

Nyaupane, G. P., Timothy, D. J., & Poudel, S. (2015). Understanding tourists in religious destinations: A social distance perspective. *Tourism Management*, *48*, 343–353. doi:10.1016/j.tourman.2014.12.009

O'Donnel, R. (1997). The Competitive Advantage of Peripheral Regions: Conceptual Issues and Research Approaches. In B. Fynes & S. Ennis (Eds.), *Competing from the Periphery* (pp. 47–82). Oak Tree.

OECD. (2020a). *OECD economic outlook, Interim report March 2020*. OECD.

Prideaux, B., Thompson, M., & Pabel, A. (2020). Lessons from COVID-19 can prepare global tourism for the economic transformation needed to combat climate change. *Tourism Geographies*, *22*(3), 667–678. doi:10.1080/14616688.2020.1762117

Ribas, J. R., Vicente, T. V. D. S., Altaf, J. G., & Troccoli, I. R. (2017). Integração de Ações na Gestão Sustentável. REAd. *Revista Eletrônica de Administração*, *23*(2), 31–57.

Saarinen, J., & Rogerson, C. M. (2014). Tourism and the millennium development goals: Perspectives beyond 2015. *Tourism Geographies*, *16*(1), 23–30. doi:10.1080/14616688.2013.851269

Sharpley, R. (2020). Tourism, sustainable development, and the theoretical divide: 20 years on. *Journal of Sustainable Tourism*, *28*(11), 1932–1946. doi:10.1080/09669582.2020.1779732

Sigala, M. (2020). Tourism and COVID-19: Impacts and implications for advancing and resetting industry and research. *Journal of Business Research*, *117*, 312–321. doi:10.1016/j.jbusres.2020.06.015 PMID:32546875

Sigala, M., Christou, E., & Gretzel, U. (Eds.). (2012). *Social media in travel, tourism and hospitality: Theory, practice and cases*. Ashgate Publishing, Ltd.

Sivam, A., & Karuppannan, S. (2012). *Density Design and Sustainable Residential Development* (Doctoral dissertation). Earthscan.

Skryl, T. V., & Gregoric, M. (2022). Tourism in the Post-COVID Age. In *Post-COVID Economic Revival* (Vol. 2, pp. 239–254). Palgrave Macmillan. doi:10.1007/978-3-030-83566-8_15

Song, J., Jai, T. M., & Li, X. (2020). Examining green reviews on TripAdvisor: Comparison between resort/luxury hotels and business/economy hotels. *International Journal of Hospitality & Tourism Administration*, *21*(2), 165–187. doi:10.1080/15256480.2018.1464418

Stepchenkova, S., & Zhan, F. (2013). Visual destination images of Peru: Comparative content analysis of DMO and user-generated photography. *Tourism Management*, *36*, 590–601. doi:10.1016/j.tourman.2012.08.006

Streimikiene, D., Svagzdiene, B., Jasinskas, E., & Simanavicius, A. (2021). Sustainable tourism development and competitiveness: The systematic literature review. *Sustainable Development*, *29*(1), 259–271. doi:10.1002d.2133

Tsartas, P. (1996). *Tourists, journeys, landscapes: Sociological approaches to tourism*. Academic Press.

UNWTO. (2018). *Press release: Tourism can and should lead sustainable development: UNWTO Secretary-General opens ITB 2018. PR 18020, 06 Mar 18*. UNWTO.

UNWTO & UNDP. (2017). *Tourism and the sustainable development goals – Journey to 2030*. UNWTO.

WTTC & McKinsey & Company. (2017). *Coping with success: Managing overcrowding in tourism destinations*. WTTC.

Zimmer, P., Camargo, M. E., Pizzoli, M. F. F., Zanandréa, G., & Bizotto, B. L. S. (2019). Consumo Consciente: O Nível de Consciência Ecológica dos Acadêmicos do Curso de Administração de uma Instituição de Ensino da Serra Gaúcha. *Desafio Online*, *7*(2), 261–277.

KEY TERMS AND DEFINITIONS

Glamping Tourism: New emerging alternative form of tourism, consisting of luxury and sustainable accommodation in nature. Its main characteristic is comfort, luxury, and direct contact of the tourist with nature.

Low Density Areas: These are areas with a low population density index. Population density is an indicator that measures the number of people living in a territorial extension.

Northern Lights: It is a striking celestial phenomenon observed in the upper layers of the atmosphere and is visible at the poles. When the phenomenon is observed at the North Pole it is called "Aurora Borealis" or "Northern Lights," and when it is observed at the South Pole it is called "Aurora Australis" or "Southern Lights".

Sustainability: Approaching the concept of sustainability is not an easy task. In fact, sustainability is a model that aims for the best possible economic results, the least possible environmental impact and, at the same time, maximum social development.

Compilation of References

Abdi, M. H. (2021). What the newcomers to transit-oriented development is confronted with? Evidence from Iranian policy and planning. *Journal of Transport Geography, 92*.

Acharya, R. P., Maraseni, T. N., & Cockfield, G. (2019). Local users and other stakeholders' perceptions of the identification and prioritization of ecosystem services in fragile mountains: A case study of Chure region of Nepal. *Forests, 10*(5), 421. Advance online publication. doi:10.3390/f10050421

Acharya, R., Maraseni, T., & Cockfield, G. (2019). Global trend of forest ecosystem services valuation – An analysis of publications. In *Ecosystem Services* (Vol. 39). Elsevier B.V. doi:10.1016/j.ecoser.2019.100979

Adelman, H., & Sorenson, J. (1994). African refugees: Development aid and repatriation. Westview Press.

Akalin & Koç. (2014). Satış personeline yönelik performans değerlendirme ölçeğinin geliştirilmesi ve psikometrik özelliklerinin incelenmesi. *Eğitim Bilimleri Araştırmaları Dergisi, 4*(2), 227-241.

Akgemci, T., & Güleş, H. K. (2009). *İşletmelerde Stratejik Yönetim*. Gazi Kitabevi.

Akın, A. (2002). İşletmelerde insan kaynakları performansını değerleme sürecinde coaching (Özel rehberlik). *C.Ü. İktisadi ve İdari Bilimler Dergisi, 3*(1).

Akse, R., Thomas, T., & Geurs, K. (2021). Mobility and accessibility paradigms in Dutch policies: An empirical analysis. *Journal of Transport and Land Use, 14*(1), 1317–1340. doi:10.5198/jtlu.2021.2097

Aktan, C. C. (2002). Performans yönetimi: Organizasyonlarda performans değerlendirme ve ölçme. *Organizasyon ve Yönetim Bilimleri Dergisi, 1*(1), 25–33.

Alamgir, M., Pert, P. L., & Turton, S. M. (2014). A review of ecosystem services research in Australia reveals a gap in integrating climate change and impacts on ecosystem services. In International Journal of Biodiversity Science, Ecosystem Services and Management (Vol. 10, Issue 2, pp. 112–127). Taylor and Francis Ltd. doi:10.1080/21513732.2014.919961

Almeida Ramos, G. M. D., & Fernandes, J. L. J. (2016). Tourism territories in low density areas: The case of Naturtejo geopark in Portugal. *Journal of Tourism, Heritage & Services Marketing, 2*(1), 14–21.

Alonso-Almeida, M. D. M., Borrajo-Millán, F., & Yi, L. (2019). Are social media data pushing overtourism? The case of Barcelona and Chinese tourists. *Sustainability, 11*(12), 3356. doi:10.3390u11123356

Altman, M. (2006). Identifying employment-creating sectors in South Africa: The role of service industries. *Development Southern Africa, 25*(5), 627–647. doi:10.1080/03768350601021871

Alvarez-garreton, C., Mendoza, P. A., Boisier, J. P., Addor, N., Galleguillos, M., Zambrano-bigiarini, M., ... Garreaud, R. (2018). The CAMELS-CL dataset : Catchment attributes and meteorology for large sample studies – Chile dataset. *Hydrol. Earth Syst, 22*(11), 5817–5846. doi:10.5194/hess-22-5817-2018

Alyavina, E., Nikitas, A., & Tchouamou Njoya, E. (2020). Mobility as a service and sustainable travel behaviour: A thematic analysis study. *Transportation Research Part F: Traffic Psychology and Behaviour*, *73*, 362–381. doi:10.1016/j.trf.2020.07.004

Apleni, L. (2017). Residents' perceptions on urban tourism as a catalyst for economic development: A case study of Buffalo City, South Africa. *African Journal of Hospitality, Tourism and Leisure*, *6*(3), 1–14.

Apleni, L., & Henama, U. S. (2020). The impact of events in boosting local economic development: A case study of Port St Johns, South Africa. *African Journal of Hospitality, Tourism and Leisure*, (1), 1–12.

Apleni, L., Vallabh, D., & Henama, U. S. (2017). Motivation for tourists' participation in religious tourism in Eastern Cape: A case study of Buffalo City, South Africa. *African Journal of Hospitality, Tourism and Leisure*, *6*(2), 1–20.

Apleni, Mxunyelwa, & Vallabh. (2017). The implications of the proposed carbon tax in South Africa's tourism industry. *African Journal of Hospitality, Tourism and Leisure*, (1), 1–15.

Apparicio, P., Gelb, J., Dubé, A.-S., Kingham, S., Gauvin, L., & Robitaille, É. (2017). The approaches to measuring the potential spatial access to urban health services revisited: Distance types and aggregation-error issues. *International Journal of Health Geographics*, *16*(1), 16. doi:10.118612942-017-0105-9 PMID:28830461

Arvor, D., Daher, F., Corpetti, T., Laslier, M., & Dubreuil, V. (2016). Monitoring of artificial water reservoirs in the Southern Brazilian Amazon with remote sensing data. *Remote Sensing for Agriculture, Ecosystems, and Hydrology XVIII*, *9998*, 999816. doi:10.1117/12.2241905

Ashley, C., De Brine, P., Lehr, A., & Wilde, H. (2007). *The role of the tourism sector in expanding economic opportunities: Opportunity, Corporate Social Responsibility Initiative*. Report No.23. Cambridge, MA: Kennedy School of Government, Harvard University.

Asian Development Bank. (2005). *Country Environmental Analysis: Azerbaijan*. Asian Development Bank.

Aslaksen, I., Nybø, S., Framstad, E., Garnåsjordet, P. A., & Skarpaas, O. (2015). Biodiversity and ecosystem services: The Nature Index for Norway. *Ecosystem Services*, *12*, 108–116. doi:10.1016/j.ecoser.2014.11.002

Auvinen, H., & Tuominen, A. (2014). Future transport systems: Long-term visions and socio-technical transitions. *European Transport Research Review*, *6*(3), 343–354. doi:10.100712544-014-0135-3

Baker, J. (1995). Refugee and labor movement in Sub-Saharan Africa. In *Studies on Emergencies and Disaster Relief, No.2* (pp. 6–28). Nordiska Afrikainstitutet.

Baral, S., Basnyat, B., Khanal, R., & Gauli, K. (2016). A Total Economic Valuation of Wetland Ecosystem Services: An Evidence from Jagadishpur Ramsar Site, Nepal. *TheScientificWorldJournal*, *2016*, 1–9. Advance online publication. doi:10.1155/2016/2605609 PMID:27830175

Barcelona Convention - Mediterranean 2017 Quality Status Report. (n.d.). *The Mediterranean Marine and Coastal Environment*. https://www.medqsr.org/mediterranean-marine-and-coastal-environment#:~:text=They%20are%20Albania%2C%20Algeria%2C%20Bosnia,scene%20of%20intense%20human%20activity

Barcik, J., & Wentkowska, A. (2014). *Prawo Unii Europejskiej, (European Union law)*. C.H. Beck.

Barichivichi, J., Gloor, E., Peylin, P., Brienen, R., Schöngart, J., Espinoza, J., & Pattnayak, K. (2018). Recent intensification of Amazon flooding extremes driven by strengthened Walker circulation. *Science Advances*, *4*(9), 8785. doi:10.1126ciadv.aat8785 PMID:30255149

Barnes, K. B., Morgan, J., Roberge, M. C., & Lowe, S. (2012). *Sprawl Development: Its patterns, consequences, and measurement*. Academic Press.

Bartczak, A., Lindhjem, H., Navrud, S., Zandersen, M., & Zylicz, T. (2008). Valuing forest recreation on the national level in a transition economy: The case of Poland. *Forest Policy and Economics*, *10*(7–8), 467–472. doi:10.1016/j.forpol.2008.04.002

Barth, N. C., & Döll, P. (2016). Assessing the ecosystem service flood protection of a riparian forest by applying a cascade approach. *Ecosystem Services*, *21*, 39–52. doi:10.1016/j.ecoser.2016.07.012

Baş, İ. M., & Atar, A. (1990). *İşletmelerde Verimlilik Denetimi: Ölçme ve Değerlendirme Modelleri*. MPM.

Basile, G., & Cavallo, A. (2020). Rural Identity, Authenticity, and Sustainability in Italian Inner Areas. *Sustainability*, *12*(3), 1272. doi:10.3390u12031272

Bastian, O., Syrbe, R. U., Slavik, J., Moravec, J., Louda, J., Kochan, B., Kochan, N., Stutzriemer, S., & Berens, A. (2017). Ecosystem services of characteristic biotope types in the Ore Mountains (Germany/Czech Republic). *The International Journal of Biodiversity Science, Ecosystem Services & Management*, *13*(1), 51–71. doi:10.1080/21513732.2016.1248865

Beck, H. E., Bruijnzeel, L. A., Van Dijk, A. I. J. M., McVicar, T. R., Scatena, F. N., & Schellekens, J. (2013). The impact of forest regeneration on streamflow in 12 mesoscale humid tropical catchments. *Hydrology and Earth System Sciences*, *17*(7), 2613–2635. doi:10.5194/hess-17-2613-2013

Behradfar, A., & Mohammadi, S. (2020). Spatio-Temporal Understanding and Representation of Transformative Urban Mobility and Trip Patterns, A Review. *J. Civil Eng. Urban.*, *10*(4), 35–41. doi:10.29252cil.2020.jceu6

Bernués, A., Rodríguez-Ortega, T., Ripoll-Bosch, R., & Alfnes, F. (2014). Socio-cultural and economic valuation of ecosystem services provided by Mediterranean mountain agroecosystems. *PLoS One*, *9*(7), e102479. Advance online publication. doi:10.1371/journal.pone.0102479 PMID:25036276

Bertram, C., & Larondelle, N. (2017). Going to the Woods Is Going Home: Recreational Benefits of a Larger Urban Forest Site — A Travel Cost Analysis for Berlin, Germany. *Ecological Economics*, *132*, 255–263. doi:10.1016/j.ecolecon.2016.10.017

Betka, T. (2011). *Wspólnoty Europejskie na arenie międzynarodowej w okresie Zimnej Wojny* [European Communities in the international arena during the Cold War]. Analyses European Union.

Bibri, S. E. (2019). On the sustainability of smart and smarter cities in the era of big data: An interdisciplinary and transdisciplinary literature review. *Journal of Big Data*, *6*(1), 25. doi:10.118640537-019-0182-7

Bibri, S. E., Krogstie, J., & Kärrholm, M. (2020). Compact city planning and development: Emerging practices and strategies for achieving the goals of sustainability. *Deve Built Environ*, *4*, 1–2. doi:10.1016/j.dibe.2020.100021

Bingham, G., Clark, E., Cooper, W., Costanza, R., Hale, T., Hayden, G., Kellert, S., Norgaard, R., Norton, B., Payne, J., Russell, C., & Suter, G. (1995). *ECOLOGICAL ECONOMICS Analysis Issues in ecosystem valuation: improving information for decision making*. Academic Press.

Bjerre, L., Helbling, M., Römer, F., & Zobel, M. (2015, Fall). Conceptualizing and Measuring Immigration Policies: A Comparative Perspective. *The International Migration Review*, *49*(3), 555–600. doi:10.1111/imre.12100

Black, R., Adger, N., Dercon, S., Geddes, A., & Thomas, D. (2011). *Migration and Global Environmental Change-Future Challenges and Opportunities*. Government Office of Science, Final Project Report. https://www.un.org/development/desa/pd/sites/www.un.org.development.desa.pd/files/unpd-cm10201202-11-1116-migration-and-global-environmental-change.pdf

Black, R. (2014). *Refugees, environment, and development*. Routledge.

Black, R., & Sessay, M. F. (1997). Forced migration, environmental change and woodfuel issues in the Senegal River Valley. *Environmental Conservation*, *24*(3), 251–260. doi:10.1017/S0376892997000337

BNC. (2016). *Geomorfología de Chile*. Biblioteca Nacional de Chile.

Bocarejo, S. (2012). Transport accessibility and social inequities: A tool for identification of mobility needs and evaluation of transport investments. *Journal of Transport Geography*, *24*(C), 142–154. doi:10.1016/j.jtrangeo.2011.12.004

Bokajło, W., & Dziubik, K. (Eds.). (2003). Unia Europejska. Leksykon integracji [European Union. Lexicon of Integration]. Europa Publishing House.

Borucki, M. (1996). *Historia powszechna (Common history) 1945-1995*. Mada.

Bostancı, S. H., & Yıldırım, S. (2021). Sustainable Communities vs. Climate Refugees: Two Opposite Results of Climate Change. In C. Popescu (Ed.), *Handbook of Research on Novel Practices and Current Successes in Achieving the Sustainable Development Goals* (pp. 298–319). IGI Global. doi:10.4018/978-1-7998-8426-2.ch015

Botelho, A., Lourenço-Gomes, L., Pinto, L., Sousa, S., & Valente, M. (2016). Using stated preference methods to assess environmental impacts of forest biomass power plants in Portugal. *Environment, Development and Sustainability*, *18*(5), 1323–1337. doi:10.100710668-016-9795-6

Bouzis, S., & Poulaki, P. (2022). Cultural Routes as a Motivation for Choosing Glamping. The case of India. In Sustainable Development, Culture, Traditions (pp. 20-31). SDCT-Journal. doi:10.26341/issn.2241-4002-2021-1b-3

Brakke, M. (2004). International tourism, demand and GDP implications: A background and empiricial analysis. *Underground Economic Review*, *1*(2), 1–38.

Breathnach, P. (2013). Regional governance and regional development: Implications of the Action Programme for Effective Local Government. *Administration*, *61*(3), 51–73.

Brelik, A. (2018). Globalization in Tourism. *Proceedings of the 2018 International Conference "Economic Science for Rural Development"*, 9-11. 10.22616/ESRD.2018.004

Britchenko, I., Zolotarov, V., Levchenko, Y., & Losonczi, P. (2022). Inter-Territorial Collaboration in the Context of Strengthening Its Economic Security. *IJCSNS International Journal of Computer Science and Network Security.*, *21*(12), 675–683.

Brown, D. O. (1998). In search of an appropriate form of tourism for Africa: Lessons from the past and suggestions for the future. *Tourism Management*, *19*(3), 237–245. doi:10.1016/S0261-5177(98)00016-8

Brussel, M., Zuidgeest, M., Pfeffer, K., & van Maarseveen, M. (2019). Access or Accessibility? A Critique of the Urban Transport SDG Indicator. *ISPRS International Journal of Geo-Information*, *8*(2), 67. doi:10.3390/ijgi8020067

Bryant, J. Jr, & Delamater, P. L. (2019). Examination of spatial accessibility at micro- and macro-levels using the enhanced two-step floating catchment area (E2SFCA) method. *Annals of GIS*, *25*(3), 219–229. doi:10.1080/19475683.2019.1641553

Burton, E., Williams, K., & Jenks, M. (1996). The compact city and urban sustainability: conflicts and complexities. In E. Burton, K. Williams, & M. Jenks (Eds.), *The compact city: a sustainable urban form* (pp. 231–247). Taylor & Francis Group.,

Butler, L., Yigitcanlar, T., & Paz, A. (2020). How Can Smart Mobility Innovations Alleviate Transportation Disadvantage? Assembling a Conceptual Framework through a Systematic Review. *Applied Sciences (Basel, Switzerland)*, *10*(18), 6306. doi:10.3390/app10186306

Camacho, R. F. (2015). *Caractererização, Simulação (à escala) e Modelação do Escoamento em Canais Artificiais. Aplicação a caso de estudo.* Dissertação para obtenção do grau de Mestre em Engenharia Civil pela Universidade da Madeira, Funchal, Portugal.

Carmona, F., & González, I. (2018). *Cartografía de la producción: dinámicas espaciales de localización de los tejidos industriales del área urbana de Madrid (1956-2014). Ciudad y formas urbanas: perspectivas transversales: Libro de resúmenes* [Production mapping: spatial dynamics of location of the industrial fabrics of the urban area of Madrid (1956-2014). City and urban forms: transversal perspectives: Book of summaries]. Academic Press.

Carvalho, F. P. (2006). Agriculture, pesticides, food security and food safety. *Environmental Science & Policy*, *9*(7–8), 685–692. doi:10.1016/j.envsci.2006.08.002

Casaca, A., & Marlow, P. (2007). The Impact of the Trans-European Transport Networks on the Development of Short Sea Shipping. *Maritime Economics & Logistics*, *9*(4), 302–323. doi:10.1057/palgrave.mel.9100184

Castanho, R. A. (2017). *Sustainable Urban Planning in Transboundary Areas: Analysis of Critical Factors for Territorial Success* [Doctoral Thesis]. University of Extremadura (UEx), Department of Vegetal Biology, Ecology, and Earth Sciences.

Castanho, R. A. (2020). The Relevance of Political Engagement and Transparency in Cross-Cooperation (CBC) Environments. Analyzing Border cities in Europe. Lex localis-Journal of Local Self-Government, 18(3), 487–502. doi:10.4335/18.3.487-502(2020)

Castanho, R. A. C., Fernández, J. C., Loures, L., & Pozo, L. F. (2017). Evolución del procedimiento de planeamiento urbano en la península ibérica y sus huellas en el paisaje urbano. Retos de futuro. *Revista Científica Monfragüe Desarrollo Resiliente, 8*(2).

Castanho, R. A., Vulevic, A., Naranjo Gómez, J., Cabezas, J., Fernández-Pozo, L., Loures, L., & Kurowska-Pysz, J. (2019). *Political Commitment and Transparency as a Critical Factor to Achieve Territorial Cohesion and Sustainable Growth. European Cross-Border Projects and Strategies.* Regional Science Policy and Practice.

Castanho, R., Vulevic, A., Fernández, J.P., Naranjo, G., & Loures, L. (2017). Accessibility andconnectivity—Movement between cities, as a critical factor to achieve success on cross-border cooperation (CBC) projects. A European analysis. Sustain. *Cities Soc.*, *2*, 181–190.

Castanho, R. A. (2019). Identifying Processes of Smart Planning, Governance andManagement in European Border Cities. Learning from City-to-City Cooperation (C2C). *Sustainability*, *2019*(11), 5476. doi:10.3390u11195476

Castanho, R. A., Naranjo Gómez, J. M., Vulevic, A., Behradfar, A., & Couto, G. (2021). Assessing Transportation Patterns in the Azores Archipelago. *Infrastructures*, *6*(1), 10. doi:10.3390/infrastructures6010010

Castanho, R., Loures, L., Cabezas, J., & Fernández-Pozo, L. (2017). Cross-Border Cooperation (CBC) in Southern Europe—An Iberian Case Study. The Eurocity Elvas-Badajoz. *Sustainability*, *9*(3), 360. doi:10.3390u9030360

Castanho, R., Lousada, S., Gómez, J. M. N., Escorcio, P., Cabezas, J., Pozo-Fernández, L., & Loures, L. (2019). *Dynamics of the Land Use Changes and the Associated Barriers and Opportunities for Sustainble. Development on Peripheral and Insular Territories*. Intechopen. doi:10.5772/intechopen.80827

Castless, S., & Miller, M. J. (1998). *The age of migration-International population movements in the modern world* (2nd ed.). Macmillan Press Limited.

CEC (Commission of European Communities). (2008). *Green Paper on territorial cohesion—turning territorial diversity into strength.* Communication from the Commission to the Council, the European Parliament, the Committee of the Regions and the European Economic and Social Committee. CEC.

Celik, A. K., Ozcan, S., Topcuoglu, A., & Yildirim, K. E. (2013). Effects of the tourism industry on the balance of payment deficits. *Anatolia*, *24*(1), 86–90. doi:10.1080/13032917.2013.772529

CELPA. (2018). *Boletim estatístico indústria Papeleira Portuguesa*. Portuguese Paper Industry Statistical Bulletin.

Cerovic, S., Vukadinovic, P. & Knezevic, M. (2015). The influence of globalization of tourism and impact on tourism on other activities with an emphasis on greenfield investments in tourism. *Sitcon*, 47-52.

Ceylan, R. (2007). Genel işletme. Anadolu Üniversitesi Yayını.

Chakrawarty, A., Ranjan, P., Klanidhi, K. B., Kaur, D., Sarkar, S., Sahu, A., ... Wig, N. (2021). Psycho-social and behavioral impact of COVID-19 on middle-aged and elderly individuals: A qualitative study. *Journal of Education and Health Promotion*, 10. PMID:34485566

Chow, V. T. (1988). Applied Hydrology. McGraw-Hill.

Chow, V. (1964). *Handbook of Applied Hydrology*. McGraw-Hill.

Christie, I. T., & Crompton, D. E. (2001). *Tourism in Africa*. The World Bank. Available at: https://www.worldbank.org/afr/wps/index.htm

Christie, I. T., Fernandes, E., Messerli, H., & Twining-Ward, L. (2013). *Tourism in Africa: Harness Tourism for Growth and Improved Livelihoods*. World Bank.

Christie, M., Hanley, N., & Hynes, S. (2007). Valuing enhancements to forest recreation using choice experiment and contingent behaviour methods. *Journal of Forest Economics*, *13*(2–3), 75–102. doi:10.1016/j.jfe.2007.02.005

Christodoulou, A., & Christidis, P. (2018). *Cross-border transport infrastructure in the EU: A methodology to assess the role of cross-border road networks, EUR 29565 EN*. Publications Office of the European Union.

Christofoletti, A. (1980). *Geomorfologia*. Edgard Blücher.

CIHEAM. (2020). *The Covid 19 Pandemic: Threats On Food Security In The Mediterranean Region*. https://www.ciheam.org/publication/the-covid-19-pandemic-threats-on-food-security-in-the-mediterranean-region/

Clarke, J. (1997). A Framework of Approaches to Sustainable Tourism. *Journal of Sustainable Tourism*, *5*(3), 224–233. doi:10.1080/09669589708667287

CNR (Comision Nacional de Riego). (2004). *Diagnóstico base de los embalses El Bato y Corrales, IV Región*. Retrieved from http://bibliotecadigital.ciren.cl/handle/123456789/9601

Condeça, J., Nascimento, J., & Barreiras, N. (2022). Monitoring the Storage Volume of Water Reservoirs Using Google Earth Engine. *Water Resources Research*, *58*(3). Advance online publication. doi:10.1029/2021WR030026

Copernicus Europe's eyes on Earth. (2022). *CORINE Land Cover*. Retrieved 01/24/2022 from https://land.copernicus.eu/pan-european/corine-land-cover

Copperthite, K. (1989). *Rigid Container Recycling: Status and Impact on the Rigid Container Industry*. U.S. Department of Commerce, International Trade Administration.

Coppola, P., Carbone, A., Aveta, C., & Stangherlin, P. (2020). Assessing transport policies for tourist mobility based on accessibility indicators. *European Transport Research Review*, *12*(1), 56. doi:10.118612544-020-00444-4

Corbau, C., Zambello, E., Rodella, I., Utizi, K., Nardin, W., & Simeoni, U. (2019). Quantifying the impacts of human activities on the evolution of Po delta territory during the last 120 years. *Journal of Environmental Management*, *232*, 702–712. doi:10.1016/j.jenvman.2018.11.096 PMID:30529412

Córdoba, R., & Morcillo, D. (2020). Marco territorial de la producción de espacio en la región funcional de Madrid [Territorial framework of the production of space in the functional region of Madrid]. *Ciudades: Revista del Instituto Universitario de Urbanística de la Universidad de Valladolid*, (23), 71–93. doi:10.24197/ciudades.23.2020.71-93

Cordova, J. (2017). *Estudio del concepto de seguridad de riego, su funciones de beneficio agrícola, caso embalse las palmas, v region, Chile*. Universidad de Chile. Departamento de ingenería civil.

Corona, P., Quatrini, V., Schirru, M., Dettori, S., & Puletti, N. (2018). Towards the economic valuation of ecosystem production from cork oak forests in sardinia (Italy). *IForest (Viterbo)*, *11*(5), 660–667. doi:10.3832/ifor2558-011

Costanza, R., de Groot, R., Farberk, S., Grasso, M., Hannon, B., Limburg, K., & Naeem, S. O, R. v, Paruelo, J., Raskin, R. G., Suttonkk, P., & van den Belt, M. (1997). The value of the world's ecosystem services and natural capital. Nature, 387.

Costanza, R., de Groot, R., Sutton, P., van der Ploeg, S., Anderson, S. J., Kubiszewski, I., Farber, S., & Turner, R. K. (2014). Changes in the global value of ecosystem services. *Global Environmental Change*, *26*(1), 152–158. doi:10.1016/j.gloenvcha.2014.04.002

Couto, G., Castanho, R. A., Santos, C., Pimentel, P., Sousa, Á., Faria, S., & Batista, M. da G. (2021). Guidelines for Tourism Sustainability in Ultra-Peripheral Territories: A Research Based on the Azores Region's Touristic Companies' Analysis. *Sustainability*, *13*(7), 3895. doi:10.3390u13073895

Croitoru, L. (2007). Valuing the non-timber forest products in the Mediterranean region. *Ecological Economics*, *63*(4), 768–775. doi:10.1016/j.ecolecon.2007.01.014

Cyberscope. (2019). *Maletsunyane Falls in Semonkong, Lesotho: the highest single-drop waterfalls in Africa?* Available at: https://cycloscope.net/maletsunyane-falls-semonkong

Czachór, Z., & Graś, A. (2006). Vademecum Europa od A do Z [Vademecum Europa from A to Z]. Vizja Press & It.

Czajkowski, M., Buszko-Briggs, M., & Hanley, N. (2009). Valuing changes in forest biodiversity. *Ecological Economics*, *68*(12), 2910–2917. doi:10.1016/j.ecolecon.2009.06.016

D'Amato, D., Rekola, M., Li, N., & Toppinen, A. (2016). Monetary valuation of forest ecosystem services in China: A literature review and identification of future research needs. In *Ecological Economics* (Vol. 121, pp. 75–84). Elsevier. doi:10.1016/j.ecolecon.2015.11.009

Daher, A. (2013). El sector inmobiliario y las crisis económicas [The real estate sector and economic crises]. *Eure (Santiago)*, *39*(118), 47-76.

Daly, J., & Gereffi, G. (2017). *Tourism global value chains and Africa*. WIDER Working Paper 2017/17. Available at: https://wider.unu.edu

David, L. M. & Carvalho, R. F. (2008). *Bacias de retenção para controlo de cheias: Reflexão sobre os métodos de dimensionamento.* 13° Encontro Nacional de Saneamento Básico.

de Graaff, J., Kessler, A., & Nibbering, J. W. (2011). Agriculture and food security in selected countries in Sub-Saharan Africa: Diversity in trends and opportunities. *Food Security*, *3*(2), 195–213. doi:10.100712571-011-0125-4

De Lange, W. J., Wise, R. W., & Nahman, A. (2010, November/December). Securing a sustainable future through a new global contract between rich and poor. *Sustainable Development (Bradford)*, *18*(6), 374–384. doi:10.1002d.413

De Lima, P., Bernabe, S., & Bubbico, R. L. (2016). *Migration and the EU, Challenges, Opportunities, the Role of EIB*. European Investment Bank.

De Souza, R. S. (2002). Evolução e condicionantes da gestão ambiental nas empresas. *READ - Revista Eletrônica de Administração, 8*(6), 1-22.

De Ville, F., & Siles-Brügge, G. (2019). The Impact of Brexit on EU Policies. *Politics and Governance, 7*(3), 1–6. doi:10.17645/pag.v7i3.2129

Demirkaya, H. (2000). *Performans ölçüm rehberi.* Sayıştay Yayın İşleri Müdürlüğü.

Deng, X., Song, C., Liu, K., Ke, L., Zhang, W., Ma, R., … Wu, Q. (2020). Remote sensing estimation of catchment-scale reservoir water impoundment in the upper Yellow River and implications for river discharge alteration. *Journal of Hydrology, 585*(December), 124791. doi:10.1016/j.jhydrol.2020.124791

Department for Environment, Food, and Rural Affairs. (2021). UK Statistics on Waste. Government Statistical Service.

Department of Tourism. (2018). *State of Tourism Report 2017/18.* National Department of Tourism.

DGA (Direccion General de Aguas). (2021a). Información pluviométrica,fluviométrica, estado de embalses y aguas subterráneas. DGA.

DGA (Direccion General de Aguas). (2021b). *Mapoteca Digital.* Retrieved from https://dga.mop.gob.cl/estudiospublicaciones/mapoteca/Paginas/default.aspx

DGA. (2013). *Estrategia nacional de recursos hídricos 2012-2025.* Retrieved from https://www.mop.cl/Documents/ENRH_2013_OK.pdf

Dhaliwal, A., & Zheng, G. (2019). Improving accessibility of EPR-insensitive tumor phenotypes using EPR-adaptive strategies: Designing a new perspective in nanomedicine delivery. *Theranostics, 9*(26), 8091–8108. doi:10.7150/thno.37204 PMID:31754383

Di Bella, A., Petino, G., & Scrofani, L. (2019). The Etna macro-region between peripheralization and innovation: Towards a smart territorial system based on tourism. *Reg Sci Policy Pract, 11*(3), 493–507. doi:10.1111/rsp3.12176

Dias, R. (2009). *Gestão ambiental: responsabilidade social e sustentabilidade* (2nd ed.). Saraiva.

Díaz-Pacheco, J., & García-Palomares, J. C. (2014). Urban Sprawl in the Mediterranean Urban Regions in Europe and the Crisis Effect on the Urban Land Development: Madrid as Study Case. *Urban Studies Research, 807381*, 1–13. Advance online publication. doi:10.1155/2014/807381

Diaz-Pacheco, J., & Gutiérrez, J. (2014). Exploring the Limitations of CORINE Land Cover for Monitoring Urban Land-Use Dynamics in Metropolitan Areas. *Journal of Land Use Science, 9*(3), 243–259. doi:10.1080/1747423X.2012.761736

Dichter, A., & Manzo, G. G. (2017). *Coping with success: Managing overcrowding in tourism destinations.* World Travel and Tourism Council.

Dingman, S. L. (2015). *Physical Hydrology* (3rd ed.). Waveland Press, Inc.

Divisekera, S. (2010). Economics of tourist's consumption behavior: Some evidence from Australia. *Tourism Management, 31*(5), 629–636. doi:10.1016/j.tourman.2009.07.001

Docherty, I., Marsden, G., & Anable, J. (2018). The governance of smart mobility. *Transportation Research Part A, Policy and Practice, 115*, 114–125. doi:10.1016/j.tra.2017.09.012

Doğan, H. (2014). *Örgütsel Adalet Algısı İle İş Performansı Arasındaki İlişki: Afyonkarahisar'da Beş Yıldızlı Termal Otel İşletmelerinde Bir Araştırma* (Master's thesis). Afyon Kocatepe Üniversitesi, Sosyal Bilimler Enstitüsü.

DOH (Direccion Obras Hidraulicas). (2017). *Actualización modelo de operación embalse el bato.* DOH.

Domingo, D., Palka, G., & Hersperger, A. M. (2021). Effect of zoning plans on urban land-use change: A multi-scenario simulation for supporting sustainable urban growth. *Sustainable Cities and Society*, *69*, 102833. doi:10.1016/j.scs.2021.102833

Donaldson, R. (2018). *Small Town Tourism in South Africa*. Springer. doi:10.1007/978-3-319-68088-0

Dowall, D., & Clarke, G. (1996). *A framework for reforming urban land policies in developing countries*. The International Bank for Reconstruction and Development.

DROTA. (2017). *Identificação das Zonas Críticas a Cartografar Considerando os Impactos da Ocorrência de Inundações (Com Risco Potencial Significativo)*. Direção Regional de Ordenamento do Território e Ambiente.

Du Preez, M., & Hosking, S. G. (2011). The value of the trout fishery at Rhodes, North Eastern Cape, South Africa: A travel cost analysis using count data models. *Journal of Environmental Planning and Management*, *54*(2), 267–282. doi:10.1080/09640568.2010.505837

Du Preez, M., & Lee, D. E. (2010). The contribution of trout fly fishing to the economy of Rhodes, North Eastern Cape, South Africa. *Development Southern Africa*, *27*(2), 241–253. doi:10.1080/03768351003740654

Dupuy, G. (1995). *Les territorios de l'automobile* [Automobile territories]. Anthropos-Economica.

Duro, J. A., Perez-Laborda, A., Turrion-Prats, J., & Fernández-Fernández, M. (2021). Covid-19 and tourism vulnerability. *Tourism Management Perspectives*, *38*, 100819. doi:10.1016/j.tmp.2021.100819 PMID:34873568

Dynia, E. (2003). *Integracja Europejska [European integration]*. LexisNexis.

Easterling, D., Kunkel, K., Wehner, M. F., & Sun, L. (2016). Detection and attribution of climate extremes in the observed record. *Weather and Climate Extremes*, *11*, 17–27. doi:10.1016/j.wace.2016.01.001

Echendu, A. J. (2022). Flooding, Food Security and the Sustainable Development Goals in Nigeria: An Assemblage and Systems Thinking Approach. *Social Science*, *2022*(11), 59. doi:10.3390ocsci11020059

Eckstein, D., Küzel, V., & Schafer, L. (2021). Global Climate Risk Index. *Germanwatch*, *1*. https://www.germanwatch.org/en/19777

ECOCE. (2019). *Plan de Manejo ECOCE.* Recovered from: https://www.ecoce.mx/plan_nacional

Economics, O. (2017). *Travel & Tourism Power and Performance Report.* Available at: https:www.oxfordeconomics.com

Edensor, T. 2010. Aurora Landscapes: Affective Atmospheres of Light and Dark. Conversations with Landscape, 227–240.

Ehrlich, P. R., Ehrlich, A. H., & Daily, G. C. (1993). Food Security, Population and Environment. *Population and Development Review*, *19*(1), 1–32. doi:10.2307/2938383

Einecker, R., & Kirby, A. (2020). Climate Change: A Bibliometric Study of Adaptation, Mitigation and Resilience. *Sustainability*, *12*(17), 6935. doi:10.3390u12176935

Engelman, M., Lagerkvist, C.-J., & Gren, I.-M. (2018). Hunters' trade-off in valuation of different game animals in Sweden. *Forest Policy and Economics*, *92*, 73–81. doi:10.1016/j.forpol.2018.04.004

Environment, U. N. (2018). *Waste Management Outlook for Latin America and the Caribbean.* United Nations Environment Programme.

EPA. (2020). *Best Practices for Solid Waste Management: A Guide for Decision-Makers in Developing Countries.* United States Environmental Protection Agency.

Erdoğdu, M., Mermod, A., & Aşkun, Y. (2016). *Social and Economic Perspectives on Sustainability*. IJOPEC Publication.

ESA. (2015). Sentinel-2 User Handbook. Sentinel 2. doi:10.1021/ie51400a018

ESPON. INTERCO. (2013). *Indicators of territorial cohesion - Final Report*. Available at: https://ise.unige.ch/isdd/IMG/pdf/interco_#nal-report_part-a_executive-summary.pdf

ESPON. TRACC. (2013). *Transport accessibility at regional/local scale and patterns in Europe*. ESPON, Final Report - Indicator Factsheets.

European Commission. (2010). *Life Cycle Thinking and Assessment for Waste Management*. Publications Office of the European Union.

European Environment Agency (EEA). (2015). *Mediterranean Sea region briefing - The European environment — state and outlook 2015*. Briefing. https://www.eea.europa.eu/soer/2015/countries/mediterranean

European Environment Agency (EEA). (2020). *CORINE land cover*. Available at: www.eea.europa.eu/publications/COR0-landcover

Eurostat. (2010). *NUTS - Nomenclature of territorial units for statistics*. http://epp.eurostat.ec.europa.eu/portal/page/portal/nuts_nomenclature/introduction

Everingham, P., & Chassagne, N. (2020). Post COVID-19 ecological and social reset: Moving away from capitalist growth models towards tourism as Buen Vivir. *Tourism Geographies*, *22*(3), 555–566. doi:10.1080/14616688.2020.1762119

Evrard, E., & Engl, A. (2018). Taking Stock of the European Grouping of Territorial Cooperation (EGTC): From Policy Formulation to Policy Implementation. In E. Medeiros (Ed.), *European Territorial Cooperation. The Urban Book Series*. Springer. doi:10.1007/978-3-319-74887-0_11

Ezebilo, E. E. (2016). Economic value of a non-market ecosystem service: An application of the travel cost method to nature recreation in Sweden. *The International Journal of Biodiversity Science, Ecosystem Services & Management*, *12*(4), 314–327. doi:10.1080/21513732.2016.1202322

Fadigas, L. (2015). *Urbanismo e Território - As Políticas Públicas*. Editorial Estampa.

Faludi, A. (2009). *Territorial cohesion under the looking glass: Synthesis paper about the history of the concept and policy background to territorial cohesion*. Retrieved from https://www.researchgate. net/profile/Andreas_Faludi/contributions

Faludi, A. (2016b), The territoriality of Cohesion policy. In S. Piattoni & L. Polverari (Eds.), Handbook on Cohesion Policy in the EU (pp. 491-505). Edward Elgar. doi:10.4337/9781784715670.00048

Faludi, A. (2016d). The Poverty of Territorialism: Revisiting European Spatial Planning. *disP – The Planning Review*, *52*(3), 73-81. doi:10.1080/02513625.2016.1235886

Faludi, A. (2004). Territorial Cohesion: Old (French) Wine in New Bottles? *Urban Studies (Edinburgh, Scotland)*, *41*(7), 1349–1363. doi:10.1080/0042098042000214833

Faludi, A. (2006). From European spatial developmentto territorial cohesion policy. *Regional Studies*, *40*(6), 667–678. doi:10.1080/00343400600868937

Faludi, A. (2010). *Cohesion, Coherence, Cooperation:European Spatial Planning Coming of Age?* Routledge. doi:10.4324/9780203842324

Faludi, A. (2013). Territorial cohesion, territorialism, territoriality and soft planning: A critical review. Environment and Planning A. *Economy and Space*, *45*(6), 1302–1317. doi:10.1068/a45299

Faludi, A. (2016a). European integration and the Territorial – Administrative Complex. *Geografiska Annaler. Series B, Human Geography*, *98*(1), 71–80. doi:10.1111/geob.12090

Faludi, A. (2016c). EU territorial cohesion, a contradiction in terms. *Planning Theory & Practice*, *17*(2), 302–313. doi:10.1080/14649357.2016.1154657

Faludi, A. (2019). New Horizons: Beyond territorialism. *Europa XXI*, *36*, 35–44. doi:10.7163/Eu21.2019.36.3

FAO, IFAD, UNICEF, WFP, & WHO. (2021). *The State of Food Security and Nutrition in the World 2021. Transforming food systems for food security, improved nutrition and affordable healthy diets for all.* FAO. doi:10.4060/cb4474en

FAO. (2018). Transforming Food and Agriculture to Achieve the SDGs: 20 interconnected actions to guide decision-makers. Technical Reference Document. FAO.

FAO. (2021). *Geo-referenced Database on Dams.* Retrieved from AQUASTAT - FAO's Global Information System on Water and Agriculture website: https://www.fao.org/aquastat/en/overview/

FAO. (2021). *World Food and Agriculture - Statistical Yearbook 2021.* doi:10.4060/cb4477en

FAOSTAT. (n.d.a). *Annual population*, https://www.fao.org/faostat/en/#data/OA

FAOSTAT. (n.d.b). *Crops and livestock products.* https://www.fao.org/faostat/en/#data/QCL

FAOSTAT. (n.d.c). *Supply Utilization Accounts.* https://www.fao.org/faostat/en/#data/SCL

FAOSTAT. (n.d.d). *Suite of Food Security Indicators.* https://www.fao.org/faostat/en/#data/FS

Fatima, K., Moridpour, S., De Gruyter, C., & Saghapour, T. (2020). Elderly Sustainable Mobility: Scientific Paper Review. *Sustainability*, *12*(18), 7319. doi:10.3390u12187319

Feldman, C. D., & Arnold, J. H. (1996). *Managing Individual and Group Behavior in Organization.* Mc.Graw-Hill International Book Company.

Felzensztein, C., Gimmon, E., & Aqueveque, C. (2013). Entrepreneurship at the Periphery: Exploring Framework Conditions in Core and Peripheral Locations. *Entrepreneurship Theory and Practice*, *37*(4), 815–835. doi:10.1111/j.1540-6520.2012.00515.x

Feranec, J., Hazeu, G., Christensen, S., & Jaffrain, G. (2007). Corine land cover change detection in Europe (case studies of the Netherlands and Slovakia). *Land Use Policy*, *24*(1), 234–247. doi:10.1016/j.landusepol.2006.02.002

Feranec, J., Jaffrain, G., Soukup, T., & Hazeu, G. (2010). Determining changes and flows in European landscapes 1990–2000 using CORINE land cover data. *Applied Geography (Sevenoaks, England)*, *30*(1), 19–35. doi:10.1016/j.apgeog.2009.07.003

Fernandes, M. J. P. (2009). *Riscos no Concelho da Ribeira Brava: Movimentos de vertente cheias rápidas e inundações.* Dissertação para obtenção do grau de Mestre em Dinâmicas Sociais e Riscos Naturais pela Universidade de Coimbra, Coimbra, Portugal.

Fernández, C., Cabezas, A., & Pinedo, P. (1998). *Urbanismo y arquitectura en el Madrid actual* [Urbanism and architecture in today's Madrid] (P. L. López Algora, Ed.). Comunidad de Madrid, Dirección General de Centros Docentes. http://www.madrid.org/bvirtual/BVCM000851.pdf

Fernández, R. (2006). *El tsunami urbanizador español y mundial: sobre sus causas y repercusiones devastadoras, y la necesidad de prepararse para el previsible estallido de la burbuja inmobiliaria* [The Spanish and global urbanizing tsunami: on its devastating causes and repercussions, and the need to prepare for the foreseeable bursting of the real estate bubble]. Virus Editorial. https://dspace-libros.metabiblioteca.com.co/jspui/bitstream/001/397/8/84-96044-74-2.pdf

Ferreira, A. F., Akasaka, Y., Greiner de Oliveira Pinheiro, M., & Chang, S. K. J. (2020). Information as the First Attribute of Accessibility: A Method for Assessing the Information Provided by Urban Rail Systems to Tourists with Reduced Mobility. *Sustainability*, *12*(23), 10185. doi:10.3390u122310185

Ferreira, A., & Papa, E. (2020). Re-enacting the mobility versus accessibility debate: Moving towards collaborative synergies among experts. *Case Studies on Transport Policy.*, *8*(3), 1002–1009. doi:10.1016/j.cstp.2020.04.006

Filion, P., Bunting, T., & Warriner, K. (1999). The Entrenchment of Urban Dispersion: Residential Preferences and Location Patterns in the Dispersed City. *Urban Studies (Edinburgh, Scotland)*, *36*(8), 1317–1347. doi:10.1080/0042098993015

Fisher, B., Turner, R. K., & Morling, P. (2009). Defining and classifying ecosystem services for decision making. *Ecological Economics*, *68*(3), 643–653. doi:10.1016/j.ecolecon.2008.09.014

Flamm, M., & Kaufmann, V. (2006). Operationalising the Concept of Motility: A Qualitative Study. *Mobilities*, *1*(2), 167–189. doi:10.1080/17450100600726563

Flavell, A., Melde, S., & Milan, A. (2020). *Migration, environment and climate change: Impacts*. Second report in the "Migration, environment and climate change" series. Texte, 43/2020, Project No. (FKZ) 3717181060, Report No. FB000245/2,ENG.

Foga, S., Scaramuzza, P. L., Guo, S., Zhu, Z., Dilley, R. D. Jr, Beckmann, T., Schmidt, G. L., Dwyer, J. L., Joseph Hughes, M., & Laue, B. (2017). Cloud detection algorithm comparison and validation for operational Landsat data products. *Remote Sensing of Environment*, *194*, 379–390. doi:10.1016/j.rse.2017.03.026

Förster, J., Schmidt, S., Bartkowski, B., Lienhoop, N., Albert, C., & Wittmer, H. (2019). Incorporating environmental costs of ecosystem service loss in political decision making: A synthesis of monetary values for Germany. *PLoS One*, *14*(2), e0211419. Advance online publication. doi:10.1371/journal.pone.0211419 PMID:30759137

Forsyth, A. (2003). Measuring density: working definitions for residential density and building intensity. *Design Brief*, *9*(1), 2-8.

Friedl, M., Gray, J., & Sulla-Menashe, D. (2015). *MCD12Q2 MODIS/Terra+Aqua Land Cover Dynamics Yearly L3 Global 500m SIN Grid V006*. Academic Press.

Fuchs, P. G., Raulino, C. E., & Guerra, J. B. O. S de A. (2020). Green business in the context of the sustainable development. In W. L. Filho, A. M. Azul, L. Brandli, A. L. Salvia, & T. Wall (Eds.), *Decent Work and Economic Growth, Encyclopedia of the UN Sustainable Development Goals* (pp. 507–518). Springer. doi:10.1007/978-3-319-71058-7_12-1

Gacka-Asiewicz, A. (2016). *Prawo Unii Europejskiej w pigułce* [European Union law in a nutshell]. Academic Press.

Gaja i Diaz, F. (2013). Tras el tsunami inmobiliario. Salir del atolladero [After the real estate tsunami. Get out of the quagmire]. In Observatorio Metropolitano de Madrid (Ed.), *Paisajes devastados después del ciclo inmobiliario: impactos regionales y urbanos de la crisis* [Landscapes devastated after the real estate cycle: regional and urban impacts of the crisis] (pp. 313-354). Traficantes de sueños. https://www.traficantes.net/sites/default/files/pdfs/Paisajes%20Devastados-TdS.pdf

Ganovel, M. F., Adger, W. N., de Campos, R. S., Boyd, E., Carr, E. R., Fabos, A., Fransen, S., Jolivet, D., Zickgraf, C., Codjoe, S. N. A., Abu, M., & Siddiqui, T. (2021). The migration-sustainability paradox: Transformations in mobile worlds. *Environmental Sustainability*, *49*, 98–109. doi:10.1016/j.cosust.2021.03.013

Gao, B. (1996). NDWI—A normalized difference water index for remote sensing of vegetation liquid water from space. *Remote Sensing of Environment, 58*(3), 257–266. doi:10.1016/S0034-4257(96)00067-3

Garcia, S., Abildtrup, J., & Stenger, A. (2018). How does economic research contribute to the management of forest ecosystem services? In Annals of Forest Science (Vol. 75, Issue 2). Springer-Verlag France. doi:10.100713595-018-0733-7

García-Palomares, J. C., & Gutiérrez, J. (2007). La ciudad dispersa: cambios recientes en los espacios residenciales de la Comunidad de Madrid [The dispersed city: recent changes in the residential spaces of the Community of Madrid]. *Anales de geografía de la Universidad Complutense, 27*(1), 45.

García-Palomares, J. C. (2010). Urban sprawl and travel to work: The case of the metropolitan area of Madrid. *Journal of Transport Geography, 18*(2), 197–213. doi:10.1016/j.jtrangeo.2009.05.012

Garreaud, R. D., Alvarez-Garreton, C., Barichivich, J., Pablo Boisier, J., Christie, D., Galleguillos, M., ... Zambrano-Bigiarini, M. (2017). The 2010-2015 megadrought in central Chile: Impacts on regional hydroclimate and vegetation. *Hydrology and Earth System Sciences, 21*(12), 6307–6327. Advance online publication. doi:10.5194/hess-21-6307-2017

Garreaud, R. D., Vuille, M., Compagnucci, R., & Marengo, J. (2009). Present-day south american climate. *Palaeogeography, Palaeoclimatology, Palaeoecology, 281*(3), 180–195. doi:10.1016/j.palaeo.2007.10.032

Garreaud, R., López, P., Minvielle, M., & Rojas, M. (2013). Large-Scale Control on the Patagonian Climate. *Journal of Climate, 26*(1), 215–230. doi:10.1175/JCLI-D-12-00001.1

Garrity, D. P., Akinnifesi, F. K., Ajayi, O. C., Weldesemayat, S. G., Mowo, J. G., Kalinganire, A., Larwanou, M., & Bayala, J. (2010). Evergreen Agriculture: A robust approach to sustainable food security in Africa. *Food Security, 2*(3), 197–214. doi:10.100712571-010-0070-7

Getzner, M., Meyerhoff, J., & Schläpfer, F. (2018). Willingness to pay for nature conservation policies in state-owned forests: An austrian case study. *Forests, 9*(9), 537. Advance online publication. doi:10.3390/f9090537

Giuliano, G., & Small, A. (1999). The determinants of growth of employment subcenters. *Journal of Transport Geography, 7*(3), 189–201. doi:10.1016/S0966-6923(98)00043-X

Gkoumas, K., dos Santos, F. L. M., Stepniak, M., & Pekar, F. (2021). Research and Innovation Supporting the European Sustainable and Smart Mobility Strategy: A Technology Perspective from Recent European Union Projects. *Applied Sciences (Basel, Switzerland), 11*(24), 1981. doi:10.3390/app112411981

Gómez, J. M. N., Vulevic, A., Couto, G., & Alexandre Castanho, R. (2021). Accessibility in European Peripheral Territories: Analyzing the Portuguese Mainland Connectivity Patterns from 1985 to 2020. *Infrastructures, 6*(6), 92. doi:10.3390/infrastructures6060092

Gonçalves, J. (2016). *Caracterização do coeficiente de rugosidade e seu efeito no escoamento em Canais Naturais. Simulação e Modelação (à escala) no laboratório de Hidráulica. Aplicação às ribeiras do Funchal.* Dissertação para obtenção do grau de Mestre em Engenharia Civil pela Universidade da Madeira, Funchal, Portugal.

Gonçalves, L. B. (2020). *Análise teórico-prática do risco de cheias no Arquipélago da Madeira – Caso de estudo dos concelhos do Funchal, Machico, Ribeira Brava e São Vicente.* Dissertação para obtenção do grau de Mestre em Engenharia Civil pela Universidade da Madeira, Funchal, Portugal.

Gonçalves, L. B., & Lousada, S. A. N. (2020). Análise Probabilística de Cheias e o Uso de Bacias de Detenção como Medida Mitigadora: Aplicação à Bacia de Santa Luzia. *Revista Científica Monfragüe Desarrollo Resiliente, 13.*

Gonçalves, L. B., Lousada, S. A. N., & Lis, M. A. (2020). Utilização da Bacia de Detenção Para o Controlo e Regularização de Caudais a Jusante, sua Sustentabilidade. In *Espacios y Sociedades em Transformación* (1st ed., Vol. 6, pp. 505–518). Thomson Reuters Aranzadi.

Gonçalves, L. B., Rodrigues, W. T., Curitiba, A. S., Torres, H., & Lousada, S. A. N. (2020). *Utilização de Bacias de Detenção Para Mitigação de Cheias em Áreas Urbanas, Espacios y Sociedades em Transfo, Rmación* (1st ed., Vol. 6). Thomson Reuters Aranzadi.

González, R. S., & Castanho, R. A. (2021). *Comparison in Recycling Processes and Methods in developed and developing countries: A Brief Preliminary Study through Literature Review-Based Research.* Online International Conference on Empirical Economics and Social Sciences (e-ICEESS'21), Turkey.

Gorelick, N., Hancher, M., Dixon, M., Ilyushchenko, S., Thau, D., & Moore, R. (2017). Google Earth Engine: Planetary-scale geospatial analysis for everyone. *Remote Sensing of Environment, 202*, 18–27. doi:10.1016/j.rse.2017.06.031

Grilli, G., Paletto, A., & de Meo, I. (2014). Economic valuation of forest recreation in an alpine valley. *Baltic Forestry, 20*(1), 167–175.

Grundlingh, A. (2011). Mutating memories and the making of a myth: Remembering The SS Mendi Disaster, 1917-2007. *South African Historical Journal, 63*(1), 20–37. doi:10.1080/02582473.2011.549372

Gudmundsson, H., & Höjer, M. (1996). Sustainable Development Principles and Their Implications for Transport. *Ecological Economics, 19*(3), 269–282. doi:10.1016/S0921-8009(96)00045-6

Gürlek, M., & Tuna, M. (2018). Reinforcing competitive advantage through green organizational culture and green innovation. *Service Industries Journal, 38*(7-8), 467–491. doi:10.1080/02642069.2017.1402889

Gutiérrez, J., García-Palomares, J. C., Romanillos, G., & Salas-Olmedo, M. H. (2017). The eruption of Airbnb in tourist cities: Comparing spatial patterns of hotels and peer-to-peer accommodation in Barcelona. *Tourism Management, 62*, 278–291. doi:10.1016/j.tourman.2017.05.003

Guzmán, S. C. (2021). *GobCDMX inaugura planta recicladora de basura, la más moderna de América Latina.* Gobierno de la Ciudad de México. Recovered from: https://www.capital21.cdmx.gob.mx/noticias/?p=25324

Hall, C. M. (2019). Constructing sustainable tourism development: The 2030 agenda and the managerial ecology of sustainable tourism. *Journal of Sustainable Tourism, 27*(7), 1044–1060. doi:10.1080/09669582.2018.1560456

Hall, C. M., Scott, D., & Gössling, S. (2020). Pandemics, transformations, and tourism be careful what you wish for. *Tourism Geographies, 22*(3), 577–598. doi:10.1080/14616688.2020.1759131

Handy, S. (2005). Planning for Accessibility: In Theory and in Practice. In Access to Destinations. Emerald Group Publishing Limited. doi:10.1108/9780080460550-007

Hardy, A., Beeton, R. J., & Pearson, L. (2002). Sustainable tourism: An overview of the concept and its position in relation to conceptualisations of tourism. *Journal of Sustainable Tourism, 10*(6), 475–496. doi:10.1080/09669580208667183

Harma, S., & Gutiérrez, J. A. (2010). An evaluation framework for viable business models for m-commerce in the information technology sector. *Electronic Markets, 20*(1), 33–52. doi:10.100712525-010-0028-9

Hay, A., & Visser, G. (2014). Socio-cultural and socio-economic features of second homes in Rosendal, *South Africa. Bulletin of Geography. Socio-Economic Series, 26*(26), 157–166. doi:10.2478/bog-2014-0051

Häyhä, T., Franzese, P. P., Paletto, A., & Fath, B. D. (2015). Assessing, valuing, and mapping ecosystem services in Alpine forests. *Ecosystem Services, 14*, 12–23. doi:10.1016/j.ecoser.2015.03.001

Heath, E. (2001). Globalisation of the Tourism Industry: Future Trends and Challenges for South Africa. *SAJEMS*, *4*(3), 542–596. doi:10.4102ajems.v4i3.2662

Hein, L. (2011). Economic Benefits Generated by Protected Areas: The Case of the Hoge Veluwe Forest, the Netherlands. *Ecology and Society*, *16*(2), art13. doi:10.5751/ES-04119-160213

Henama, U. S. (2013). Attracting Indian Outbound Tourists to South Africa: A BRICS Perspective. *India Quarterly*, *69*(3), 229–247. doi:10.1177/0974928413489466

Henama, U. S. (2017). International Migration and National Tourism Industry: The Case of South Africa. *The EUrASE-ANs: The Journal of Global Socio-Economic Dynamics*, *2*(3), 72–82.

Hern, J. (2013). Urban landscape sustainability and resilience: The promise and challenges of integrating ecology with urban planning and design. *Landscape Ecology*, *28*(6), 1203–1212. doi:10.100710980-012-9799-z

Hersperger, A. M., Oliveira, E., Pagliarin, S., Palka, G., Verburg, P., Bolliger, J., & Grădinaru, S. (2018). Urban land-use change: The role of strategic spatial planning. *Global Environmental Change*, *51*, 32–42. doi:10.1016/j.gloenvcha.2018.05.001

Hidayati, I., Yamu, C., & Tan, W. (2019). The Emergence of Mobility Inequality in Greater Jakarta, Indonesia: A Socio-Spatial Analysis of Path Dependencies in Transport–Land Use Policies. *Sustainability*, *11*(18), 5115. doi:10.3390u11185115

Hill, T. (2003). *Reduce, Reuse, Recycle*. Ready-Ed Publications.

Hix, S., Noury, A., & Roland, G. (2005). Power to the Parties: Cohesion and Competition in the European Parliament, 1979–2001. *British Journal of Political Science*, *35*(2), 209–234. doi:10.1017/S0007123405000128

Hlaváčková, P., & Šafařík, D. (2016). Quantification of the utility value of the recreational function of forests from the aspect of valuation practice. *Journal of Forest Science*, *62*(8), 345–356. doi:10.17221/50/2016-JFS

Honkanen, A., & Mustonen, P. (2007). Tourism consumption revisited-an empirical study of Finnish consumers. *Research on Finnish Society*, *1*, 43–58.

Hoogendoorn, G., & Visser, G. (2015). Focusing on the 'blessing' and not the 'curse' of second homes: notes from South Africa. *Urban Forum*, *47*(2), 179-184.

Hoogendoorn, G., Mellett, R., & Visser, G. (2005). Second Home Tourism in Africa: Reflections on the South African Experience. *Urban Forum*, *16*(2-3), 112-154.

Hoogendoorn, G. (2011). Low-income earners as second home tourists in South Africa? *Tourism Review International*, *15*(1), 37–50. doi:10.3727/154427211X13139345020219

Hoogendoorn, G., & Fitchett, J. (2018). Perspectives on Second Homes, Climate Change and Tourism in South Africa. *African Journal of Hospitality, Tourism and Leisure*, *7*(2), 1–18.

Hoogendoorn, G., & Visser, G. (2004). Second home and small-town (re) development: The case of Clarens. *Journal of Family Ecology and Consumer Sciences*, *31*, 105–115.

Hoogendoorn, G., & Visser, G. (2010). The role of second homes in local economic development in five small South African towns. *Development Southern Africa*, *27*(4), 547–562. doi:10.1080/0376835X.2010.508585

Hoogendoorn, G., & Visser, G. (2011). Economic Development through Second Home Development: Evidence from South Africa. *Tijdschrift voor Economische en Sociale Geografie*, *102*(3), 275–289. doi:10.1111/j.1467-9663.2011.00663.x

Hoogendoorn, G., Visser, G., & Marais, L. (2009). Changing Countrysides, Changing Villages: Second Homes in Rhodes, South Africa. *The South African Geographical Journal*, *91*(2), 75–83. doi:10.1080/03736245.2009.9725334

Hornberger, G., & Wiberg, P. (2014). *Elements of physical hydrology*. Johns Hopkins University Press.

Hossain, M. S., Bujang, J. S., Zakaria, M. H., & Hashim, M. (2015). Assessment of Landsat 7 Scan Line Corrector-off data gap-filling methods for seagrass distribution mapping. *International Journal of Remote Sensing*, *36*(4), 1188–1215. doi:10.1080/01431161.2015.1007257

Hougner, C., Colding, J., & Söderqvist, T. (2005). *Economic valuation of a seed dispersal service in the Stockholm National Urban Park*., doi:10.1016/j.ecolecon.2005

ICMDP. (2021). *Regional Migration Outlook 2021-Turkey & Western Balkans*. ICMDP Turkey & Western Balkans Regional Coordination Office. https://www.icmpd.org/file/download/51375/file/Turkey0and0Western0Balkans0Regional0Migration0Outlook.pdf

ICNF, Instituto da Conservação da Natureza e das Florestas. (2006). *Estratégia nacional para as florestas* [National strategy for forests]. Author.

ICNF, Instituto da Conservação da Natureza e das Florestas. (2019a). *IFN 6*. Author.

ICNF, Instituto da Conservação da Natureza e das Florestas. (2019b). *Programa Regional de Ordenamento Florestal, PROF da Região de Trás-os-Montes e Alto Douro. Documento Estratégico. CAP. A B e C* [Regional Program for Forest Management, PROF of the Trás-os-Montes and Alto Douro Region. STRATEGIC DOCUMENT. Chap. A B and C]. Author.

IGM. (2018). *Mapa Geomorfológico de Chile*. Instituto Geográfico Militar.

ILO. (2021). *Impact of COVID-19 on nexus between climate change and labour migration in selected South Asian countries: An exploratory study*. https://www.ilo.org/wcmsp5/groups/public/---ed_protect/---protrav/---migrant/documents/publication/wcms_822838.pdf

INE. (2019). *Instituto Nacional de Estatísticas* [National Statistical Institute]. Estatísticas Agrícolas.

Ingaramo, R., & Salizzoni, E. (2017). Valuing Forest Ecosystem Services for Spatial and Landscape Planning and Design. *Journal Valori e Valutazioni*, *19*, 65–78.

International Eco-Energy Association. (2002). *Proceedings of the 6th Baku International Congress "Energy, Ecology, Economy."* Author.

IOM. (2008). *Migration, development, and environment*. IOM International Organization of Migration, No. 35. https://publications.iom.int/system/files/pdf/mrs_35_1.pdf

IOM. (2020). *World migration report 2020*. https://publications.iom.int/system/files/pdf/wmr_2020.pdf

IPCC. (2014). *Climate Change 2014: Synthesis Report. Contribution of Working Groups I, II and III to the Fifth Assessment Report of the Intergovernmental Panel on Climate Change*. IPCC.

IPCC. (2018). Summary for Policymakers. In *Global Warming of 1.5°C. An IPCC Special Report on the impacts of global warming of 1.5°C above pre-industrial levels and related global greenhouse gas emission pathways, in the context of strengthening the global response to the threat of climate change, sustainable development, and efforts to eradicate poverty*. World Meteorological Organization.

Irvine, P. M., Kepe, D. T., & Hamunime, N. P. (2016). Whose Mecca? Divergent experiences of post-productisvism and tourism in Nieu Bethesda, South Africa. *The South African Geographical Journal*, *98*(2), 386–401. doi:10.1080/03736245.2015.1052843

Islam, T., & Chandrasekaran, U. (2016). Religiosity and ecologically conscious consumption behaviour. *Asian Journal of Business Research*, *5*(1), 1–18.

Jabar, A. S. A., Sulong, G., & George, L. E. (2014). Survey on gap filling algorithms in Landsat 7 ETM+ images. *Journal of Theoretical and Applied Information Technology*, *63*(1), 136–146.

Jacobsen, J. K. S., & Munar, A. M. (2012). Tourist information search and destination choice in a digital age. *Tourism Management Perspectives*, *1*, 39–47. doi:10.1016/j.tmp.2011.12.005

Jaraíz-Cabanillas, F. J., Mora-Aliseda, J., Jeong, J. S., & Garrido-Velarde, J. (2018). Methodological proposal to classify and delineate natural protected áreas. Study case: Region of Extremadura, Spain. *Land Use Policy*, *79*, 310–319. doi:10.1016/j.landusepol.2018.08.034

Järv, O., Tenkanen, H., Salonen, M., Ahas, R., & Toivonen, T. (2018). Dynamic cities: Location-based accessibility modelling as a function of time. *Applied Geography (Sevenoaks, England)*, *95*, 101–110. doi:10.1016/j.apgeog.2018.04.009

Jean-Paul Rodrigue. (2020). *The Geography of Transport Systems. FIFTH EDITION*. Routledge.

Jednolity Akt Europejski – podpisany 17 lutego 1986 r. (Single European Act - signed on 17 February 1986) (Luksemburg) oraz 28 lutego 1986 r. (Haga) (wszedł w życie 1 lipca 1987 r.) (Dz.U. z 2004 r. Nr 90, poz. 864/5).

Jeon, M. M., Kang, M. M., & Desmarais, E. (2016). Residents' perceived quality of life in a cultural-heritage tourism destination. *Applied Research in Quality of Life*, *11*(1), 105–123. doi:10.100711482-014-9357-8

Jóhannesdóttir, G. R. (2010). Landscape and Aesthetic Values: Not Only in the Eye of the Beholder. In K. Benediktsson & K. Lund (Eds.), *Conversations with Landscape* (pp. 107–123). Ashgate.

Johnson, M. (2001). Environmental Impacts of Urban Sprawl: A Survey of the Literature and Proposed Research Agenda. *Environment and Planning A. Economy and Space*, *33*(4), 717–735. doi:10.1068/a3327

Jones, S. K., Fremier, A. K., DeClerck, F. A., Smedley, D., Pieck, A. O., & Mulligan, M. (2017). Big data and multiple methods for mapping small reservoirs: Comparing accuracies for applications in agricultural landscapes. *Remote Sensing*, *9*(12), 1307. Advance online publication. doi:10.3390/rs9121307

Jopson, T. (2018). *Rhodes Village-up where you belong*. Available at: https://www.getaway.co.za/travel-ideas/destinations-travel-ideas/rhodes-village/

Jordová, R., & Brůhová-Foltýnová, H. (2021). Rise of a New Sustainable Urban Mobility Planning Paradigm in Local Governance: Does the SUMP Make a Difference? *Sustainability*, *13*(11), 5950. doi:10.3390u13115950

Kammerlander, M., Schanes, K., Hartwig, F., Jäger, J., Omann, I., & O'Keeffe, M. (2015). A resource-efficient and sufficient future mobility system for improved well-being in Europe. *Eur J Futures Res*, *3*(1), 8. doi:10.100740309-015-0065-x

Khalfaoui, M., Daly-Hassen, H., Stiti, B., & Jebari, S. (2020). Toward decision-making support: Valuation and mapping of new management scenarios for tunisian cork oak forests. *Forests*, *11*(2), 197. Advance online publication. doi:10.3390/f11020197

Khan, S. A. R., Sharif, A., Golpira, H., & Kumar, A. (2019, November/December). A green ideology in Asian emerging economies: From environmental policy and sustainable development. *Sustainable Development*, *27*(6), 1063–1075. doi:10.1002d.1958

Kibreab, G. (1996, December). Eritrean and Ethiopian urban refugees in Khartoum: What the eye refuses to see. *African Studies Review*, *39*(3), 131–178. doi:10.2307/524946

Kibria, A. S. M. G., Behie, A., Costanza, R., Groves, C., & Farrell, T. (2017). The value of ecosystem services obtained from the protected forest of Cambodia: The case of Veun Sai-Siem Pang National Park. *Ecosystem Services*, *26*, 27–36. doi:10.1016/j.ecoser.2017.05.008

Kinigadner, J., & Büttner, B. (2021). How accessibility instruments contribute to a low carbon mobility transition: Lessons from planning practice in the Munich region. *Transport Policy*, *111*, 157–167. doi:10.1016/j.tranpol.2021.07.019

Kişi, N. (2019). A Strategic Approach to Sustainable Tourism Development Using the A'WOT Hybrid Method: A Case Study of Zonguldak, Turkey. *Sustainability*, *11*(4), 964. doi:10.3390u11040964

Klein, I., Mayr, S., Gessner, U., Hirner, A., & Kuenzer, C. (2021). Water and hydropower reservoirs: High temporal resolution time series derived from MODIS data to characterize seasonality and variability. *Remote Sensing of Environment, 253*(November), 112207. doi:10.1016/j.rse.2020.112207

Klein, I., Gessner, U., Dietz, A. J., & Kuenzer, C. (2017). Global WaterPack – A 250 m resolution dataset revealing the daily dynamics of global inland water bodies. *Remote Sensing of Environment*, *198*, 345–362. doi:10.1016/j.rse.2017.06.045

Kmec, S., Hesse, M., Wille, C., & Reckinger, R. (Eds.). (2016). *Spaces and Identities in Border Regions: Politics - Media – Subjects*. Transcript.

Kogo, B. K., Kumar, L., & Koech, R. (2021). Climate change and variability in Kenya: A review of impacts on agriculture and food security. *Environment, Development and Sustainability*, *23*(1), 23–43. doi:10.100710668-020-00589-1

Kokkossis, H., Tsartas, P., & Grimba, E. (2011). *Special and alternative forms of tourism*. Kritiki.

Kolossov, V., & Scott, J. (2013). Selected conceptual issues in border studies. *Belgeo, 1*.

Kondolf, G. M., Gao, Y., Annandale, G. W., Morris, G. L., Jiang, E., Zhang, J., Cao, Y., Carling, P., Fu, K., Guo, Q., Hotchkiss, R., Peteuil, C., Sumi, T., Wang, H.-W., Wang, Z., Wei, Z., Wu, B., Wu, C., & Yang, C. T. (2014). Sustainable sediment management in reservoirs and regulated rivers: Experiences from five continents. *Earth's Future*, *2*(5), 256–280. doi:10.1002/2013EF000184

Kountouris, Y., & Remoundou, K. (2011). *Valuing the Welfare Cost of Forest Fires: a Life Satisfaction Approach*. Academic Press.

Kowalski, J., & Ślusarczyk, Z. (2006). *Unia Europejska. Proces integracji europejskiej i zarys proble-matyki instytucjonalno-prawnej [European Union. The process of European integration and the outline of institutional and legal issues]*. Polskie Wydawnictwo Prawnicze.

Kucsicsa, G., Popovici, E. A., Bălteanu, D., Grigorescu, I., Dumitraşcu, M., & Mitrică, B. (2019). Future land use/cover changes in Romania: Regional simulations based on CLUE-S model and CORINE land cover database. *Landscape and Ecological Engineering*, *15*(1), 75–90. doi:10.100711355-018-0362-1

Kuligowski, R. (2014). *Unia Europejska po Traktacie z Lizbony, Wydanie II – poprawione i rozszerzone* [The European Union after the Treaty of Lisbon, Edition II - Revised and extended]. Academic Press.

Kurowska-Pysz, J., Castanho, R., & Loures, L. (2018). Sustainable Planning of Cross-Border Cooperation: A Strategy for Alliances in Border Cities. *Sustainability*, *10*(5), 1416. doi:10.3390u10051416

Kuźniar, R. (1996). Porządek międzynarodowy –faza nierównowagi [International order - phase of imbalance, International affairs]. *Sprawy międzynarodowe, 2*.

Kwan, M.-P. (1998). Space-Time and Integral Measures of Individual Accessibility: A Comparative Analysis Using a Point-based Framework. *Geographical Analysis*, *30*(3), 191–216. doi:10.1111/j.1538-4632.1998.tb00396.x

Landau, L. B., Segatti, A., & Misago, J. P. (2013). Planning and participation in cities that move: Identifying obstacles to municipal mobility management. Public Admin. *Dev.*, *33*, 113–124.

Lang, T. (2015). Socio-economic and political responses to regional polarisation and socio-spatial peripheralisation in Central and Eastern Europe: A research agenda. *Hungarian Geographical Bulletin*, *64*(3), 171–185. doi:10.15201/hungeobull.64.3.2

Lankia, T., Neuvonen, M., Pouta, E., Sievänen, T., & Torvelainen, J. (2020). Outdoor recreation in ecosystem service accounting: Pilot accounts from Finland. *Scandinavian Journal of Forest Research*, *35*(3–4), 186–197. doi:10.1080/02827581.2020.1760342

Lanzini, P., & Stocchetti, A. (2021). From techno-centrism to socio-centrism: The evolution of principles for urban sustainable mobility. *International Journal of Sustainable Transportation*, *15*(11), 815–825. doi:10.1080/15568318.2020.1827315

Larsson, A., & Olsson, J. (2017). Potentials and limitations for the use of accessibility measures for national transport policy goals in freight transport and logistics: Evidence from Västra Götaland County, Sweden. *REGION*, *4*(1), 71–92. doi:10.18335/region.v4i1.172

Łastawski, K. (2006). *Historia integracji europejskiej [History of European integration]*. Wydawnictwo Adam Marszałek.

Laurenti, R., Singh, J., Cotrim, J. M., Toni, M., & Sinha, R. (2019). Characterizing the Sharing Economy State of the Research: A Systematic Map. *Sustainability*, *11*(20), 5729. doi:10.3390u11205729

Law and your Environment. (2017). *Law Search.* Recovered from: http://www.environmentlaw.org.uk/rte.asp?id=88

Le Clercq, F., & Bertolini, L. (2003). Achieving Sustainable Accessibility: An Evaluation of Policy Measures theAmsterdam in Area. *Built Environment*, *29*(1), 36–47. doi:10.2148/benv.29.1.36.53949

Lebo, J., & Schelling, D. (2001). Design and Appraisal of Rural Transport Infrastructure: Ensuring Basic Access for Rural Communities. World Bank Technical Paper; No. 496. Washington, DC: World Bank.

Lee, S. E. (1966). A Theory of Migration. *Demography*, *3*(1), 47–57. doi:10.2307/2060063

Legislation.gov.uk. (2013). *The Waste Electrical and Electronic Equipment Regulations 201.* Recovered from https://www.legislation.gov.uk/uksi/2013/3113/contents/made

Lehner, B., Liermann, C. R., Revenga, C., Vörösmarty, C., Fekete, B., Crouzet, P., Döll, P., Endejan, M., Frenken, K., Magome, J., Nilsson, C., Robertson, J. C., Rödel, R., Sindorf, N., & Wisser, D. (2011). High-resolution mapping of the world's reservoirs and dams for sustainable river-flow management. *Frontiers in Ecology and the Environment*, *9*(9), 494–502. doi:10.1890/100125

Levesque, J.F., Harris, M.F., & Russell, G. (2013). Patient-centered access to health care: conceptualising access at the interface of health systems and populations. *Int J Equity Health, 11*, 12-18.

Levine, J. (2020). A century of evolution of the accessibility concept. *Transportation Research Part D: Transport and Environment.*

Levine, J., & Garb, Y. (2002). Congestion pricing's conditional promise: Promotion of accessibility or mobility? *Transport Policy, Elsevier*, *9*(3), 179–188. doi:10.1016/S0967-070X(02)00007-0

Lew, A. A., Cheer, J. M., Haywood, M., Brouder, P., & Salazar, N. B. (2020). Visions of travel and tourism after the global COVID-19 transformation of 2020. *Tourism Geographies*, *22*(3), 455–466. doi:10.1080/14616688.2020.1770326

Liang, Y., Song, Q., Wu, N., Li, J., Zhong, Y., & Zeng, W. (2021). Repercussions of COVID-19 pandemic on solid waste generation and management strategies. *Frontiers of Environmental Science & Engineering*, *15*, 115.

Lienhoop, N., & Brouwer, R. (2015). Agri-environmental policy valuation: Farmers' contract design preferences for afforestation schemes. *Land Use Policy*, *42*, 568–577. doi:10.1016/j.landusepol.2014.09.017

Lipper, L., Thornton, P., Campbell, B., Baedeker, T., Braimoh, A., Bwalya, M., Caron, P., Cattaneo, A., Garrity, D., Henry, K., Hottle, R., Jackson, L., Jarvis, A., Kossam, F., Mann, W., McCarthy, N., Meybeck, A., Neufeldt, H., Remington, T., ... Torquebiau, E. F. (2014). Climate-smart agriculture for food security. *Nature Climate Change*, *4*(12), 1068–1072. doi:10.1038/nclimate2437

Liu, Q., Liu, Y., Zhang, C., An, Z., & Zhao, P. (2021). Elderly mobility during the COVID-19 pandemic: A qualitative exploration in Kunming, China. *Journal of Transport Geography*, *96*(C).

Liu, R., Chen, Y., Wu, J., Xu, T., Gao, L., & Zhao, X. (2018). Mapping spatial accessibility of public transportation network in an urban area – A case study of Shanghai Hongqiao Transportation Hub. *Transportation Research Part D, Transport and Environment*, *59*, 478–495. doi:10.1016/j.trd.2018.01.003

Liu, Y., Cao, X., & Li, T. (2020). Influence of Accessibility on Land Use and Landscape Pattern Based on Mapping Knowledge Domains: Review and Implications. *Journal of Advanced Transportation*, *2020*, 7985719–12.

Li, Z., Li, J. J., & Shi, X. P. (2020). A Two-Stage Multisite and Multivariate Weather Generator. *Journal of Environmental Informatics*, *35*(2), 148–159.

Long, D., & Hoogendoorn, G. (2014). Second home owner perceptions of their environmental impacts: The case of Hartebeespoort. *Urban Forum*, *25*, 517-530.

Lopes, L. F. G., dos Santos Bento, J. M. R., Arede Correia Cristovão, A. F., & Baptista, F. O. (2015). Exploring the effect of land use on ecosystem services: The distributive issues. *Land Use Policy*, *45*, 141–149. doi:10.1016/j.landusepol.2014.12.008

López-Moreno, J. I., Vicente-Serrano, S. M., Angulo-Martínez, M., Beguería, S., & Kenawy, A. (2010). Trends in daily precipitation on the northeastern Iberian Peninsula, 1955–2006. *International Journal of Climatology*, *30*(7), 1026–1041. doi:10.1002/joc.1945

Loucks, D. P., & van Beek, E. (2017). Water Resources Planning and Management: An Overview. In *Water Resource Systems Planning and Management*. Springer. doi:10.1007/978-3-319-44234-1_1

Loukaitou-Sideris, A. (2013). New Rail Hubs along High-Speed Rail Corridor in California: Urban Design Challenges. *Transportation Research Record: Journal of the Transportation Research Board*, *2350*(1), 1–8. doi:10.3141/2350-01

Loures, L. (2011). *Planning and Design in Post-industrial Landscapes: East Bank Arade River – Lagoa, Case Study* [Ph.D Thesis]. University of Algarve, Portugal.

Lousada, S. A. N., & Camacho, R. (2018). *Hidrologia, recursos hídricos e ambiente - Aulas Teóricas* (Vol. 1). Universidade da Madeira.

Lousada, S. A. N., Gonçalves, L. B., & Velarde, J. G. (2020). *Controlo e regularização de caudais excedentes por meio de bacia de detenção: simulação para bacia hidrográfica de João Gomes, Funchal. Espacios y Sociedades em Transformación* (1st ed., Vol. 6). Thomson Reuters Aranzadi.

Lousada, S., & Castanho, R. A. (2021). GIS-based Assessment of Morphological and Hydrological Parameters of Ribeira dos Socorridos and Ribeira do Vigário Basins, Madeira Island, Portugal. *Current World Environment*, *16*(2), 408–426. doi:10.12944/CWE.16.2.08

Lousada, S., Gonçalves, L., & Atmaca, A. (2022). Hydraulic Planning in Insular Urban Territories: The Case of Madeira Island—São Vicente. *Water (Basel)*, *2022*(14), 112. doi:10.3390/w14010112

Lozano-Parra, J., Schnabel, S., & Ceballos-Barbancho, A. (2015). The role of vegetation covers on soil wetting processes at rainfall event scale in scattered tree woodland of Mediterranean climate. *Journal of Hydrology (Amsterdam)*, *529*, 951–961. doi:10.1016/j.jhydrol.2015.09.018

Ludeman, K. (2000). How to conduct self-directed 360. *Training & Development*, *54*(7), 44–44.

Lund, K. A. (2016). Chasing the lights: Darkness, tourism and the northern lights. In *Green Ice* (pp. 49–71). Palgrave Macmillan. doi:10.1057/978-1-137-58736-7_3

Luukkonen, J. (2010). Territorial cohesion policy in the light of peripherality. *The Town Planning Review*, *81*(4), 445–466. doi:10.3828/tpr.2010.12

Madureira, L., Nunes, L. C., Borges, J. G., & Falcão, A. O. (2011). Assessing forest management strategies using a contingent valuation approach and advanced visualisation techniques: A Portuguese case study. *Journal of Forest Economics*, *17*(4), 399–414. doi:10.1016/j.jfe.2011.04.001

Maity, R. (2018). *Statistical Methods in Hydrology and Hydroclimatology*. Springer. doi:10.1007/978-981-10-8779-0

Makkonen, T., & Williams, A. M. (2016). Border region studies: The structure of an "offbeat" field of regional studies. *Regional Studies, Regional Science*, *3*(1), 355–367. doi:10.1080/21681376.2016.1209982

Malinowski, L. (2002). Unia Europejska: geneza, współczesność [European Union: genesis, modernity]. SIG Management School.

Małuszyńska, E., & Gruchman, B. (Eds.). (2005). *Kompendium wiedzy o Unii Europejskiej [Compendium of knowledge on the European Union]*. PWN.

Malý, J. (2019). Polycentric Urban Systems and Territorial Cohesion. In E. Medeiros (Ed.), *Territorial Cohesion. The Urban Book Series*. Springer. doi:10.1007/978-3-030-03386-6_4

Mammadov, G. S., Khalilov, M. Y., & Mammadova, S. Z. (2009). Ecology Atlas. Academic Press.

Mammadov, G., & Khalilov, M. (2005). *Ecology and Protection of the Environment. Elm* Press.

Maphanga, P. M., & Henama, U. S. (2019). The Tourism Impact of Ebola in Africa: Lessons on Crisis Management. *African Journal of Hospitality, Tourism and Leisure*, *8*(3), 1–11.

Marenin, O. (2010). *Challenges for Integrated Border Management in the European Union*. Geneva Centre for the Democratic Control of Armed Forces Occasional Paper No. 17: 28-45.

Marini Govigli, V., Górriz-Mifsud, E., & Varela, E. (2019). Zonal travel cost approaches to assess recreational wild mushroom picking value: Trade-offs between online and onsite data collection strategies. *Forest Policy and Economics*, *102*, 51–65. doi:10.1016/j.forpol.2019.02.003

Marszałek, A. (1996). Członkostwo krajów Europy Środkowo-Wschodniej we Wspólnotach Europejskich, a przesłanki międzynarodowej integracji [Membership of Central and Eastern European countries in the European Communities and the rationale for international integration]. In J. Fiszer (Ed.), Państwa narodowew euroatlantyckich strukturach, Kom [Nation states in Euro-Atlantic structures, Kom]. The Political Sciences Department of PAN.

Marszałek-Kawa, J. (2003). *Procesy integracyjne w Europie po II wojnie światowej [Integration processes in Europe after World War II]*. Wydawnictwo Adam Marszałek.

Marta-Pedroso, C., Laporta, L., Proença, V., Azevedo, J. C., & Domingos, T. (2014). Changes in the ecosystem services provided by forests and their economic valuation: A review. In Forest Landscapes and Global Change: Challenges for Research and Management (pp. 107–137). Springer. doi:10.1007/978-1-4939-0953-7_5

Marta-Pedroso, C., Laporta, L., Gama, I., & Domingos, T. (2018). Economic valuation and mapping of ecosystem services in the context of protected area management (Natural park of Serra de São Mamede, Portugal). *One Ecosystem*, *3*, e26722. Advance online publication. doi:10.3897/oneeco.3.e26722

Martin, A. (2005, May). Environmental Conflict between Refugee and Host Communities. *Journal of Peace Research*, *42*(3), 329–346. doi:10.1177/0022343305052015

Martínez Arroyo, A., Ruíz Suárez, L. G., Gavilán García, A., Ramírez Muñoz, T., & Huerta Colosia, D. (2020). *Panorama de la generación y manejo de residuos sólidos y médicos durante la emergencia sanitaria por COVID-19*. Instituto Nacional de Ecología y Cambio Climático INECC.

Martínez, J. N. (2021). *A la pandemia se suma otro problema… la generación de desperdicios*. Dirección General de Comunicación Social. Recovered from: https://www.dgcs.unam.mx/boletin/bdboletin/2021_080.html#:~:text=Con%20el%20confinamiento%20en%20M%C3%A9xico,Regional%20de%20Investigaciones%20Multidisciplinarias%20(CRIM

Martínez, M., Rojas, C., Condeço-Melhorado, A., & Carrasco, J. A. (2021). Accessibility Indicators for the Geographical Assessment of Transport Planning in a Latin American Metropolitan Area. *Geographies*, *1*(2), 124–142. doi:10.3390/geographies1020008

Martín, R., Martín, M., Fernández, S., Mejía, Z., & Bedriñana, A. (2019). A Spatial Analysis of the Achievements, in Terms of Regional Development, accomplished by the Initial EU-Member Cohesion Fund Beneficiaries Using a Synthetic Indicator. *Sustainability*, *11*(8), 2343. doi:10.3390u11082343

Martín-Uceda, J., & Vicente Rufí, J. (2021). Territorial Development and Cross-Border Cooperation: A Review of the Consequences of European INTERREG Policies on the Spanish–French Border (2007–2020). *Sustainability*, *13*(21), 12017. doi:10.3390u132112017

Martin-Vide, J. (2004). Spatial distribution of a daily precipitation concentration index in peninsular Spain. *International Journal of Climatology*, *24*(8), 959–971. doi:10.1002/joc.1030

Martín-Vide, J., & Gómez, L. (1999). Regionalization of peninsular Spain based on the length of dry spells. *International Journal of Climatology*, *19*(5), 537–555. doi:10.1002/(SICI)1097-0088(199904)19:5<537::AID-JOC371>3.0.CO;2-X

Masiero, M., Pettenella, D. M., & Secco, L. (2016). From failure to value: Economic valuation for a selected set of products and services from Mediterranean forests. *Forest Systems*, *25*(1), 051. Advance online publication. doi:10.5424/fs/2016251-08160

Mathis, R. L. (2015). *Human resource management: Essential perspectives*. Cengage Learning.

Matsiori, S., Anagnos, N., & Soutsas, K. (2012). Economic valuation of forest recreation: The case of the University Forest of Pertouli in Greece. *Journal of Food Agriculture and Environment*, *10*(2), 866–870.

Mayer, H., Habersetzer, A., & Meili, R. (2016). Rural–Urban Linkages and Sustainable Regional Development: The Role of Entrepreneurs in Linking Peripheries and Centers. *Sustainability*, *8*(8), 745. doi:10.3390u8080745

McFeeters, S. (1996). The use of the Normalized Difference Water Index (NDWI) in the delineation of open water features. *International Journal of Remote Sensing*, *17*(7), 1425–1432. doi:10.1080/01431169608948714

McKinsey & World Travel & Tourism Council. (2017). *Coping with Success: Managing Overcrowding in Tourism Destinations*. WTTC.

Medeiros, E. (2016a). Territorial Cohesion: An EU concept. *European Journal of Spatial Development*, 60. Available from: http://www.nordregio.se/Global/EJSD/Refereed articles/refereed60.pdf

Medeiros, E. (2016b). EU Cohesion Policy in Spain. *Regional Studies.* doi:10.1080/00343404.2016.1187719

Medeiros, E. (2013a). Assessing territorial impacts of the EU Cohesion Policy: The Portuguese case. *European Planning Studies*, *22*(9), 1960–1988. doi:10.1080/09654313.2013.813910

Medeiros, E. (2013b). Euro-Meso-Macro:The new regions in Iberian and European Space. *Regional Studies*, *47*(8), 249–1266. doi:10.1080/00343404.2011.602336

Medeiros, E. (2014a). The Europeanization of Spatial Planning processes in Portugal within the EU Cohesion policy Strategies (1989-2013). *Geographyand Spatial Planning Journal*, *6*, 201–222. doi:10.17127/got/2014.6.012

Medeiros, E. (2019). Cross-border transports and cross-border mobility in EU border regions. *Case Studies on Transport Policy.*, *7*(1), 1–12. doi:10.1016/j.cstp.2018.11.001

Medeiros, E., Ferreira, R., Boijmans, P., Verschelde, N., Spisiak, R., Skoniezki, P., Dietachmair, J., Hurnaus, K., Ebster, M., Madsen, S., Ballaguy, R.-L., Volponi, E., Isinger, E., Voiry, P., Markl-Hummel, L., Harster, P., Sippel, L., Nolte, J., Maarfield, S., ... Berzi, M. (2021). Boosting cross-border regions through better cross-border transport services. The European case. *Case Stud Transp Policy*, *9*(1), 291–301. doi:10.1016/j.cstp.2021.01.006

Meehl, G., Karl, T., Easterling, D. R., Changnon, S., Pielke, R. Jr, Changnon, D., Evans, J., Groisman, P. Y., Knutson, T. R., Kunkel, K. E., Mearns, L. O., Parmesan, C., Pulwarty, R., Root, T., Sylves, R. T., Whetton, P., & Zwiers, F. (2000). An Introduction to Trends in Extreme Weather and Climate Events: Observations, Socioeconomic Impacts, Terrestrial Ecological Impacts, and Model Projections. *Bulletin of the American Meteorological Society*, *81*(3), 413–416. doi:10.1175/1520-0477(2000)081<0413:AITTIE>2.3.CO;2

Melo, P. C., Graham, D. J., & Brage-Ardao, R. (2013). The productivity of transport infrastructure investment: A meta-analysis of empirical evidence. *Regional Science and Urban Economics*, *43*(5), 695–706. doi:10.1016/j.regsciurbeco.2013.05.002

Mendes, A. M. S. C., Madureira, L., Sottomayor, M., Alves, R., & Rosário, J. V. C. (2021). *ECOFOR.PT – Valorização Económica dos Bens e Serviços dos Ecossistemas Florestais de Portugal. Relatório Científico. Operação 20.2.3 – Assistência Técnica RRN – Área 3.* Candidatura Nº PDR 2020-2023-045913. https://www.researchgate.net/publication/358646726_ECOFORPT_-_Valorizacao_Economica_dos_Bens_e_Servicos_dos_Ecossistemas_Florestais_de_Portugal_Relatorio_Cientifico

Mendes, A. M. S. C. (2005). Portugal. In T. E. V. Towards, M. Merlo, & L. Croitoru (Eds.), *Valuing Mediterranean Forests* (pp. 331–352). CAB International. doi:10.1079/9780851999975.0331

Meng, L., Zhang, Z., Zhang, W., Ye, J., Wu, C., Chen, D., & Song, C. (2019). An Automatic Extraction Method for Lakes and Reservoirs Using Satellite Images. *IEEE Access: Practical Innovations, Open Solutions*, *7*, 62443–62456. doi:10.1109/ACCESS.2019.2916148

Mercado, R. G. (2002). *Regional Development in the Philippines: A Review of Experience.* State of the Art and Agenda for Research and Action.

Merlo, M. L., Croitoru, Merlo, M., Croitoru, L. (Eds.). (2005). Valuing Mediterranean forests_ towards total economic value-CABI. CABI Publishing.

Meya, J. N. (2020). Environmental Inequality and Economic Valuation. *Environmental and Resource Economics*, *76*(2–3), 235–270. doi:10.100710640-020-00423-2

Migration Data Portal. (2020). *What you need to know about the impact of environmental change on migration.* https://www.migrationdataportal.org/blog/what-you-need-know-about-impact-environmental-change-migration

Migration Data Portal. (2021). *Migration policies and governance.* https://www.migrationdataportal.org/ar/themes/migrationspolitik-und-regierungsfuehrung

Militino, A. F., Montesino-SanMartin, M., Pérez-Goya, U., & Ugarte, M. D. (2020). Using RGISTools to estimate water levels in reservoirs and lakes. *Remote Sensing*, *12*(12), 1934. Advance online publication. doi:10.3390/rs12121934

Millennium Ecosystem Assessment. (2005). *Ecosystems and human well-being : synthesis.* Island Press.

Milohnić, I., Bonifačić, J. C., & Licul, I. (2019). Transformation of camping into glamping–trends and perspectives. *Tourism in Southeast Europe.*, *5*, 457–473. doi:10.20867/tosee.05.30

Ministerio de Obras Publicas - Chile. (2013). *Decreto 50.* Author.

Ministry of Environment and Natural Resources. (2003). *National Caspian Action Plan of Azerbaijan.* Author.

Ministry of Interior Presidency of Migration Management. (2022). *Temporary protection.* https://www.goc.gov.tr/gecici-koruma5638

Ministry of the Environment. (2018). *Japan's Resource Circulation Policy for Plastics.* Author.

Ministry of the Environment. (2021). *Municipal solid waste emissions and disposal in FY2019.* Author.

Mogale, P. T., & Odeku, K. O. O. (2019). Perspectives on transformative legislative and policy framework promoting tourism for poverty alleviation in South Africa. *African Journal of Hospitality, Tourism and Leisure*, *8*(2), 1–14.

Montero, M. J., López-Casares, S., García, L., & Hernández, J. (2005). Visual Impact on Wetlands: Consequence of Building Sprawls in Rural Areas of the West of Spain. *MODSIM Intnl Cong on Modelling and Simulation; Modelling and Simulation Society of Australia and New Zealand: Melbourne, Australia*, 170–176.

Mora Aliseda, J., & Castellano Álvarez, F. J. (2005). Evolución y redefinición del concepto de desarrollo rural: de la sectorización hacia el enfoque holístico. Junta de Extremadura. Mérida [Evolution and redefinition of the concept of rural development: from sectorization to the holistic approach. Regional Government of Extremadura]. In Políticas urbanas y territoriales en la Península Ibérica. Tomo II [Urban and territorial policies in the Iberian Peninsula. Volume II]. Academic Press.

Mora, J., Garrido Velarde, J., & Mora, C. (2018). Gobernanza de los recursos hídricos transfronterizos: Una Propuesta [Governance of transboundary water resources: A Proposal]. *Vertentes do Direito*, *5*(2), 1–15. doi:10.20873/uft.2359-0106.2018.v5n2.p1-15

Morrissey, J. W. (2013, December). Understanding the relationship between environmental change and migration: The development of an effects framework based on the case of northern Ethiopia. *Global Environmental Change*, *23*(6), 1501–1510. doi:10.1016/j.gloenvcha.2013.07.021

Moullin, J. C., Dickson, K. S., Stadnick, N. A., Albers, B., Nilsen, P., Broder-Fingert, S., Mukasa, B., & Aarons, G. A. (2020). Ten recommendations for using implementation frameworks in research and practice. *Implement Sci Commun*, *1*(1), 42. doi:10.118643058-020-00023-7 PMID:32885199

Moura, A., Silva, A., Gonçalves, L., & Lousada, S. (2020). Numerical modelling of the flow rate in artificial water channels: Application to Ribeira Brava's stream. *Revista Brasileira de Planejamento e Desenvolvimento, Curitiba*, *9*(1), 39–59. doi:10.3895/rbpd.v9n1.10974

Mrabet, R., Savé, R., Toreti, A., Caiola, N., Chentouf, M., Llasat, M. C., Mohamed, A. A. A., Santeramo, F. G., Sanz-Cobena, A., & Tsikliras, A. (2020). Food. In Climate and Environmental Change in the Mediterranean Basin – Current Situation and Risks for the Future. First Mediterranean Assessment Report. Union for the Mediterranean, Plan Bleu, UNEP/MAP.

Mtsiliza, N. S. (2017). Seeking strategies for sustainability in Tourism Entrepreneurship in South Africa. *African Journal of Hospitality, Tourism and Leisure*, *6*(4), 1–10.

Mulligan, M., van Soesbergen, A., & Sáenz, L. (2020). GOODD, a global dataset of more than 38,000 georeferenced dams. *Scientific Data*, *7*(1), 1–8. doi:10.103841597-020-0362-5 PMID:31964896

Naldini, A. (2018). Improvements and risks of the proposed evaluation of Cohesion Policy in the 2021–27 period: A personal reflection to open a debate. *Evaluation*, *24*(4), 496–504. doi:10.1177/1356389018804261

Naranjo, J. M., Castanho, R. A., & Vulevic, A. (2021). Analyzing Transportation Logistics and Infrastructure Sustainability in the Iberian Peninsula: The Case of Portugal Mainland. *European Planning Studies*, 1–23. Advance online publication. doi:10.1080/09654313.2021.2014789

Naredo, J. M. (2014). El modelo inmobiliario español y sus consecuencias [The Spanish real estate model and its consequences]. *Boletín CF+S*, (44), 13-28. http://habitat.aq.upm.es/boletin/n44/ajnar.html

NASA Global Climate Change. (n.d.). *Scientific Consensus: Earth's Climate Is Warming*. https://climate.nasa.gov/scientific-consensus/

Ngai, P. B., & Koehn, P. H. (2002). *Organizational communication in refugee camp situations*. UNHCR The UN Refugee Agency, New Issues in Refugee Research, Working Paper No. 71.

Ninan, K. N., & Kontoleon, A. (2016). Valuing forest ecosystem services and disservices - Case study of a protected area in India. *Ecosystem Services*, *20*, 1–14. doi:10.1016/j.ecoser.2016.05.001

Notaro, S., & Paletto, A. (2012). The economic valuation of natural hazards in mountain forests: An approach based on the replacement cost method. *Journal of Forest Economics*, *18*(4), 318–328. doi:10.1016/j.jfe.2012.06.002

NovemberB. (2020). *About us*. Available at: http://www.berlinnovember.co.za/index.php/about berlin-november

Nuestra esfera. (2014). *¿Cómo se clasifican los residuos?* Accessed at http://nuestraesfera.cl/zoom/como-se-clasifican-los-residuos/#:~:text=Los%20residuos%20pueden%20ser%20clasificados,cartones%2C%20vidrios%2C%20por%20ejemplo)

Nunes, J. R., Ramos-Miras, J., Lopez-Pineiro, A., Loures, L., Gil, C., Coelho, J., & Loures, A. (2014). Concentrations of available heavy metals in Mediterranean agricultural soils and their relation with some soil selected properties: A case study in typical Mediterranean soils. *Sustainability (Switzerland)*, *6*(12), 9124–9138. doi:10.3390u6129124

Nyaupane, G. P., Timothy, D. J., & Poudel, S. (2015). Understanding tourists in religious destinations: A social distance perspective. *Tourism Management*, *48*, 343–353. doi:10.1016/j.tourman.2014.12.009

O'Donnel, R. (1997). The Competitive Advantage of Peripheral Regions: Conceptual Issues and Research Approaches. In B. Fynes & S. Ennis (Eds.), *Competing from the Periphery* (pp. 47–82). Oak Tree.

O'Lear, S. (1997, June). Migration and the Environment: A Review of Recent Literature. *Social Science Quarterly*, *78*(2), 606–618.

OECD. (2014). *OECD Tourism Trends and Policies, 2014*. OECD Publishing. doi:10.1787/tour-2014-

OECD. (2016). *States of fragility 2016: Undestanding violence.* . https://www.oecd-ilibrary.org/sites/9789264267213-5-en/index.html?itemId=/content/component/9789264267213-5-en#:~:text=Fragility%20is%20defined%20as%20the,absorb%20or%20mitigate%20those%20risks doi:10.1787/9789264267213-5-en

OECD. (2017). *Analysing megatrends to better shape the future of tourism: Issue Paper.* OECD Publishing.

OECD. (2017). *Interrelations between Public Policies, Migration and Development.* OECD Development Center, OECD Publishing. https://www.oecd-ilibrary.org/docserver/9789264265615-en.pdf?expires=1651525981&id=id&accname=guest&checksum=580CCE6E6B161AA7868D7D49F3B0DDE6

OECD. (2018). Analysing megatrends to better shape the future of tourism. OECD Tourism Papers, No. 2018/02. OECD Publishing.

OECD. (2020a). *OECD economic outlook, Interim report March 2020.* OECD.

Okraszewska, R., Romanowska, A., Wołek, M., Oskarbski, J., Birr, K., & Jamroz, K. (2018). Integration of a Multilevel Transport System Model into Sustainable Urban Mobility Planning. *Sustainability*, *10*(2), 479. doi:10.3390u10020479

Oliveira, F., Pintassilgo, P., Pinto, P., Mendes, I., & Silva, J. A. (2017). Segmenting visitors based on willingness to pay for recreational benefits: The case of Leiria National Forest. *Tourism Economics*, *23*(3), 680–691. doi:10.5367/te.2015.0526

Olsson, L. E., Friman, M., & Lättman, K. (2021). Accessibility Barriers and Perceived Accessibility: Implications for Public Transport. *Urban Science*, *5*(3), 63. doi:10.3390/urbansci5030063

Om, K.-C., Ren, G., Li, S., & Kang-Chol, O. (2018). Climatological characteristics and long-term variation of rainy season and torrential rain over DPR Korea. *Weather and Climate Extremes*, *22*, 48–58. doi:10.1016/j.wace.2018.09.003

Önder, B. (2000). Kamu Yayın Kurumlarında Personel Yönetimi. TRT.

Oppido, S., & Ragozino, S. (2019). Unbalanced development and peripheralisation processes: atesting phase to map studies.*AESOP Annual Congress Venice 2019 Planning for Transition*, 3381–3393.

Páez, A., Scott, D. M., & Morency, C. (2012). Measuring accessibility: Positive and normative implementations of various accessibility indicators. *Journal of Transport Geography*, *25*, 141–153. doi:10.1016/j.jtrangeo.2012.03.016

Paiva, D. M. B., Freire, A. P., & de Mattos Fortes, R. P. (2021). Accessibility and Software Engineering Processes: A Systematic Literature Review. *Journal of Systems and Software*, *171*, 110819. doi:10.1016/j.jss.2020.110819

Paletto, A., de Meo, I., Grilli, G., & Nikodinoska, N. (2017). Effects of different thinning systems on the economic value of ecosystem services: A case-study in a black pine peri-urban forest in Central Italy. *Annals of Forest Research*, *60*(2), 313–326. doi:10.15287/afr.2017.799

Paletto, A., Geitner, C., Grilli, G., Hastik, R., Pastorella, F., & Garcìa, L. R. (2015). Mapping the value of ecosystem services: A case study from the Austrian Alps. *Annals of Forest Research*, *58*(1), 157–175. doi:10.15287/afr.2015.335

Parzymies, S. (2012). *Unia Europejska od Maastricht do Lizbony. Polityczne aspekty aktywności [European Union from Maastricht to Lisbon. Political aspects of activity].* DIALOG.

Paskoff, R. (1996). Atlas de las formas de relieve de Chile. Instituto Geográfico Militar de Chile.

Pawlak, K., & Kołodziejczak, M. (2020). The Role of Agriculture in Ensuring Food Security in Developing Countries: Considerations in the Context of the Problem of Sustainable Food Production. *Sustainability*, *12*(13), 5488. doi:10.3390u12135488

Pécoud, A., & de Guchteneire, P. (2006). International migration, border controls and human rights: Assessing the relevance of a right to mobility. *Journal of Borderlands Studies*, *21*(1), 69–86. doi:10.1080/08865655.2006.9695652

Peña, S., & Durand, F. (2022). Mobility planning in cross-border metropolitan regions: The European and North American experiences. *Territory, Politics, Governance*, *10*(2), 219–236. doi:10.1080/21622671.2020.1769716

Penga, Y., Wua, P., Schartupb, A., & Zhanga, Y. (2021). Plastic waste release caused by COVID-19 and its fate in the global ocean. *Proceedings of the National Academy of Sciences of the United States of America*.

Pérez-Escamilla, R. (2017, July). Food Security and the 2015–2030 Sustainable Development Goals: From Human to Planetary Health: Perspectives and Opinions. *Current Developments in Nutrition*, *1*(7), e000513. doi:10.3945/cdn.117.000513 PMID:29955711

Petheram, L., Zander, K. K., Campbell, B. M., High, C., & Stacey, N. (2010). Strange Changes: Indigenous perspectives of climate change and adaptation in NE Arnhem Land (Australia). *Global Environmental Change*, *20*(4), 681–692. doi:10.1016/j.gloenvcha.2010.05.002

Petrauskiene, K., Dvarioniene, J., Kaveckis, G., Kliaugaite, D., Chenadec, J., Hehn, L., Pérez, B., Bordi, C., Scavino, G., Vignoli, A., & Erman, M. (2020). Situation Analysis of Policies for Electric Mobility Development: Experience from Five European Regions. *Sustainability*, *12*(7), 2935. doi:10.3390u12072935

Petrie, H., & Bevan, N. (2009). *The Evaluation of Accessibility, Usability, and User Experience*. The Universal Access Handbook. doi:10.1201/9781420064995-c20

Pham, D. H., Shimizu, T., & Van Nguyen, T. (2021). A Literature Review on Interactions Between Stakeholders Through Accessibility Indicators Under Mobility as a Service Context. Int. J. *Intelligent Transportation Systems Research*, *19*(2), 468–476. doi:10.100713177-021-00257-2

Piekermann, K., & Neubauer, J. (2002). *European Accessibility and Peripherality: Concepts, Models, and Indicators*. Nordregio Working Paper.

Pienaar, J. J., & Visser, G. (2009). The Thorny Issue of Identifying Second Homes in South Africa. *Urban Forum*, *20*, 455 469.

Pinna, F., Masala, F., & Garau, C. (2017). Urban Policies and Mobility Trends in Italian Smart Cities. *Sustainability*, *9*(4), 494. doi:10.3390u9040494

Plata, W., Gómez, M., & Bosque, J. (2008). Análisis de factores explicativos del crecimiento urbano en la Comunidad de Madrid a través de métodos estadísticos y sig [Analysis of explanatory factors of urban growth in the Community of Madrid through statistical methods and sig]. In *Tecnologías de la Información Geográfica para el desarrollo territorial: XIII Congreso Nacional de Tecnologías de la Información Geográfica* [Geographic Information Technologies for Territorial Development: XIII National Congress of Geographic Information Technologies]. Academic Press.

Plieninger, T., Höchtl, F., & Spek, T. (2006). Traditional land-use and nature conservation in European rural landscape. *Environmental Science & Policy*, *9*(9), 317–321. doi:10.1016/j.envsci.2006.03.001

Popa, B., Borz, S. A., Nita, M. D., Ioras, F., Iordache, E., Borlea, F., Pache, R., & Abrudan, I. V. (2016). Forest ecosystem services valuation in different management scenarios: A case study of the maramures mountains. *Baltic Forestry*, *22*(2), 327–340.

Porru, S., Misso, F. E., Pani, F. E., & Repetto, C. (2020). Smart mobility and public transport: Opportunities and challenges in rural and urban areas. *Journal of Traffic and Transportation Engineering*, *7*(1), 88–97. doi:10.1016/j.jtte.2019.10.002

Porto, R. L., Filho, K. Z., Tucci, C. E. M., & Bidone, F. (1993). *Drenagem urbana. Hidrologia: Ciência e Aplicação. 2ª Edição.* Editora da Universidade UFRGS.

Prada, S., Gaspar, A., Sequeira, M. M., & Nunes, A. (2005). Disponibilidades Hídricas na Ilha da Madeira. In C. de Lanzarote & C. I. de Aguas de Lanzarote (Eds.), *AQUAMAC – Técnicas e métodos para a gestão sustentável da água na Macaronésia. Disponibilidades Hídricas na Ilha da Madeira, Publisher: Instituto Tecnológico das Canarias* (pp. 261–294). Instituto Tecnológico de Canarias.

Prideaux, B., Thompson, M., & Pabel, A. (2020). Lessons from COVID-19 can prepare global tourism for the economic transformation needed to combat climate change. *Tourism Geographies*, *22*(3), 667–678. doi:10.1080/14616688.2020.1762117

Prus, P., & Sikora, M. (2021). The Impact of Transport Infrastructure on the Sustainable Development of the Region—Case Study. *Agriculture*, *11*(4), 279. doi:10.3390/agriculture11040279

Puga, D. (2002, October 1). European regional policies in light of recent location theories. *Journal of Economic Geography*, *2*(4), 373–406. doi:10.1093/jeg/2.4.373

Pugh, S. D. (1991). *Organizational Behaviour*. Prentice Hall.

Pullano, G., Valdano, E., Scarpa, N., Rubrichi, S., & Colizza, V. (2020). Evaluating the effect of demographic factors, socioeconomic factors, and risk aversion on mobility during the COVID-19 epidemic in France under lockdown: A population-based study. *The Lancet. Digital Health*, *2*(12), e638–e649. doi:10.1016/S2589-7500(20)30243-0 PMID:33163951

Qing, Y., Zhu-Guo, M., & Liang, C. (2011). A Preliminary Analysis of the Relationship between Precipitation Variation Trends and Altitude in China. *Atmospheric and Oceanic Science Letters*, *4*(1), 41–46. doi:10.1080/16742834.2011.11446899

Qin, J., Luo, S., Yi, D., Jiang, H., & Zhang, J. (2022). Measuring Cluster-Based Spatial Access to Shopping Stores under Real-Time Travel Time. *Sustainability*, *14*(4), 2310. doi:10.3390u14042310

Ramukumba, T., Mmbengwa, M., Mwamayi, K. A., & Groenewald, J. A. (2012). Analysis of the socio-economic impacts of tourism for emerging tourism entrepreneurs: The case of George municipality in the Western Cape Province, South Africa. *Journal of Hospitality Management and Tourism*, *3*(3), 39–45. doi:10.5897/JHMT11.038

Ratisurakarn, T. (2019). The Valuation of Forest Ecological Services: A Meta-Analysis. *Southeast Asian Journal of Economics*, *7*(2), 61–84.

Raum, S. (2018). Reasons for Adoption and Advocacy of the Ecosystem Services Concept in UK Forestry. *Ecological Economics*, *143*, 47–54. doi:10.1016/j.ecolecon.2017.07.001

Ribas, J. R., Vicente, T. V. D. S., Altaf, J. G., & Troccoli, I. R. (2017). Integração de Ações na Gestão Sustentável. REAd. *Revista Eletrônica de Administração*, *23*(2), 31–57.

Riccioli, F., Fratini, R., Fagarazzi, C., Cozzi, M., Viccaro, M., Romano, S., Rocchini, D., Diaz, S. E., & Tattoni, C. (2020). Mapping the recreational value of coppices' management systems in Tuscany. *Sustainability (Switzerland)*, *12*(19), 1–18. doi:10.3390u12198039

Roberts, W., Williams, G. P., Jackson, E., Nelson, E. J., & Ames, D. P. (2018). Hydrostats: A Python package for characterizing errors between observed and predicted time series. *Hydrology*, *5*(4), 66. Advance online publication. doi:10.3390/hydrology5040066

Rodrigues, L. N., Sano, E. E., Steenhuis, T. S., & Passo, D. P. (2012). Estimation of small reservoir storage capacities with remote sensing in the Brazilian Savannah region. *Water Resources Management*, *26*(4), 873–882. doi:10.100711269-011-9941-8

Rodríguez-Pose, A. (2008). The Rise of the "City-region" Concept and its Development Policy Implications. *European Planning Studies*, *16*(8), 1025–1046. doi:10.1080/09654310802315567

Rogerson, C. M. (2014). The Uneven Geography of Tourism in South Africa. *African Journal of Hospitality, Tourism and Leisure*, *3*(1), 1–14.

Rogerson, C. M. (2015). Revisiting VFR tourism in South Africa. *The South African Geographical Journal*, *97*(2), 139–157. doi:10.1080/03736245.2015.1028981

Rogerson, C. M. (2016). Outside the cities: Tourism pathways in South Africa's small towns and rural areas. *African Journal of Hospitality, Tourism and Leisure*, *6*(3), 1–10.

Rogerson, C. M. (2017). Visiting friends and relatives travel matters for sub-Saharan Africa. *African Journal of Hospitality, Tourism and Leisure*, *5*(3), 1–16.

Rogerson, C. M., & Hoogendoorn, G. (2014). VFR Travel and Second Home Tourism: The Missing Link: The Case of South Africa. *Tourism Review International*, *18*(3), 167–178. doi:10.3727/154427214X14101901317156

Rogerson, C. M. I, & Kiambo, W. (2007). The growth and promotion of regional tourism in the developing world: The South African experience. *Development Southern Africa*, *24*(3), 505–521. doi:10.1080/03768350701445608

Rogerson, J., & Rogerson, C. M. (2014a). Agritourism and local economic development in South Africa. *Bulletin of Geography. Socio-Economic Series*, *26*(26), 93–106. doi:10.2478/bog-2014-0047

Rogerson, J., & Rogerson, C. M. (2014b). Maximising the local development potential of Nature Tourism accommodation establishments in South Africa. *African Journal of Hospitality, Tourism and Leisure*, *3*(1), 1–20.

Romero González, J. (2005). El Gobierno del Territorio en España. Balance de iniciativas de coordinación y cooperación territorial [The Territorial Government in Spain. Balance of coordination and territorial cooperation initiatives]. *Boletín de la Asociación de Geógrafos Españoles*, (39), 59–86.

Romero, J., Brandis, D., & Melo, C. (2015). El giro neoliberal de las políticas para la ciudad en España. Balance a partir de los ejemplos de Madrid y Valencia [The neoliberal turn of policies for the city in Spain. Balance based on the examples of Madrid and Valencia]. *Boletín de la Asociación de Geógrafos Españoles*.

Rosegrant, M. W., & Ringler, C. (2000). Impact on food security and rural development of transferring water out of agriculture. *Water Policy*, *1*(6), 567–586. doi:10.1016/S1366-7017(99)00018-5

Roy, P. K., Mohanty, A., Wagner, A., Sharif, S., Khalil, H., & Misra, M. (2021). *Impacts of COVID-19 Outbreak on the Municipal Solid Waste Management: Now and beyond the Pandemic*. ACS Environmental.

Rubel, F., & Kottek, M. (2010). Observed and projected climate shifts 1901–2100 depicted by world maps of the Koppen-Geiger climate classification. *Meteorologische Zeitschrift (Berlin)*, *19*(2), 135–141. doi:10.1127/0941-2948/2010/0430

Ruszkowski, J., Górnicz, E., & Żurek, M. (2004). *Leksykon integracji europejskiej [Lexicon of European integration]*. PWN.

Rutkowska, I. (1998). *Od wspólnot europejskich do Unii Europejskiej [From the European Communities to the European Union]*. Wydawnictwo Zachodnio-pomorskiej Szkoły Biznesu w Szczecinie.

Saarinen, J., & Rogerson, C. M. (2014). Tourism and the millennium development goals: Perspectives beyond 2015. *Tourism Geographies*, *16*(1), 23–30. doi:10.1080/14616688.2013.851269

Saayman, A. & Saayman, M. (2010). Forecasting tourist arrivals in South Africa. *Acta Commercii*, 281-293.

Sabuncuoğlu, Z. (2013). İnsan kaynakları yönetimi (uygulamalı). *BETA Bulletin of Experimental Treatments for AIDS*.

Sassen, S. (2013). When territory deborders territoriality. *Territory, Politics, Governance*, *1*(1), 21–45. doi:10.1080/21622671.2013.769895

Schermerhorn, J. R., Jr. (1989). *Management for productivity*. https://aydancag.com/geleneksel-performans-degerlendirme-yontemleri/

Schnabel, S. (1998). La precipitación como factor en los procesos hidrológicos y erosivos. Análisis de datos de Cáceres capital. *Revija za Geografijo*, *X*, 137–152.

Schwatke, C., Scherer, D., & Dettmering, D. (2019). Automated extraction of consistent time-variable water surfaces of lakes and reservoirs based on Landsat and Sentinel-2. In Remote Sensing (Vol. 11). doi:10.3390/rs11091010

Scola, L. A., Takahashi, R. H. C., & Cerqueira, S. A. A. G. (2014). Multipurpose Water Reservoir Management: An Evolutionary Multiobjective Optimization Approach. *Mathematical Problems in Engineering*, *2014*(2). doi:10.1155/2014/638259

Secretaría de Medio Ambiente y Recursos Naturales. (2020). *Diagnóstico Básico para la Gestión Integral de los Residuos*. Author.

Sessitsch, A., & Birgit Mitter, B. (2015, January). 21st century agriculture: Integration of plant microbiomes for improved crop production and food security. *Microbial Biotechnology*, *8*(1), 32–33. doi:10.1111/1751-7915.12180 PMID:25545820

Sevelius, J. M., Gutierrez-Mock, L., Zamudio-Haas, S., McCree, B., Ngo, A., Jackson, A., Clynes, C., Venegas, L., Salinas, A., Herrera, C., Stein, E., Operario, D., & Gamarel, K. (2020). Research with Marginalized Communities: Challenges to Continuity During the COVID-19 Pandemic. *AIDS and Behavior*, *24*(7), 2009–2012. doi:10.100710461-020-02920-3 PMID:32415617

Sharpley, R. (2020). Tourism, sustainable development, and the theoretical divide: 20 years on. *Journal of Sustainable Tourism*, *28*(11), 1932–1946. doi:10.1080/09669582.2020.1779732

Sharpley, R., & Telfer, D. J. (2004). *Tourism and Development: Concepts and Issues*. Channel View Publishers.

Sheppard, S. R. J. (1989). *Visual Simulation: A User's Guide for Architects, Engineers and Planners*. Van Nostrand Reinhold.

Shrestha, N. K., & Wang, J. (2020). Water Quality Management of a Cold Climate Region Watershed in Changing Climate. *Journal of Environmental Informatics*, *35*(1), 56–80.

Siddiq & Taylor. (2021). Tools of the Trade? *Journal of the American Planning Association*, *87*(4), 497–511. doi:10.1080/01944363.2021.1899036

Sifolo, P. P. S. (2016). A reflection on tourism and related security implications on the African continent. *African Journal of Hospitality, Tourism and Leisure*, *5*(3), 1–10.

Sigala, M. (2020). Tourism and COVID-19: Impacts and implications for advancing and resetting industry and research. *Journal of Business Research*, *117*, 312–321. doi:10.1016/j.jbusres.2020.06.015 PMID:32546875

Sigala, M., Christou, E., & Gretzel, U. (Eds.). (2012). *Social media in travel, tourism and hospitality: Theory, practice and cases*. Ashgate Publishing, Ltd.

Signe, L. (2018). *Africa's tourism potential: Trends, drivers, opportunities and strategies. African Growth Initiative*. Brookings Institute.

Sil, Â., Fernandes, P. M., Rodrigues, A. P., Alonso, J. M., Honrado, J. P., Perera, A., & Azevedo, J. C. (2019). Farmland abandonment decreases the fire regulation capacity and the fire protection ecosystem service in mountain landscapes. *Ecosystem Services*, *36*, 100908. Advance online publication. doi:10.1016/j.ecoser.2019.100908

Sil, Â., Fonseca, F., Gonçalves, J., Honrado, J., Marta-Pedroso, C., Alonso, J., Ramos, M., & Azevedo, J. C. (2017). Analysing carbon sequestration and storage dynamics in a changing mountain landscape in Portugal: Insights for management and planning. *The International Journal of Biodiversity Science, Ecosystem Services & Management*, *13*(2), 82–104. doi:10.1080/21513732.2017.1297331

Silva, A. R. F. & Lousada, S. A. N. & Moura, A. D. S. & Gonçalves, L. B. (2020). Modelação numérica do escoamento em canais artificiais: aplicação à ribeira da Ribeira Brava. *Revista Científica Monfragüe Desarrollo Resiliente, 13.*

Silva, C., & Larsson, A. (2018). *Challenges for Accessibility Planning and Research in the context of Sustainable Mobility.* Discussion Paper, International Transport Forum, Paris.

Sippel, S., Mitchell, D., Black, M. T., Dittus, A. J., Harrington, L., Schaller, N., & Otto, F. E. L. (2015). Combining large model ensembles with extreme value statistics to improve attribution statements of rare events. *Weather and Climate Extremes*, *9*, 25–35. doi:10.1016/j.wace.2015.06.004

Sivam, A., & Karuppannan, S. (2012). *Density Design and Sustainable Residential Development* (Doctoral dissertation). Earthscan.

Skryl, T. V., & Gregoric, M. (2022). Tourism in the Post-COVID Age. In *Post-COVID Economic Revival* (Vol. 2, pp. 239–254). Palgrave Macmillan. doi:10.1007/978-3-030-83566-8_15

SNIRH. (2021). *Sistema Nacional de Informação de Recursos Hídricos*. Disponível em: https://snirh.apambiente.pt/index.php?idMain=2&idItem=1&objCover=920123704&objSite=920685506

Soares, A. L., Rego, F. C., McPherson, E. G., Simpson, J. R., Peper, P. J., & Xiao, Q. (2011). Benefits and costs of street trees in Lisbon, Portugal. *Urban Forestry & Urban Greening*, *10*(2), 69–78. doi:10.1016/j.ufug.2010.12.001

Soliman, A. M. (2004). Regional planning scenarios in South Lebanon: The challenge of rural-urban interactions in the era of liberation and globalization. *Habitat International*, *28*(3), 385–408. doi:10.1016/S0197-3975(03)00039-0

Soliman, T., Mourits, M. C. M., van der Werf, W., Hengeveld, G. M., Robinet, C., & Lansink, A. G. J. M. O. (2012). Framework for Modelling Economic Impacts of Invasive Species, Applied to Pine Wood Nematode in Europe. *PLoS One*, *7*(9), e45505. Advance online publication. doi:10.1371/journal.pone.0045505 PMID:23029059

Solís, E. (2008). El horizonte urbano madrileño: Más allá de la región político-administrativa [The Madrilenian Urban Horizont: Beyond the Administrative Region]. *Anales de Geografía de la Universidad Complutense*, *28*(1), 133.

Solon, J. (2009). Spatial context of urbanization: Landscape pattern and changes between 1950 and 1990 in the Warsaw metropolitan area, Poland. *Landscape and Urban Planning*, *93*(3), 250–261. doi:10.1016/j.landurbplan.2009.07.012

Song, J., Jai, T. M., & Li, X. (2020). Examining green reviews on TripAdvisor: Comparison between resort/luxury hotels and business/economy hotels. *International Journal of Hospitality & Tourism Administration*, *21*(2), 165–187. doi:10.1080/15256480.2018.1464418

Spanish Statistical Office. (2021). *Municipal Register*. Retrieved from https://www.ine.es/jaxiT3/Tabla.htm?t=2881

Spiertz, J. H. J. (2009). Nitrogen, Sustainable Agriculture and Food Security: A Review. In E. Lichtfouse, M. Navarrete, P. Debaeke, S. Véronique, & C. Alberola (Eds.), *Sustainable Agriculture*. Springer. doi:10.1007/978-90-481-2666-8_39

Squires, G. (2002). *Urban sprawl: Causes, consequences, & policy responses*. The Urban Institute.

Stanković, M., Gladović, P., Popović, V., & Lukovac, V. (2018). Selection Criteria and Assessment of the Impact of Traffic Accessibility on the Development of Suburbs. *Sustainability*, *10*(6), 1977. doi:10.3390u10061977

Stepchenkova, S., & Zhan, F. (2013). Visual destination images of Peru: Comparative content analysis of DMO and user-generated photography. *Tourism Management*, *36*, 590–601. doi:10.1016/j.tourman.2012.08.006

Strahler, A. N. (1964). Quantitative geomorphology of drainage basins and channel networks. McGraw-Hill.

Strange, T., & Bayley, A. (2008). *Sustainable development-Linking economy, society, environment*. OECD Publication. doi:10.1787/9789264055742-en

Streimikiene, D., Svagzdiene, B., Jasinskas, E., & Simanavicius, A. (2021). Sustainable tourism development and competitiveness: The systematic literature review. *Sustainable Development*, *29*(1), 259–271. doi:10.1002d.2133

Sundberg, J. (2011). Diabolic *Caminos* in the Desert and Cat Fights on the Río: A Posthumanist Political Ecology of Boundary Enforcement in the United States–Mexico Borderlands. *Annals of the Association of American Geographers*, *101*(2), 318–336. doi:10.1080/00045608.2010.538323

Sverige, A. (2020). *Swedish Waste Management 2019*. Recovered from: https://www.avfallsverige.se/fileadmin/user_upload/Publikationer/SAH_2019_publ20_eng.pdf

Szymańska, E., Panfiluk, E., & Kiryluk, H. (2021). Innovative Solutions for the Development of Sustainable Transport and Improvement of the Tourist Accessibility of Peripheral Areas: The Case of the Białowieża Forest Region. *Sustainability*, *13*(4), 2381. doi:10.3390u13042381

Tang, M., Liao, H., Wan, Z., Herrera-Viedma, E., & Rosen, M. A. (2018). Ten years of Sustainability (2009 to 2018): A bibliometric overview. *Sustainability (Switzerland)*, *10*(5), 1655. Advance online publication. doi:10.3390u10051655

Tech, B. (2019). *These 6 towns in South Africa are attracting a new wave of semigrants*. Available at: https://businesstech.co.za/news/property/344518/these-6-towns-in-south-africa-are-attracting-a-new-wave-of-semigrants/

Tencaliec, P., Favre, A., Prieur, C., & Mathevet, T. (2015). Reconstruction of missing daily streamflow data using dynamic regression models. *Water Resources Research*, *51*(12), 9447–9463. doi:10.1002/2015WR017399

Tennøy, A., Hansson, L., Lissandrello, E., & Næss, P. (2016). *How planners' use and non-use of expert knowledge in land use and transport planning affect the goal achievement potential of plans? Experiences from three Scandinavian cities*. Academic Press.

Teodoro, A. C., & Duarte, L. (2013). Forest fire risk maps: A GIS open source application - a case study in Norwest of Portugal. *International Journal of Geographical Information Science*, *27*(4), 699–720. doi:10.1080/13658816.2012.721554

The Economist and Intelligence Unite. (2017). *Fixing Food: The Mediterranean Region Building Sustainable Food Systems Through Capacity-Building And Co-Operation*. https://www.barillacfn.com/m/publications/bcfn-fixingfood-themediterraneanregion2017.pdf

The World Bank Data. (2022). *Personal remittances, received (% GDP)*. https://data.worldbank.org/indicator/BX.TRF.PWKR.DT.GD.ZS?locations=HT-AM-GE-PH-MA

The World Bank. (2021, November 17). Remittance Flows Register Robust 7.3 Percent. *Growth*, *2021*. https://www.worldbank.org/en/news/press-release/2021/11/17/remittance-flows-register-robust-7-3-percent-growth-in-2021

Tiboni, M., Rossetti, S., Vetturi, D., Torrisi, V., Botticini, F., & Schaefer, M. D. (2021). Urban Policies and Planning Approaches for a Safer and Climate Friendlier Mobility in Cities: Strategies, Initiatives and Some Analysis. *Sustainability*, *13*(4), 1778. doi:10.3390u13041778

Tilman, D., Fargione, J., Wolf, B., Dantonio, C., Dobson, A., Owarth, R., & Schindler, D. (2001). Forecasting Agriculturally Driven Global Environmental Change. *Science*, *292*(5515), 281–284. doi:10.1126cience.1057544 PMID:11303102

Traktat o funkcjonowaniu Unii Europejskiej – tekst skonsolidowany uwzględniający zmiany wprowadzone Traktatem z Lizbony (Dz.U.2004.90.864/2) (Treaty on the Functioning of the European Union - consolidated text taking into account the changes introduced by the Treaty of Lisbon).

Traktat o funkcjonowaniu Unii Europejskiej, (Treaty on the Functioning of the European Union) art. 13.

Traktat o Unii Europejskiej (tzw. Traktat z Maastricht) (The Treaty on European Union (the so-called,, Treaty on European Union"). (Maastricht Treaty)) – podpisany 7 lutego 1992 r. (Dz.U. z 2004 r. Nr 90, poz. 864/30).

Traktat o utworzeniu Europejskiej Wspólnoty Energii Atomowej (tzw. Traktat Rzymski) (Treaty establishing the European Atomic Energy Community) – podpisany 25 marca 1957 r. (Dz.U. z 2004 r. Nr 90, poz. 864/3); J. Kowalski, Z. Ślusarczyk, Unia Europejska (…) (European Union (…)).

Traktat o utworzeniu Europejskiej Wspólnoty Energii Atomowej. (Treaty establishing the European Atomic Energy Community).

Traktat o utworzeniu Europejskiej Wspólnoty Gospodarczej (tzw. Traktat Rzymski) (The Treaty establishing the European Economic Community (the so-called,, Treaty of Lisbon"). Treaty of Rome)) – podpisany (signed) 25 marca 1957 r. (Dz.U. z 2004 r. Nr 90, poz. 864/2).

Traktat o utworzeniu Europejskiej Wspólnoty Gospodarczej. (Treaty establishing the European Economic Community).

Traktat o utworzeniu Europejskiej Wspólnoty Węgla i Stali (tzw. Traktat Paryski) – podpisany 18 kwietnia 1951 r. (wszedł w życie 23 lipca 1952 r.; wygasł 23 lipca 2002 r.) (Treaty on the establishment of the European Coal and Steel Community (the so-called,, Treaty of Lisbon"). Treaty of Paris) - signed on 18 April 1951 (entered into force on 23 July 1952; expired on 23 July 2002).

Traktat ustanawiający jednolitą Radę i jednolitą Komisję Wspólnot Europejskich (tzw. traktat o fuzji) – podpisany 8 kwietnia 1965 r. (Treaty establishing a single Council and a single Commission of the European Communities (the so-called,, Single Council"). the Merger Treaty) - signed on 8 April 1965.

Traktat ustanawiający Konstytucję dla Europy (Treaty establishing a Constitution for Europe) (Dz.Urz. UE, C 310, 16 grudzień 2004).

Traktat z Amsterdamu zmieniający Traktat o Unii Europejskiej, traktaty ustanawiające Wspólnoty Europejskie i niektóre związane z nimi akty (tzw. Traktat Amsterdamski) (The Treaty of Amsterdam amending the Treaty on European Union, the Treaties establishing the European Communities and certain related acts (the so-called,, Amsterdam Treaty")) – podpisany 2 października 1997 r. (wszedł w życie 1 maja 1999 r.) (Dz.U. z 2004 r. Nr 90, poz. 864/31).

Traktat z Lizbony zmieniający Traktat o Unii Europejskiej i Traktat ustanawiający Wspólnotę Europejską podpisany w Lizbonie dnia 13 grudnia 2007 r. (Treaty of Lisbon amending the Treaty on European Union and the Treaty establishing the European Community signed in Lisbon on 13 December 2007.) (Dz.U. 2009 nr 203 poz. 1569).

Traktat z Nicei zmieniający Traktat o Unii Europejskiej, traktaty ustanawiające Wspólnoty Europejskie i niektóre związane z nimi akty (tzw. Traktat Nicejski) –przyjęty 11 grudnia 2000 r. i podpisany 26 lutego 2001 r. (wszedł w życie 1 lutego 2003 r.) (Dz.U. z 2004 r. Nr 90, poz. 864/32). (The Treaty of Nice amending the Treaty on European Union, the Treaties establishing the European Communities and certain related acts (the so-called,, Nice Treaty") Treaty of Nice) - Adopted on 11 December 2000 and signed on 26 February 2001 (entered into force on 1 February 2003) (OJ L 338, 30. 12. 2000, p. 1. U. of 2004, No. 90, item. 864/32).

Trenberth, K. E., Dai, A., Rasmussen, R. M., & Parsons, D. B. (2003). *The Changing Character of Precipitation*. American Meteorological Society. doi:10.1175/BAMS-84-9-1205

Tsartas, P. (1996). *Tourists, journeys, landscapes: Sociological approaches to tourism*. Academic Press.

Tucci, C. E. M. (1993). *Controle de Enchentes. Hidrologia: Ciência e Aplicação. 2ª Edição*. Editora da Universidade UFRGS.

Tucci, C. E. M., Porto, R. L., & Barros, M. T. (1995). *Drenagem urbana. Universidade Federal do Rio Grande do Sul*. UFRGS.

Turienzo, J., Cabanelas, P., & Lampón, J. F. (2022). The Mobility Industry Trends Through the Lens of the Social Analysis: A Multi-Level Perspective Approach. *SAGE Open*, *12*(1). Advance online publication. doi:10.1177/21582440211069145

Tyranowski, J. (1999). *Prawo Europejskie. Zagadnienia instytucjonalne z uwzględnieniem Traktatu Amsterdamskiego [European law. Institutional issues in the light of the Amsterdam Treaty]*. Properium.

U.S. Department of Agriculture. (n.d.). *Climate Change, Global Food Security, and the U.S. Food System*. https://www.usda.gov/oce/energy-and-environment/food-security#:~:text=Climate%20change%20is%20likely%20to,food%20safety%2C%20among%20other%20causes

Ukko, J., Saunila, M., Rantala, T., & Havukainen, J. (2019, May/June). Sustainable development: Implications and definition for open sustainability. *Sustainable Development (Bradford)*, *27*(3), 321–336. doi:10.1002d.1904

UN. (2015). *Transforming our world: the 2030 Agenda for Sustainable Development- Resolution adopted by the General Assembly on 25 September 2015. 17th Session Ageda items 15 and 116. A/RES/70/1*. https://www.un.org/ga/search/view_doc.asp?symbol=A/RES/70/1&Lang=E

UNEP. (2015). *Global Waste Management Outlook*. United Nations Environment Programme.

UNHCR. (2021). *Refugee data finder*. https://www.unhcr.org/refugee-statistics/

UNHCR. (2022). *Operational Data Portal-Refugee situations*. https://data2.unhcr.org/en/situations/syria

United Nations Conference on Trade and Development. (2017). *Economic Development in Africa Report 2017. Tourism for Transformative and Inclusive Growth*. United Nations.

United Nations Department of Economic and Social Affairs. (n.d.). *The 17 Goals*. https://sdgs.un.org/goals

UNWTO & UNDP. (2017). *Tourism and the sustainable development goals – Journey to 2030*. UNWTO.

UNWTO. (2018). *Press release: Tourism can and should lead sustainable development: UNWTO Secretary-General opens ITB 2018. PR 18020, 06 Mar 18*. UNWTO.

Vale, D. S., Saraiva, M., & Pereira, M. (2015). Active accessibility: A review of operational measures of walking and cycling accessibility. *Journal of Transport and Land Use*, *9*(1). Advance online publication. doi:10.5198/jtlu.2015.593

Valenzuela, R., & Garreaud, R. (2019). Extreme Daily Rainfall in Central-Southern Chile and Its Relationship with Low-Level Horizontal Water Vapor Fluxes. *Journal of Hydrometeorology*, *20*(9), 1829–1850. doi:10.1175/JHM-D-19-0036.1

Van Den Hoek, J., Getirana, A., Jung, H. C., Okeowo, M. A., & Lee, H. (2019). Monitoring reservoir drought dynamics with landsat and radar/lidar altimetry time series in persistently cloudy eastern Brazil. *Remote Sensing*, *11*(7), 827. Advance online publication. doi:10.3390/rs11070827

van Eldijk, J., Gil, J., & Marcus, L. (2022). Disentangling barrier effects of transport infrastructure: Synthesising research for the practice of impact assessment. *European Transport Research Review*, *14*(1), 1. doi:10.118612544-021-00517-y

Vandenbulcke, G., Steenberghen, T., & Thomas, I. (2009). Mapping accessibility in Belgium: A tool for land-use and transport planning? *Journal of Transport Geography*, *17*(1), 39–53. doi:10.1016/j.jtrangeo.2008.04.008

Vargues, P., & Loures, L. (2008). Using Geographic Information Systems in Visual and Aesthetic Analysis: The case study of a golf course in Algarve. *WSEAS Transactions on Environment and Development*, *4*(9), 774–783.

Vecchiato, D., & Tempesta, T. (2013). Valuing the benefits of an afforestation project in a peri-urban area with choice experiments. *Forest Policy and Economics*, *26*, 111–120. doi:10.1016/j.forpol.2012.10.001

Velaga, N. R., Nelson, J. D., Wright, S. D., & Farrington, J. H. (2012). The Potential Role of Flexible Transport Services in Enhancing Rural Public Transport Provision. *Journal of Public Transportation*, *15*(1), 111–131. doi:10.5038/2375-0901.15.1.7

Vendemmia, B., Pucci, P., & Beria, P. (2021). An institutional periphery in discussion. Rethinking the inner areas in Italy. *Applied Geography (Sevenoaks, England)*, *135*, 102537.

Véron, J. (2014). *How are migration and the environment connected?* https://www.ined.fr/en/everything_about_population/demographic-facts-sheets/focus-on/migrations_environment/

Vettorazzi, S. (2018). Establishing the Connecting Europe Facility 2021-2027. European Parliamentary Research Service.

Vieira, I. L. S., Barreto, V., Figueira, C., Lousada, S., & Prada, S. (2016). The use of detention basins to reduce flash flood hazard in small and step volcanic watersheds – a simulation from Madeira Island. *Journal of Flood Risk Management*. Advance online publication. doi:10.1111/jfr3.12285

Vilar-Compte, M., Burrola-Méndez, S., Lozano-Marrufo, A., Isabel Ferré-Eguiluz, I., Flores, D., Gaitán-Rossi, P., Teruel, G., & Rafael Pérez-Escamilla, R. (2021). Urban poverty and nutrition challenges associated with accessibility to a healthy diet: A global systematic literature review. *International Journal for Equity in Health*, *20*(1), 40. doi:10.118612939-020-01330-0 PMID:33472636

Visser, G. (2004). Second homes and local development: Issues arising from Cape Town's De Waterkant. *GeoJournal*, *60*(3), 259–271. doi:10.1023/B:GEJO.0000034733.80648.88

Visser, G., & Hoogendoorn, G. (2015). A decade of second home tourism in South Africa: Research prospects for the developing world? *The South African Geographical Journal*, *97*(2), 111–222. doi:10.1080/03736245.2015.1028976

Vitale Brovarone, E., & Cotella, G. (2020). Improving Rural Accessibility: A Multilayer Approach. *Sustainability*, *12*(7), 2876.

von Essen, M., do Rosário, I. T., Santos-Reis, M., & Nicholas, K. A. (2019). Valuing and mapping cork and carbon across land use scenarios in a Portuguese montado landscape. *PLoS One*, *14*(3), e0212174. Advance online publication. doi:10.1371/journal.pone.0212174 PMID:30845222

Vulevic, A., Obradovic, V., Castanho, R.A., & Djordjevic, D. (2020). Cross-Border Cooperation (CBC) in a Multi-Level Governance System in Southeastern Europe Territories: How to Manage Territorial Governance Processes in Serbia-Romania Border Space. In *Cross-Border Cooperation (CBC) Strategies for Sustainable Development*. IGI Global. doi:10.4018/978-1-7998-2513-5.ch004

Vulevic, A. (2016). *Accessibility concepts and indicators in transportation strategic planning issues: theoretical framework and literature review.* Logistics & Sustainable Transport. doi:10.1515/jlst-2016-0006

Vulevic, A., Castanho, R. A., Naranjo Gómez, J. M., Loures, L., Cabezas, J., Fernández-Pozo, L., & Martín Gallardo, J. (2020). Accessibility Dynamics and Regional Cross-Border Cooperation (CBC) Perspectives in the Portuguese—Spanish Borderland. *Sustainability*, *12*(5), 1978. doi:10.3390u12051978

Wageningen University. (2017a). *Rural migration and environmental degradation: A vicious cycle?* NOW VENI. https://www.wur.nl/en/project/Rural_migration_and_environmental_degradation.htmq

Wageningen University. (2017b). *Innovation to improve the fate of refugees and their hosts.* https://www.wur.nl/en/Research-Results/Research-Institutes/Environmental-Research/Programmes/Green-Cities/Migration-and-settlements.htm

Wang, X., Shi, R., & Zhou, Y. (2020). Dynamics of urban sprawl and sustainable development in China. *Socio-Economic Planning Sciences, 70*, 100736. doi:10.1016/j.seps.2019.100736

Wang, Z., Han, Q., & De Vries, B. (2012). Land Use/Land Cover and Accessibility: Implications of the Correlations for Land Use and Transport Planning. *Applied Spatial Analysis and Policy, 12*, 923–940.

Watson, D., Elander, M., Gylling, A., Andersson, T., & Heikkilä, P. (2017). *Stimulating Textile-to-Textile Recycling.* Nordic Council of Ministers.

Weber, H. (2017). Politics of 'Leaving No One Behind': Contesting the 2030 Sustainable Development Goals Agenda. *Globalizations, 14*(3), 399-414. doi:10.1080/14747731.2016.1275404

Wetherald, R. T., & Manabe, S. (2007). Climate Change. IPCC.

Whitaker, B. E. (1999). *Changing opportunities: Refugees and host communities in western Tanzania.* New Issues In Refugee Research, Working Paper No. 11.

Wiesenthal, T., Condeço-Melhorado, A., & Leduc, G. (2015). Innovation in the European transport sector: A review. *Transport Policy, Elsevier, 42*(C), 86–93.

Wojtaszczyk, K. (Ed.). (2006). Integracja europejska [European integration]. Poltext.

Wong Villanueva, J. L., Kidokoro, T., & Seta, F. (2020). Cross-Border Integration, Cooperation and Governance: A Systems Approach for Evaluating "Good" Governance in Cross-Border Regions. *Journal of Borderlands Studies.*

World Atlas. (n.d.). *Mediterranean Countries.* https://www.worldatlas.com/articles/mediterranean-countries.html

World Economic Forum. (2018). *Travel and Tourism Competitiveness Report, 2018.* WEF.

World Meteorological Organization (WMO). (2021). *WMO: Climate change threatens sustainable development.* Press Release Number: 22092021. https://public.wmo.int/en/media/press-release/wmo-climate-change-threatens-sustainable-development

Worrell, E., & Reuter, M. A. (2014). *Handbook of Recycling.* Elsevier.

Wulfhorst, G., Büttner, B., & Ji, C. (2017). The TUM Accessibility Atlas as a tool for supporting policies of sustainable mobility in metropolitan regions. *Transportation Research Part A: Policy and Practice, Elsevier., 104*, 121–136.

Xu, H. (2006). Modification of normalised difference water index (NDWI) to enhance open water features in remotely sensed imagery. *International Journal of Remote Sensing, 27*(14), 3025–3033. doi:10.1080/01431160600589179

Xu, Q., Yang, R., Dong, Y.-X., Liu, Y.-X., & Qiu, L.-R. (2016). The influence of rapid urbanization and land use changes on terrestrial carbon sources/sinks in Guangzhou, China. *Ecological Indicators, 70*, 304–316. doi:10.1016/j.ecolind.2016.05.052

Yan, X., Zhao, X., Han, Y., Hentenryck, P. V., & Dillahunt, T. (2021). Mobility-on-demand versus fixed-route transit systems: An evaluation of traveler preferences in low-income communities. *Transportation Research Part A: Policy and Practice, Elsevier., 148*, 481–495.

Yao, F., Wang, J., Wang, C., & Crétaux, J. F. (2019). Constructing long-term high-frequency time series of global lake and reservoir areas using Landsat imagery. *Remote Sensing of Environment*, *232*(July), 111210. doi:10.1016/j.rse.2019.111210

Ye, E. M., Du, J. T., Hansen, P., Ashman, H., Sigala, M., & Huang, S. S. (2021). Understanding roles in collaborative information behaviour: A case of Chinese group travelling. *Information Processing & Management*, *58*(4), 102581.

Yıldırım, S., & Kaplan, M. (2022). The Threat of Invasive Alien Marine Species to Blue Economy: The Mediterranean Case. In Implications for Entrepreneurship and Enterprise Development in the Blue Economy. IGI Global. doi:10.4018/978-1-6684-3393-5.ch004

Yıldırım, S., Yıldırım, D. Ç., & Gedikli, A. (2018). Sustainable Consumption Trends in the World in the Context of Green Economy and Sustainability. In Sustainable Development: Concepts, Methodologies, Tools, and Applications (pp. 1605-1624). IGI Global. doi:10.4018/978-1-5225-3817-2.ch071

Yıldırım, S., & Yıldırım, D. C. (2021). Achieving Seafood Security in the Mediterranean Region: A Case of Turkey. In R. Castanho & J. Martín Gallardo (Eds.), *Management and Conservation of Mediterranean Environments* (pp. 175–195). IGI Global. doi:10.4018/978-1-7998-7391-4.ch011

Yu, B. Y., Wu, P., Sui, J., Ni, J., & Whitcombe, T. (2020). Variation of Runoff and Sediment Transport in the Huai River–A Case Study. *Journal of Environmental Informatics*, *35*(2), 138–147. doi:10.3808/jei.202000429

Zeder, J., & Fischer, E. (2020). Observed extreme precipitation trends and scaling in Central Europe. *Weather and Climate Extremes*, *29*, 100266. doi:10.1016/j.wace.2020.100266

Zetter, R. (1994). Shelter provision and settlement policies for refugees: A state of the art review. In *Studies on Emergencies and Disaster Relief, No.2* (pp. 29–98). Nordiska Afrikainstitutet.

Zhang, S., Gao, H., & Naz, B. (2014). Monitoring reservoir storage in South Asia from multisatellite remote sensing. *Water Resources Research*, *50*(11), 5375–5377. doi:10.1002/2014WR015829

Zhang, T. (2001). *Public participation in China's urban development*. Taylor & Francis Group.

Zhang, X., Lu, H., & Holt, J. B. (2011). Modeling spatial accessibility to parks: A national study. *International Journal of Health Geographics*, *10*, 31.

Zhao, G., & Gao, H. (2018). Automatic Correction of Contaminated Images for Assessment of Reservoir Surface Area Dynamics. *Geophysical Research Letters*, *45*(12), 6092–6099. doi:10.1029/2018GL078343 PMID:35095126

Zimmer, P., Camargo, M. E., Pizzoli, M. F. F., Zanandréa, G., & Bizotto, B. L. S. (2019). Consumo Consciente: O Nível de Consciência Ecológica dos Acadêmicos do Curso de Administração de uma Instituição de Ensino da Serra Gaúcha. *Desafio Online*, *7*(2), 261–277.

Zoido, F. (2001). La Convención Europea del Paisaje y su aplicación en España [The European Landscape Convention and its application in Spain]. *Ciudad y Territorio, Estudios Territoriales, 33*(128), 275-281.

Zoido, F. (2000). Proteger y realzar el paisaje [Protect and enhance the landscape]. *Boletín de la Asociación de Geógrafos Profesionales de Andalucía*, (7), 7–15.

Zolin, M. B., Ferretti, P., & Grandi, M. (2020). Sustainability in Peripheral and Ultra-Peripheral Rural Areas through a Multi-Attribute Analysis: The Case of the Italian Insular Region. *Sustainability*, *12*(22), 9380.

Zwiers, F., Zhang, X., & Feng, Y. (2011). Anthropogenic Influence on Long Return Period Daily Temperature Extremes at Regional Scales. *Journal of Climate*, *24*(3), 881–892. doi:10.1175/2010JCLI3908.1

About the Contributors

Rui Alexandre Castanho holds an International Ph.D. on Sustainable Planning in Borderlands. Nowadays, he's a Professor at the WSB University, Poland, and the Coordinator of International Relations on the Research Institute on Territorial Cooperation and Inter-Organizational Cooperation. Also, he is a pos-doc researcher on the GREAT Project, University of Azores, Ponta Delgada, Portugal. Furthermore, he collaborates with the Environmental Resources Analysis Research Group (ARAM) - University of Extremadura (UEx); VALORIZA - Research Center for the Enhancement of Endogenous Resources, Polytechnic Institute of Portalegre (IPP), Portugal; CITUR - Madeira - Centre for Tourism Research, Development and Innovation, Madeira, Portugal; and AQUAGEO Research Group - University of Campinas (UNICAMP), Brazil. Moreover, he has Graduated in Green Spaces Engineering and Agricultural Engineering and holds a Master's Degree in Planning, Audit, and Control in Urban Green Spaces.

* * *

Ignacio Aguirre Belmar holds a geography degree from the Pontificia Universidad Católica de Chile. In 2021, he began graduate studies in the Department of Geography and Planning and the Centre for Hydrology at the University of Saskatchewan, Canada. He has worked as a water resources consultant on several projects. His scientific interests include ecohydrology, wetlands, hydrological modelling, hydroinformatics, process-based modelling, remote sensing, and nature-based solutions. He has presented at the European Geosciences Union (EGU) General Assembly, Biohydrology meeting and published two book chapters.

Lwazi Apleni teaches tourism at the University of Zululand, in KwaZulu-Natal in the Republic of South Africa. He holds a Master's Degree in Management Sciences from University of Science & Technology Beijing (China) obtained in 2015. In his Master's journey, He focused on "The implications of carbon tax in South Africa's tourism industry". Lwazi is the recipient of Chinese Scholarship Council (2013-2015). As part of his career growth and development, he is currently reading his PhD (specialising in Local Economic Development) at the University of Johannesburg having registered for the programme in June 2020. His research interests include sustainable development, tourism development and tourism management.

Stylianos Bouzis is a PhD Candidate at the International Hellenic University. Certified adult educator (included in the register of educators of Panteion University). Holder of Bachelor in Russian language, literature and culture and Master of science in Strategic Management of Tourism Destinations and Hospitality Enterprises. With passion in designing and developing new emerging alternative forms of tourism.

José Cabezas has experience in research applied to sustainable management, conservation and recovery of Natural Ecosystems (environment: water, soil, fauna, flora...), Land Use Planning, Environmental Impact, Remote Sensing and Geographic Information Systems, as well as in Cartographic Modelling and Spatial Data Infrastructure (SDI), aspects that are fundamentally oriented towards the conservation of transboundary Mediterranean ecosystems and their sustainability.

Fatma Erdogan completed her doctorate at Tekirdağ Namık Kemal University. she as several publications including SSCI, ESCI, Scopus and other international indexed journals. She mostly studies on economics, sustainable development, and sustainability.

Alexandros Filiopoulos is a graduate student at University of Aegean (strategic management of tourism destinations and hospitality enterprises), Bachelor in History and Archaeology at University of Ioannina, with passion in designing and developing new alternative forms of tourism. Well rounded young professional, team player, with excellent communication skills, ready to respond in challenging projects and work environments, which are operating with strict deadlines.

Leonardo Bazilio Gonçalves is a Bachelor in Civil Engineering from University Center São Camilo (CUSC-ES, Brazil), and holds a Master's Degree from University of Madeira (UMa, Portugal). Futhermore, he has a line of research focused on water resources management and urban planning, collaborating with the University of Madeira's Department of Civil Engineering and Geology (DCEG). Nowadays, he works as a visiting assistant professor in the Higher Professional Technical Course (CTeSP) in Civil Construction from University of Madeira.

Unathi Sonwabile Henama is a Senior Lecturer in the Department of Tourism and Events Management. He was formerly employed by the Department of Tourism Management at the Tshwane University of Technology. He has a PhD from Mid Sweden University and a Masters from the University of the Free State. He has written numerous articles and presented research papers at both local and international conferences. He is the leading tourism commentator in South Africa, and his views are highly sought after by TV, print and online news outlets.

Sérgio António Neves Lousada holds an International Ph.D. in Civil Engineering - Hydraulics. He teaches at the University of Madeira, field of Hydraulics, Environment and Water Resources and Construction. Furthermore, he collaborates with the Environmental Resources Analysis Research Group (ARAM) - University of Extremadura (UEx); VALORIZA - Research Center for the Enhancement of Endogenous Resources, Polytechnic Institute of Portalegre (IPP), Portugal; CITUR - Madeira - Centre for Tourism Research, Development and Innovation, Madeira, Portugal; and Institute of Research on Territorial Governance and Inter-Organizational Cooperation, Dąbrowa Górnicza, Poland. Moreover, he holds an International Master's Degree in Ports and Coasts Engineering.

Lívia Madureira is currently associate professor at the UTAD (University of Trás-os-Montes e Alto Douro) and senior researcher at the CETRAD Centre of Transdisciplinary Development Studies). Research interests comprise: Sustainability, Environmental Valuation; Rural and Agricultural Knowledge and Innovation Systems Mixed-methods, comparative studies and multi-level analysis are areas of methodological research interest.

José Martin Gallardo holds a PhD in Sciences (Biology) at the University of Extremadura. Interested in freshwater ecology.

Jankie Kgalabi Phale affectionately as JK is a BTech: Tourism Management graduate at the Department of Tourism at the Tshwane University of Technology. He is currently employed as a Student development practitioner. JK has great interest in collaborating towards tourism research whenever opportunities present themselves.

José Rosário is currently a PhD student of Agribusiness and sustainability at University of Trás-os-Montes e Alto Douro (UTAD). He finished Master study in 2010 on Rural Sustainable Development in Czech University Of Life Sciences Prague, Faculty of Tropical AgriSciences.

Ana Vulevic is a Ph.D. in Spatial Planning (University of Belgrade) and scientific researcher at Institute of Transportation CIP, Department of Architecture and Urban planning (from 1996. year) engaged in fields of work covered by the Institute include the so-called public sector - urban planning, spatial planning, transport planning, environmental planning, analyses for adapting spatial and urban plans). Thesis title: Infrastructural Corridors and their Influence on Spatial Development Example Pan European Corridor VII - Danube, in Serbia. Her research interests are spatial and transports analysis, empirical research and applications (e.g., accessibility indicators, spatial and statistical analyses) bridging the gap between theory and practice.

Durmuş Çağrı Yıldırım was born in 1981. After graduating from Kocaeli University, Department of Economics in 2005, he started to study for a master's degree in Economics. In 2007, he received a master's degree in Economic Policy. In 2011, he received his PhD in Economics. While working as a doctoral student, he worked at the University of Almeria-Spain for a while. He worked as a research assistant at Kocaeli University between 2006-2012. He worked as an Assistant Professor at the Department of Banking and Finance at Recep Tayyip Erdoğan University between 2013-2015. Since April 2015, he has been working as Professor at Tekirdag Namık Kemal University. His current research interests are current and emerging issues such as sustainable development, energy, health, employment and Islamic economics. https://www.researchgate.net/profile/Durmus-Yildirim-2, https://publons.com/researcher/3067692/durmus-cagr-yildirim/.

Seda Yıldırım currently works at Tekirdag Namık Kemal University in Turkey. She studies on sustainable development, 2030 sustainable development goals, sustainable consumption patterns, Islamic finance and banking, consumption patterns, environmental quality and food security. She is married and has one child. https://orcid.org/0000-0003-4367-6652, https://twitter.com/sdayildr?s=08 https://www.researchgate.net/profile/Seda-Yildirim-10, https://publons.com/researcher/2986513/seda-yldrm/.

Index

M

N

P

R

S

T

U

V

W

www.ingramcontent.com/pod-product-compliance
Lightning Source LLC
LaVergne TN
LVHW060417300926
849162LV00006B/1453
* 9 7 8 1 6 6 8 4 4 5 4 8 8 *